Biodegradable Polymers and Plastics

Biodegradable Polymers and Plastics

Edited by

Emo Chiellini and Roberto Solaro
University of Pisa
Pisa, Italy

Kluwer Academic / Plenum Publishers
New York, Boston, Dordrecht, London, Moscow

Library of Congress Cataloging-in-Publication Data

World Conference on Biodegradable Polymers & Plastics (7th : 2002 : Tirrenia, Italy)
Biodegrable polymers and plastics / edited by Emo Chiellini, Roberto Solaro.
p. cm.
Includes bibliographical references and index.
ISBN 0-306-47884-6
1. Polymers--Biodegradation--Congresses. 2. Biodegradable plastics--Congresses. 3. Plastic scrap--Congresses. I. Chiellini, Emo. II. Solaro, Roberto III. Title.

QP801.P64W67 2002
668.4'192--dc22

2003058934

Proceedings of the 7th World Conference on Biodegradable Polymers & Plastics organized by the European Degradable Polymer Society in conjunction with the Bioenvironmental Polymer Society and the Biodegradable Plastics Society, under the auspices of ICS-UNIDO (Italy) and INSTM Consortium (Italy) and under the patronage of IUPAC—International Union of Pure and Applied Chemistry (USA) and Ministero dell'Ambiente e della Tutela del Territorio (Italy), held on June 4–8, 2002, in Terrenia (Pisa), Italy.

ISBN 0-306-47884-6

233 Spring Street, New York, New York 10013

http://www.wkap.nl/

10 9 8 7 6 5 4 3 2 1

A C.I.P. record for this book is available from the Library of Congress

Printed in the United States of America

Preface

Synthetic and semi-synthetic polymeric materials were originally developed for their durability and resistance to all forms of degradation including biodegradation. Special performance characteristics are achieved in items derived therefrom through the control and maintenance of their molecular weight and functionality during the processing and under service conditions. Polymeric materials were and are currently widely accepted because of their ease of processability and amenability to provide a large variety of cost effective items that help enhancing the comfort and quality of life in the modern industrial society.

The widespread utilization of plastics in various mercantile segments that make the polymeric materials so convenient and useful to the human life, has contributed however to create a serious plastic waste burden, sometimes unfairly oversized by media because of the visible dispersion of plastic litter in the environment and the heavy contribution to landfill depletion due to the unfavorable weight to volume ratio of plastic items.

On the other hand, the expectations in the 21st century for polymeric material demand are in favor of 2 to 3 fold production increase in the next couple of decades, thus overcoming the world-wide annual production of paper as a consequence of the increase of plastic consumption in developing countries and countries in transition. Indeed, the magnitude of the indicated jump of plastic consumption with respect to the present annual level of 2-15 kg pro-capita can be easily envisaged for those countries once they will approach the living standards of industrialized countries with an annual average consumption pro-capita of 100 kg.

The design, production and consumption of polymeric materials for commodity and specialty plastic items have certainly to face all the constraints and regulations already in place or to be issued in the near future, dealing with the management of primary and post-consume plastic waste. Therefore, the formulation of environmentally sound degradable polymeric materials and relevant plastic items will constitute a key option among those available for the management of plastic waste.

In this connection the 7th World Conference on "Biodegradable Polymers & Plastics" held in Tirrenia (Pisa) - Italy in June 2002, as renamed continuation of the series of six former International Scientific Workshops on Biodegradable Polymers and Plastics started in 1989, was focused on the following topics and issues comprising:

- Environmentally Degradable Polymeric Materials (EDPs)
- Water-soluble/Swellable Biodegradable Polymers
- EDPs from Renewable Resources
- Biopolymers
- Bioresorbable Materials for Biomedical Applications
- Biorelated Polymers
- Standards and Regulations on EDPs.

The building of a common understanding background for a more rational utilization of resources in the fabrication and consumption of plastic items and in approaching issues bound to plastic waste management, consitute one of the major goal of the Conference. In order to avoid misuse of some fundamental concepts and for a fair appreciation of EDPs, it is useful to provide some general definitions that has been amply debated and basically accepted on a ground of common consensus.

Fundamental concepts and provision of general guidelines aimed at meeting the sustainability criteria of the modern industrial development constitute a focal point of the Conference. Polymeric materials and plastics deriving from fossil and renewable feedstocks, meeting the environmental degradability and biodegradability criteria, were considered in relation to their specific segments of applications including commodity and specialty areas.

The present book comprises some of the major contributions given at the Conference. For convenience they have been grouped in four sections even though they have a common backbone encompassing the synthesis and characterization of polymeric materials meant to be qualified as environmentally compatible and degradable with ultimate propensity to biodegradation.

The Editors
Emo Chiellini and Roberto Solaro

Acknowledgements

The Editors wish to acknowledge the continuous and tireless dedication provided by Ms Maria G. Viola who managed to transform all contributions into a camera-ready format and to handle all the correspondence with the authors to the present volume as well as with the publisher's contact person, Ms Joanna Lawrence.
The sponsorship provided by the following Institutions and Companies is also gratefully acknowledged: IUPAC-International Union of Pure and Applied Chemistry, University of Pisa, Kedrion SpA, Novamont SpA, EPI – Environmental Plastics Inc., Rheometric Scientific Italy, Belotti Strumenti Srl, Mettler Toledo Italy, and Idroplast Srl.

Contributors

ABE HIDEKI, Department of Innovative and Engineered Materials and the SORST Group of Japan Science and Technology Corporation (JST), Tokyo Institute of Technology, 4259 Nagatsuta, Midori-ku, Yokohama 226-8502, Japan, and Polymer Chemistry Laboratory, RIKEN Institute, 2-1 Hirosawa, Wako-shi, Saitama 351-0198, Japan
ASANO YASUHIRO, Department of Applied Physics and Chemistry, Fukui University of Technology, Fukui, Japan
BARBAUD CHRISTEL, LRP, UMR 7581, CNRS, Université Paris 12 - Val de Marne, 2/8 rue Henry Dunant 94320 Thias, France
BILLINGHAM NORMAN C., Department of Chemistry, University of Sussex, Brighton, BN1 9QJ, UK and EPI (Europe) Ltd., Unit 7, Dunston Place, Duston Road, Chesterfield, Derbyshire, S41 8NL, UK
BOESEL LUCIANO F., Department of Polymer Engineering, University of Minho, Campus de Azurém, 4800-058 - Guimarães, Portugal
BONORA MICHELA, Ciba Specialty Chemicals SpA, via Pila 6/3, I-40044 Sasso Marconi, Italy
CHEN GUO-QIANG, Department of Biological Science and Biotechnology, Tsinghua University, Beijing 100084, China
CHIELLINI EMO, Department of Chemistry & Industrial Chemistry, University of Pisa, via Risorgimento 35, 56126 Pisa, Italy
CHIELLINI FEDERICA, Department of Chemistry & Industrial Chemistry, University of Pisa, via Risorgimento 35, 56126 Pisa
CINELLI PATRIZIA, Department of Chemistry & Industrial Chemistry, University of Pisa, via Risorgimento 35, 56126 Pisa
CONNOR ERIC F., IBM Almaden research Center, San Jose, California 95120 (USA)
CORTI ANDREA, Department of Chemistry & Industrial Chemistry, University of Pisa, via Risorgimento 35 - 56126 Pisa, Italy
CRISTIANO FEDELE, Department of Chemistry & Industrial Chemistry, University of Pisa, via Risorgimento 35 - 56126 Pisa, Italy
DE CORTE DAVID, Ciba Specialty Chemicals, CH-4002 Basel, Switzerland
DEGLI INNOCENTI FRANCESCO, Novamont SpA, via Fauser 8, I-28100 Novara

DI MAIO ERNESTO, Institute of Composite Materials Technology (ITMC-CNR) & Department of Materials and Production Engineering, University of Naples "Federico II", Piazzale Tecchio 80, 80125 Naples, Italy

DI YINGWEY W., Institute of Composite Materials Technology (ITMC-CNR) & Department of Materials and Production Engineering, University of Naples "Federico II", Piazzale Tecchio 80, 80125 Naples, Italy

DIJKSTRA PIETER J., Department of Chemical Technology, University of Twente, P.O. Box 217, 7500 AE Enschede, The Netherlands

DOI YOSHIHARU, Department of Innovative and Engineered Materials and the SORST Group of Japan Science and Technology Corporation (JST), Tokyo Institute of Technology, 4259 Nagatsuta, Midori-ku, Yokohama 226-8502, Japan, and Polymer Chemistry Laboratory, RIKEN Institute, 2-1 Hirosawa, Wako-shi, Saitama 351-0198, Japan

DUBOIS PHILIPPE, Laboratory of Polymeric and Composite Materials, University of Mons-Hainaut, 20 Place du Parc, 7000 Mons, Belgium

ELVIRA CARLOS, Institute of Science and Technology of Polymers, CSIC, c/Juan de la Cierva, 3, 28006, Madrid, Spain

FEIJEN JAN, Department of Chemical Technology, University of Twente, P.O. Box 217, 7500 AE Enschede, The Netherlands

GAN ZHIHUA, Department of Innovative and Engineered Materials and the SORST Group of Japan Science and Technology Corporation (JST), Tokyo Institute of Technology, 4259 Nagatsuta, Midori-ku, Yokohama 226-8502, Japan

GLAUSER T., IBM Almaden research Center, San Jose, California 95120 (USA)

GUÉRIN PHILIPPE, LRP, UMR 7581, CNRS, Université Paris 12 - Val de Marne, 2/8 rue Henry Dunant 94320 Thias, France

HATAKEYAMA HYOE, Department of Applied Physics and Chemistry, Fukui University of Technology, Fukui, Japan

HATAKEYAMA TATSUKO, Department of Textile Science, Otsuma Women's University, Tokyo, Japan

HEDRICK JAMES L., IBM Almaden research Center, San Jose, California 95120 (USA)

HIGO SACHIKO, Biodegradable Plastics Society, Grande bldg., 26-9, Hatchobori 2 Chome, Chuo-ku, Tokyo 104-0032 Japan

HOSHINO AKIRA, Biodegradable Plastics Society, Grande bldg., 26-9, Hatchobori 2 Chome, Chuo-ku, Tokyo 104-0032 Japan

HUANG SAMUEL J., Institute of Materials Science, University of Connecticut, Storrs, Connecticut 06269-3136, USA

IANNACE SALVATORE, Institute of Composite Materials Technology (ITMC-CNR) & Department of Materials and Production Engineering, University of Naples "Federico II", Piazzale Tecchio 80, 80125 Naples, Italy

ILIEVA IVANOVA VASSILKA, Department of Chemistry & Industrial Chemistry, University of Pisa, via Risorgimento 35, 56126 Pisa

IM SEOUNG SOON, Department of Textile & Polymer Engineering, Hanyang University, 17 Haengdang-Dong, Seongdong-Gu, Seoul, 133-791, Korea

ITO MICHIO, Biodegradable Plastics Society, Grande bldg., 26-9, Hatchobori 2 Chome, Chuo-ku, Tokyo 104-0032 Japan

KIM SOO HYUN, Biomaterials Research Center, Korea Institute of Science and Technology, P.O. Box 131, Gheongryang, Seoul, 130-650 Korea

KIM YOUNG HA, Biomaterials Research Center, Korea Institute of Science and Technology, P.O. Box 131, Gheongryang, Seoul, 130-650 Korea

KOBERSTEIN JEFFREY T., Department of Chemical Engineering and Applied Chemistry, Columbia Univeristy, 500 West 120 Street, New York, USA
KUWABARA KAZUHIRO, Polymer Chemistry Laboratory, RIKEN Institute, 2-1 Hirosawa, Wako-shi, Saitama 351-0198, Japan
LANGLOIS VALÉRIE, LRP, UMR 7581, CNRS, Université Paris 12 - Val de Marne, 2/8 rue Henry Dunant 94320 Thias, France
LEE SOO HONG, Biomaterials Research Center, Korea Institute of Science and Technology, P.O. Box 131, Gheongryang, Seoul, 130-650 Korea
LI YAN, Kaufert Laboratory, University of Minnesota, 2004 Folwell Avenue, St. Paul, MN, USA
LIPS PRISCILLA A.M., Department of Chemical Technology, University of Twente, P.O. Box 217, 7500 AE Enschede, The Netherlands
MANO JOÃO F., Department of Polymer Engineering, University of Minho, Campus de Azurém, 4800-058 - Guimarães, Portugal, and 3B's Research Group, University of Minho, Campus de Gualtar, 4710-057, Braga, Portugal
MARÉCHAL FREDDY, APME Technical & Environmental Centre
MENSITIERI GIUSEPPE, Institute of Composite Materials Technology (ITMC-CNR) & Department of Materials and Production Engineering, University of Naples "Federico II", Piazzale Tecchio 80, 80125 Naples, Italy
MIERTUS STANISLAV, International Centre for Science and High Technology of the United Nations Industrial Development Organization (ICS-UNIDO), AREA Science Park, 34012 Trieste, Italy
MIZUTANI AKIKO, Biodegradable Plastics Society, Grande bldg., 26-9, Hatchobori 2 Chome, Chuo-ku, Tokyo 104-0032 Japan
MÖCK ANDREAS, IBM Almaden research Center, San Jose, California 95120 (USA)
MOMOCHI MASANORI, Biodegradable Plastics Society, Grande bldg., 26-9, Hatchobori 2 Chome, Chuo-ku, Tokyo 104-0032 Japan
NICOLAIS LUIGI, Institute of Composite Materials Technology (ITMC-CNR) & Department of Materials and Production Engineering, University of Naples "Federico II", Piazzale Tecchio 80, 80125 Naples, Italy
NYCE GREGORY W., IBM Almaden research Center, San Jose, California 95120 (USA)
PACK JI WON, Biomaterials Research Center, Korea Institute of Science and Technology, P.O. Box 131, Gheongryang, Seoul, 130-650 Korea
PARK JUN WUK, Department of Textile & Polymer Engineering, Hanyang University, 17 Haengdang-Dong, Seongdong-Gu, Seoul, 133-791, Korea
PATEL MARTIN, Department of Science, Technology and Society, Utrecht University, Padualaan 14, 3584 CH Utrech, The Netherlands
PAUL MARIE-AMÉLIE, Laboratory of Polymeric and Composite Materials, University of Mons-Hainaut, 20 Place du Parc, 7000 Mons, Belgium
POLLET ERIC, Laboratory of Polymeric and Composite Materials, University of Mons-Hainaut, 20 Place du Parc, 7000 Mons, Belgium
REIS RUI L., Department of Polymer Engineering, University of Minho, Campus de Azurém, 4800-058 - Guimarães, Portugal, and 3B's Research Group, University of Minho, Campus de Gualtar, 4710-057, Braga, Portugal
REN XIN, International Centre for Science and High Technology of the United Nations Industrial Development Organization (ICS-UNIDO), AREA Science Park, 34012 Trieste, Italy
RENARD ESTELLE, LRP, UMR 7581, CNRS, Université Paris 12 - Val de Marne, 2/8 rue Henry Dunant 94320 Thias, France
RESKE JÖRAN, INTERSEROH GmbH, Stollwerckstr. 94, D-51149 Köln, Germany

SAN ROMÁN JÚLIO, Institute of Science and Technology of Polymers, CSIC, c/Juan de la Cierva, 3, 28006, Madrid, Spain
SARKANEN SIMO, Kaufert Laboratory, University of Minnesota, 2004 Folwell Avenue, St. Paul, MN, USA
SAWADA HIDEO, Biodegradable Plastics Society, Grande bldg., 26-9, Hatchobori 2 Chome, Chuo-ku, Tokyo 104-0032 Japan
SCOTT GERALD, Aston University, Birmingham, UK
SIGNORI FRANCESCA, Department of Chemistry & Industrial Chemistry, University of Pisa, via Risorgimento 35, 56126 Pisa, Italy
SMITH DAWN A., Institute of Materials Science, University of Connecticut, Storrs, Connecticut 06269-3136, USA
SOLARO ROBERTO, Department of Chemistry & Industrial Chemistry, University of Pisa, via Risorgimento 35, 56126 Pisa, Italy
SWIFT GRAHAM, GS Polymer Consultants, 1078 Eastchurch, Chapel Hill, North Carolina 27517, USA
TAKAKUWA KYOHEI, Biodegradable Plastics Society, Grande bldg., 26-9, Hatchobori 2 Chome, Chuo-ku, Tokyo 104-0032 Japan
TSUJI MASAO, Biodegradable Plastics Society, Grande bldg., 26-9, Hatchobori 2 Chome, Chuo-ku, Tokyo 104-0032 Japan
UEMATSU SHOGO, School of Pharmaceutical Sciences, University of Shizuoka, 52-1, Yada, Shizuoka-shi 422-8526, Japan

Contents

PART 1

STANDARDS AND POLICIES

Science and Standards

GERALD SCOTT
Aston University, Birmingham, UK

1. WHY ARE STANDARDS NECESSARY?

In the 1980s there was a great deal of public interest in claims by industrial manufacturers of plastics for "environmentally friendly" polyethylene containing starch, which was claimed to be biodegradable. This led to the setting up of a task force by the Attorneys General of the USA to investigate "green marketing" and the publication in 1990 of the Green Report[1]. This contained very severe strictures against such "deceptive" claims without associated scientific evidence and later led to legal proceedings against companies who continued to make false or unsupportable claims.

The 'Green Report' introduced the concept of a "level playing field" through life cycle assessment for the degradable plastics industries. The following are excerpts.

"Environmental claims must be uniform and supported by competent and reliable scientific evidence"

"Product life cycle assessment involves consideration of environmental effects at every stage in the product's life cycle, including the natural resources and energy consumed and the waste created in the manufacture, distribution and disposal of a product and its packaging...Such assessments will only provide useful comparative information about how to reduce environmental problems associated with products if they are conducted using uniform and consistent assumptions"

2. LIFE CYCLE ASSESSMENT OF BIODEGRADABLE POLYMERS

It is sometimes assumed in that polymers from renewable resources are by definition 'environmentally friendly', or in modern parlance, 'sustainable'. One definition of sustainable suggests that the development of new products for the benefit of society should not have an unacceptable effect on resource depletion and environmental pollution. However, 'unacceptable' is a relative term and invites comparison of one material with another by life-cycle assessment (LCA)[2]. Companies engaged in the development of degradable polymers from renewable resources have initiated life-cycle assessment comparisons of their products with the commodity synthetic polymers, notably polyethylene. It has not so far been shown unambiguously that bio-based polymers are more environmentally sustainable than the present range of commodity polymers[3,4]. This results from the same reason that led to concern in the 1980s; namely lack of consistency and uniformity of the assumptions made. In some cases they actually contradict one another.

The most important measures of sustainable biodegradable plastics are energy utilised and environmental pollution generated during manufacture, since the production of waste involves further energy input to disperse potentially toxic waste products to an acceptable level in the environment[2]. The ability to utilise the waste in one of the main recycling procedures (i.e. mechanical recycling, energy recovery, and composting or biological recycling) is crucially important and land utilisation during manufacture and disposal should be taken into account[2].

2.1 Energy Utilisation During the Manufacture of Degradable Plastics

It is commonly assumed in life cycle assessments that the production of bio-based plastics uses no fossil resources because they are synthesised and returned to the carbon cycle by biological processes. However, all chemical (including biochemical) manufacturing operations involve energy input. The use of fossil carbon begins when the ground is tilled before planting the seed. It continues in the use of fertilisers that are synthesised in chemical plants and oil is used when the crop is harvested and transported to the factory for conversion to plastics. Organic wastes are major products of biopolymer manufacture and again these have to be disposed of with fossil energy input. Finally, the manufacturing processes used in the production of modified biopolymers to give useful products always require the use of fossil fuels.

LCAs of degradable polymers published by bioplastic producers[5-10] generally assume that there is no fossil carbon input in the manufacture of bioplastics, whereas that of polyolefins is positive. Further, it is assumed that fossil-based plastics can only be disposed of in landfill or by incineration without energy generation. There appears to be no recognition that PE can be "recovered" from the waste stream by pyrolysis to give monomers and fuels[2], or as compost for the benefit of agriculture. In practice, since the calorific value of PE (43 GJ/t) is almost identical to that of the oil from which it was manufactured, the carbon content of the plastic is ecologically neutral since it replaces fossil fuels in waste-to-energy incineration. The total non-recoverable energy used in the manufacture of PE is 21.6 GJ/t, compared with the estimates given for the manufacture of starch-based products, which vary between 25.4 GJ/t and 52.5 GJ/t, depending on the co-agent in the formulations. Blends with fossil-based polymers or additives are of course energy positive during manufacture[11]. Furthermore, the energy produced is considerably less than that from PE if biopolymers are used as a source of fuel

2.2 Land Resource Utilisation

None of the LCA studies have so far considered land utilisation in the ecological balance. At present bio-based polymers such as PHA, PLA and starch are produced from food crops. This does not present a problem in the short-term if the polymers are to be used in specialised 'niche' applications on the basis of a temporary surplus of food crops but it cannot be used as the basis of long-term sustainable development of bio-based plastics if the intention is to completely replace polyolefins in packaging. For example the anticipated scale of production of PLA during the present decade (not more than 500,000 tonnes/annum worldwide) is less than 1% of the worldwide production of polyolefins and, if the production of food-based biodegradable polymers was to increase toward the level of the fossil-based polymers, there would be serious competition between polymer feedstocks and food production, particularly in developing countries[12]. On the other hand, if bio-based feedstocks could be based on biological wastes or on crops grown on marginal land, then the situation could be quite different. This objective and the efficient use of biomass as fuel represent a major challenge to the bioengineering industries.

3. DEGRADATION OF CARBON-CHAIN POLYMERS

Carbon-chain polymers are normally associated with synthetic polymers made by the polymerisation of vinyl compounds but many naturally occurring polymers contain uninterrupted sequences of carbon atoms. The most studied and best understood of these is natural *cis*-polyisoprene (NR), synthesised by the rubber tree *Hevea Braziliensis,* which is chemically identical to synthetic *cis*-IR (*cis*-polyisoprene rubber).

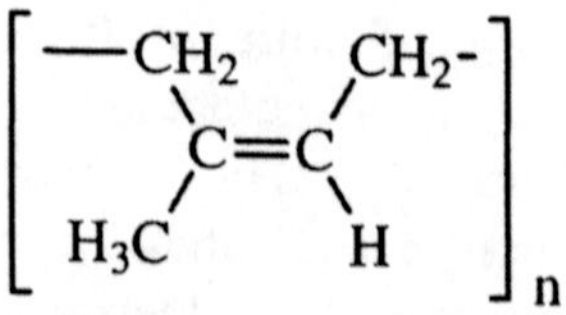

cis-polyisoprene, NR

Natural rubber was one of the earliest industrial polymers to be developed commercially and it was recognised even before it reached the industrialised countries that it very rapidly lost its initial useful properties in the environment. Furthermore, rubber latex products were rapidly attacked by microorganisms, leading to more general loss of mechanical properties and to eventual bioassimilation in the soil environment[13-17].

The synthetic polyolefins, although more environmentally stable than the polydiene rubbers, are nevertheless much less resistant to the environment than might have been expected on the basis of their formal structures[18]. There are several reasons for this, which are now well documented[18-20]. The first is the presence of a small amount of unsaturation in the polymers as they are manufactured. It will be seen below that these markedly decrease the stability of saturated polymers out of all proportion to their concentration. The second reason is that the processing operations (extrusion, injection moulding, etc.) are very damaging to the polymer due to the high shearing forces induced in the polymer molecules. In the highly viscous state of the molten polymer, the chemical bonds of the polymer chains are broken to give free radicals, which immediately react with the oxygen that is always present in commercial operations to give peroxyl radicals and hydroperoxides. The latter play a key role in subsequent polymer degradation, since they are initiators for further peroxidation[21].

$$\text{P-P'} \xrightarrow{\text{Shear}} \text{P}^{\bullet} + {}^{\bullet}\text{P'} \xrightarrow{O_2} \text{POO}^{\bullet} + \text{P'OO}^{\bullet} \xrightarrow{\text{2PH}} \text{POOH} + \text{P'OOH} + 2\text{P}^{\bullet} \qquad (1)$$

P, P' are long-chain alkyl groups.

The third reason is the inadvertent incorporation of transition metal ions during manufacture. or during the conversion of polymers to industrial materials. This chemistry can be turned to an advantage in the induced acceleration of polymers to lower molar mass materials, leading to their bioassimilation in the environment.

4. HYDROPEROXIDES AND THE PEROXIDATION CHAIN MECHANISM

Hydroperoxides and their breakdown products are of fundamental importance to polymer biodegradation. Not only are they the cause of the free radical chain reaction by rapid dissociation caused by thermal (Δ) and UV (hν) energy to give initiating radicals (reaction 2)[21,22], but PO is also the source of the ultimate low molar mass degradation products that are readily bioassimilated by microorganisms.

$$\text{POOH} \xrightarrow{\Delta,\, h\nu} \text{PO}^{\bullet} + {}^{\bullet}\text{OH} \xrightarrow{2\text{PH}} 2\text{P}^{\bullet} + \text{POH} + \text{H}_2\text{O} \qquad (2)$$

The following chain reaction is the essential peroxidation sequence, which continues as long as oxygen is present in the system.

$$\text{P}^{\bullet} + \text{O}_2 \rightarrow \text{POO}^{\bullet} \qquad (3)$$

$$\text{POO}^{\bullet} + \text{PH} \rightarrow \text{POOH} + \text{P}^{\bullet} \qquad (4)$$

Reaction (4) is rate controlling. The kinetics of the peroxidation chain reaction has been discussed in many reviews and standard texts and the reader is directed to these for further information[15,19,20].

The oxidative stabilities of hydrocarbon polymers in the absence of added initiators differs by two orders of magnitude, from the relatively stable unbranched polyethylenes through the branched polypropylene to the unsaturated polydienes[18];

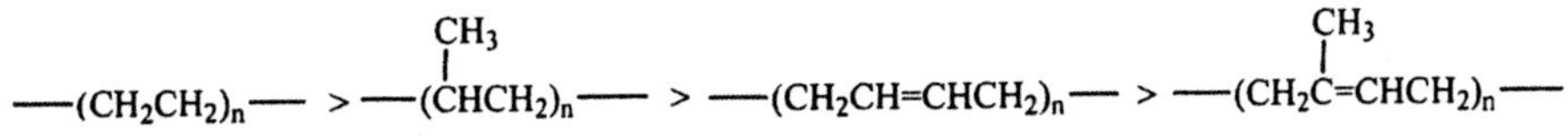

PE PP *cis*-PB *cis*-PI

This order reflects the increasing ease of abstraction of the weakest P-H bonds (namely the methylenic hydrogens) in reaction (4)[19]. Consequently blends of saturated polymers with rubbers or co-polymers of saturated and unsaturated polymers peroxidise more rapidly than the saturated polymers themselves and this kind of modification has sometimes been used to increase the rate of bioassimilation of polymers through peroxidation.

In the absence of antioxidants and stabilisers, the concentration of hydroperoxides increases rapidly in an 'autooxidising' polymer by reactions (2)-(4) until the rate of decomposition of hydroperoxides is equal to the rate of their formation. The rate at which this state is reached in the environment is normally determined by the influence of external factors that promote the decomposition of hydroperoxides.

4.1 Promoters of Hydroperoxide Decomposition

Hydroperoxides are relatively stable at ambient temperatures in the absence of promoters of decomposition, the most important of which are UV light and transition metal ions. Activation of hydroperoxide decomposition by UV light, reaction[2], is the main cause of polyolefin peroxidation in sunlight. Consequently peroxidation and physical degradation of hydrocarbon polymers is very much faster in sunlight than in the dark at the same temperature, although the relative rates depend very much on the presence or absence of photosensitisers. In the modern synthesis of polymers from vinyl monomers using organometallic catalysts, the metal ions are not normally removed and this can also give rise to oxidative instability during service[23]. Transition metal catalysts react with polymer hydroperoxides as follows;

$$M^{n+} + POOH \rightarrow M^{(n+1)+} + PO^{\bullet} + OH^{-} \qquad (5)$$

$$M^{(n+1)+} + POOH \rightarrow M^{n+} + POO^{\bullet} + H^{+} \qquad (6)$$

$$POO^{\bullet} + PH \longrightarrow POOH + P^{\bullet} \xrightarrow{O_2/PH} POOH \qquad (7)$$

$$PO^{\bullet} + PH \longrightarrow POH + P^{\bullet} \xrightarrow{O_2/PH} POOH \qquad (8)$$

The reaction sequence (5)-(8) leads to the rapid accumulation of hydroperoxides and the attainment of the stationary peroxide concentration. Ionic Cu, Mn, Fe, Cr, and Co are the most effective promoters of peroxidation whereas Ni, Ce, V, Ti, and Zn are less effective[24]. Transition metal ions are also initiators of photooxidation.

4.2 Products of Hydroperoxide Decomposition

The most important physical effect of peroxidation in hydrocarbon polymers is to reduce the molar mass of the polymer leading to deterioration in mechanical properties. Peroxidation, whether abiotically or biologically initiated, leads to the formation of hydrophilic chemical species such as carboxylic acids and alcohols particularly in the surface layers of the polymer[22]. This enables microorganisms to colonise the polymer surface and utilise the low molar mass oxygenated species as nutrients in the absence of other sources of carbon[25].

Abiotic peroxidation of the polyolefins gives rise to vicinal hydroperoxides and this process is particularly favoured in the α-olefins, such as polypropylene due to the susceptibility of the *tertiary* carbon atom to hydrogen abstraction via a hydrogen-bonded intermediate. A major proportion of the peroxidic products are hydrogen-bonded vicinal hydroperoxides that break down to small biodegradable molecules such as carboxylic acids, alcohols and ketones[26] as well as longer chain oxygen-modified breakdown products, which oxo-biodegrade more slowly. The decomposition of the vicinal hydroperoxides is also facilitated by internal hydrogen bonding and the low molar mass products of this self-induced degradation are small biodegradable molecules such as acetic and formic acids.

In polyethylene the proportion of vicinal hydroperoxides is lower than in PP and random chain scission is initially the dominant process. However, a variety of low molar mass oxidation products are formed via vicinal hydroperoxides (Scheme 1)[26]. The alkoxyl radicals formed by decomposition of the hydroperoxides contain weak carbon-carbon bonds in the α positions to the hydroperoxide groups which lead to the formation of low molecular weight aldehydes and alcohols that further oxidise rapidly to abiotically stable carboxylic acids. These are biodegradable species, similar to those formed by hydrolysis of aliphatic polyesters and they provide an environment for rapid biofilm formation (Section 5.5).

5. MICROBIAL DEGRADATION OF CARBON-CHAIN POLMERS

5.1 *Cis*-Polyisoprene

Abiotic peroxidation of cis-PI occurs at ambient temperatures so long as oxygen is present in the system. However, oxo-biodegradation also proceeds in parallel in microbially active environments. It has been demonstrated experimentally[27,28] that pure strains of bacteria (in particular *actinomycetes*) and fungi cause up to 55% loss of mass of rubber sheets in 70 days. The actinomycete, *Nocardia* (sp. St835A) was found by Tsuchii and co-workers[29] to be particularly effective in degrading NR rubber gloves in the absence of any other source of carbon. A mass loss of 75% was achieved in two weeks and the same strain in laboratory fermenters led to complete degradation of NR in 45 days[30].

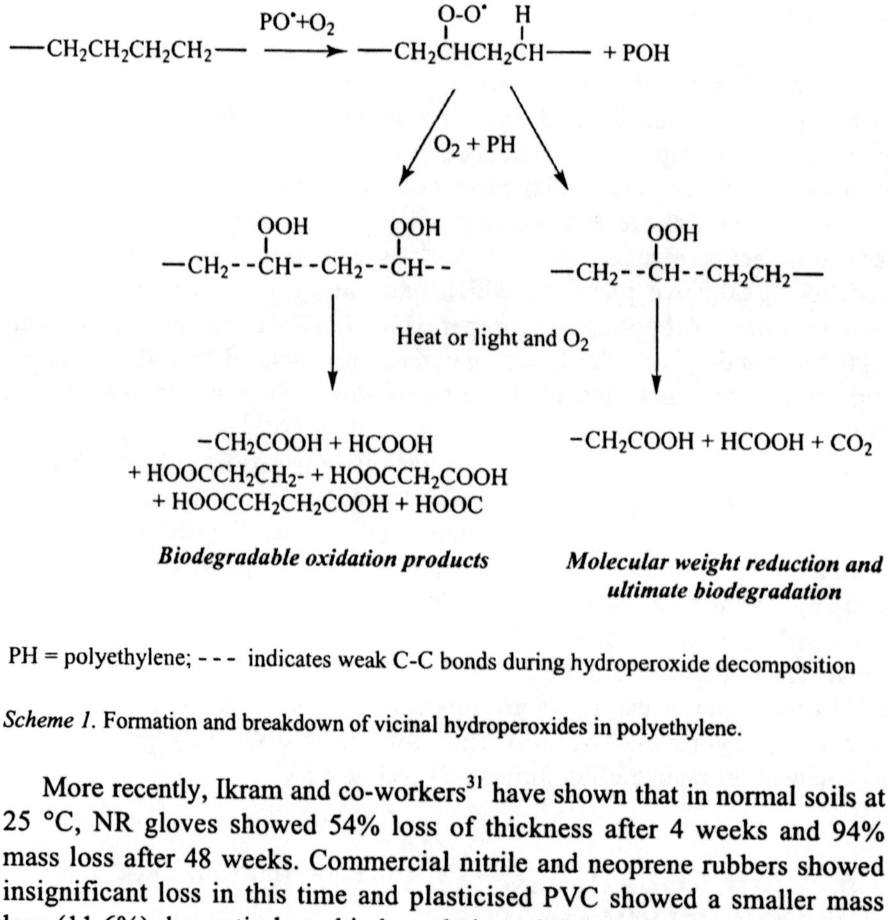

Scheme 1. Formation and breakdown of vicinal hydroperoxides in polyethylene.

More recently, Ikram and co-workers[31] have shown that in normal soils at 25 °C, NR gloves showed 54% loss of thickness after 4 weeks and 94% mass loss after 48 weeks. Commercial nitrile and neoprene rubbers showed insignificant loss in this time and plasticised PVC showed a smaller mass loss (11.6%) due entirely to biodegradation of the plasticiser and not to the biodegradation of the polymer itself. Bacterial populations on the NR gloves (12317/mg) were higher than for fungi (441/mg), which were in turn significantly higher than actinomycetes (297/mg). Nevertheless, Heisey and Papadatos[32] isolated 10 actinomycetes (seven strains of *Streptomycetes*, two strains of *Amycolatopsis* and one strain of *Nocardia*) from soil that reduced the mass of NR gloves from 10-18% in 6 weeks.

Ikram has subsequently shown[33] that mass loss of NR is highly dependent upon the nutrients in the soil; particularly nitrogen and phosphorus. After 24 weeks NR in the high N (100mg/l), P (150mg/l) system had lost 61.5% of its mass whereas in the low N (10mg/l), P (15mg/l) system, only 23.6% mass was lost. Control (unfertilised) soil produced least mass loss (17.3%). Microbial growth rates measured on the rubber pieces were in decreasing order as expected (Table 1).

Table 1. Effect of added soil nutrients on the mass loss of rubber and plastic films (%) after 40 weeks in soil[33]

Polymer	Nutrient treatment*		
	High	Low	Control
NR	-82.4	-38.5	-29.7
Neoprene	+0.3	-13.0	-1.1
Nitrile	-4.3	-3.2	-3.5
Plasticised PVC	-26.1	-13.4	-11.1

* Nutrients added: High 100 mg/l N and 150 mg/l P; Low 10 mg/l N. 15 mg/l P; Control nil

Steinbüchel and co-workers[34,35], using rubbers as the sole source of carbon, found that NR and IR (synthetic polyisoprene rubber) biodegrade at a similar rate in the presence of *Pseudomonas aeruginosa.* NR gloves were 26% mineralised in 6 week compared with 21% for IR gloves. This slight difference may well have been due to the difference in the antioxidants used in the formulation, although these were not identified. It is clear, however that, contrary to the views of some environmentalists[36], there is no intrinsic difference between natural and synthetic polymers.

It has been pointed out that some actinomyctes can utilise CO_2 as a source of carbon[37]. It is therefore necessary to equate microbial growth and associated formation of protein to loss of weight of the substrate. Table 2, taken from the work of Heisey[32] shows that there is indeed a broad correlation between mass loss and protein formation and Delort and co-workers[38] have shown that loss of carboxylic acids formed during abiotic peroxidation of PE correlates with the formation of protein and polysaccharides, almost certainly associated with the cross-linked bacterial cell wall structure.

Nocardia and *P. aeruginosa*[34,35] were shown to break the *cis*-PI chain by an oxidative mechanism since aldehyde groups were found to accumulate during microbial degradation. This is always the first product formed during the abiotic peroxidation of *cis*-PI and the evidence suggests that the bacteria initiate a radical-chain peroxidation. This will be discussed further in the context of polyolefin biodegradation.

Chlorinated polymers are much more resistant to abiotic peroxidation than pure hydrocarbon polymers[18] and nitrile rubbers, although susceptible to peroxidation are normally highly stabilised by extraction-resistant antioxidants[20,39].

Table 2. Mass changes of NR strips and protein concentration produced by rubber metabolising microorganisms[32]

Isolate	Mass change of rubber strips (%)	Protein concentration* (mg/g of rubber)
Control	1 ± 1	1 ± 0
1	1 ± 0	2 ± 0
2	0 ± 2	2 ± 0
3	-8 ± 1	26 ± 11
4	-9 ± 1	35 ± 3
5	-11 ± 2	27 ± 9
6	-11 ± 0	29 ± 3
7	-12 ± 3	27 ± 4
8	-12 ± 1	28 ± 7
9	-13 ± 1	32 ± 2
10	-14 ± 2	40 ± 3
11	-16 ± 4	39 ± 4
12	-16 ±4	46 ± 9
13	-16 ± 2	45 ± 3
14	18 ± 2	46 ± 2

* Total in the culture broth and on the rubber strips.

Fully formulated tyre rubbers, in contrast to latex rubbers used in domestic products, are highly resistant to peroxidadation and hence biodegradation. Tyres survive almost unchanged in the outdoor environment when discarded and have to be disposed of by some other means. Antioxidants added during processing are thus the controlling factor in the biodegradation of technological rubber products.

5.2 Lignin and Lignocellulose

Lignin is another polymer that, like *cis*-polyisoprene, bridges the gap between natural and synthetic polymers. Lignin is a cross-linked polymer containing benzene rings (see below). It is formed in chemical association with cellulose (lignocellulose) and constitutes the tough cell wall structure of plants[40]. The aromatic structures contain alkoxy and hydrocarbon substituents that link the basic unit below into a macromolecular structure through carbon-carbon and carbon-oxygen bonds. Both the chemical and physical properties of lignin resemble those of the synthetic phenol-formaldehyde (PF) resins and it acts as an adhesive for cellulose fibre in a manner that anticipates the synthetic fibreglass composites in modern polymer technology[41]. Like the PF resins, lignocellulose is strong and tough and provides physical protection to the growing plant. In addition it provides chemical and biological protection to wood, straw, husks, etc. A further similarity to PF resins is the relative resistance of lignin to peroxidation due to the presence in the polymer of many antioxidant-active phenolic groups which act as protective agents against abiotic peroxidation and biological attack by peroxidase enzymes (see below).

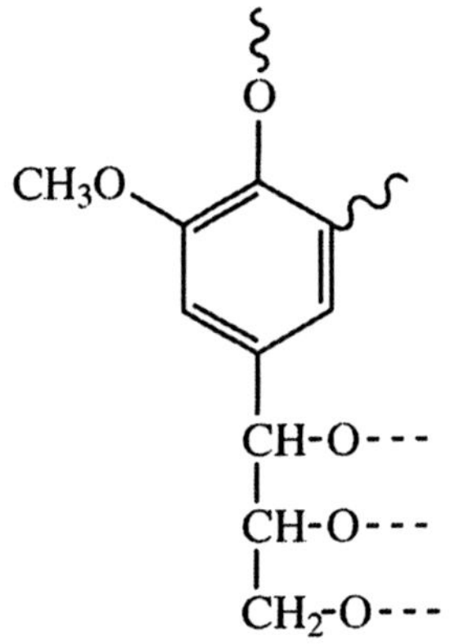

Lignin monomeric unit

∿∿ indicates potential sites through which dehydropolymerization and cross-linking may occur.

---- indicates sites through which attachment to cellulose may occur.

Cellulose is almost always found in natural products in combination with lignin (25-30% in most woods). The crystalline structure of the former provides reinforcement and tensile strength for an otherwise rather weak material[41]. Relatively small amounts of lignin inhibit attack by the hydrolytic microorganisms that degrade pure cellulose, for example in compost[42]. Lignocellulose, due to its physical (hydrophobic) and chemical inertness, does not readily degrade either abiotically or biotically and when it does occur, the lignin tends to accumulate[43]. However, lignin does biodegrade slowly under composting conditions. Lignin in grass, hay and straw were found to biodegrade to the extent of 17-53% in 100 days[44]. In laboratory incubation studies, thermophilic composting of grass straw showed 45% degradation in 45 days[45] but the process tends to slow down at more extended times, probably due to the accumulation of lignin. Janssen[46] has estimated by carbon labelling that the time for complete conversion of straw to carbon dioxide is about ten years and as will be seen in Section 3.3 a considerable proportion of the lignin is converted to humus (Section 5.4).

There has been intense interest in recent years in the selective removal of lignin from wood pulp during papermaking. Since lignin cannot biodegrade by a hydrolytic process, the lignin component of lignocellulose biodegrades slowly by oxidative attack due to extra-cellular peroxidases formed from actinomycetes such as *Streptomyces viridosporus*[47-49]. A number of peroxidases have been isolated that remove lignin from lignocellulose without affecting cellulose itself. Manganese peroxidase (MnP) in particular has been implicated as an important enzyme formed by white rot fungi during the delignification of Kraft pulps[50-52].

Unlike the peroxidation of the hydrocarbon polymers, the oxidation of lignin occurs by a stoichiometric process and not a chain reaction. Because phenols are antioxidants, the phenoxyl radicals formed are too stable to participate in a peroxidation chain reaction and the aromatic system is converted to quinoid compounds and ultimately humus. Both abiotic transition metal ion catalysed peroxidation and biological oxidation are involved in the conversion of lignin to humus.

5.3 Tannins

Tannins are derived from gallic acid and its derivative by dehydropolymerisation (see below). The role of redox reactions involving iron, hydrogen peroxide and hydroperoxides is well understood[53]. However, extra-cellular peroxidases (ferriprotoporphorins) are the biological equivalents, which act many times faster than the inorganic system[54]. Both carbon-carbon and carbon-oxygen linkages are formed in the dehydropolymerisation of simple phenols and these are also present in lignin and the tannins.

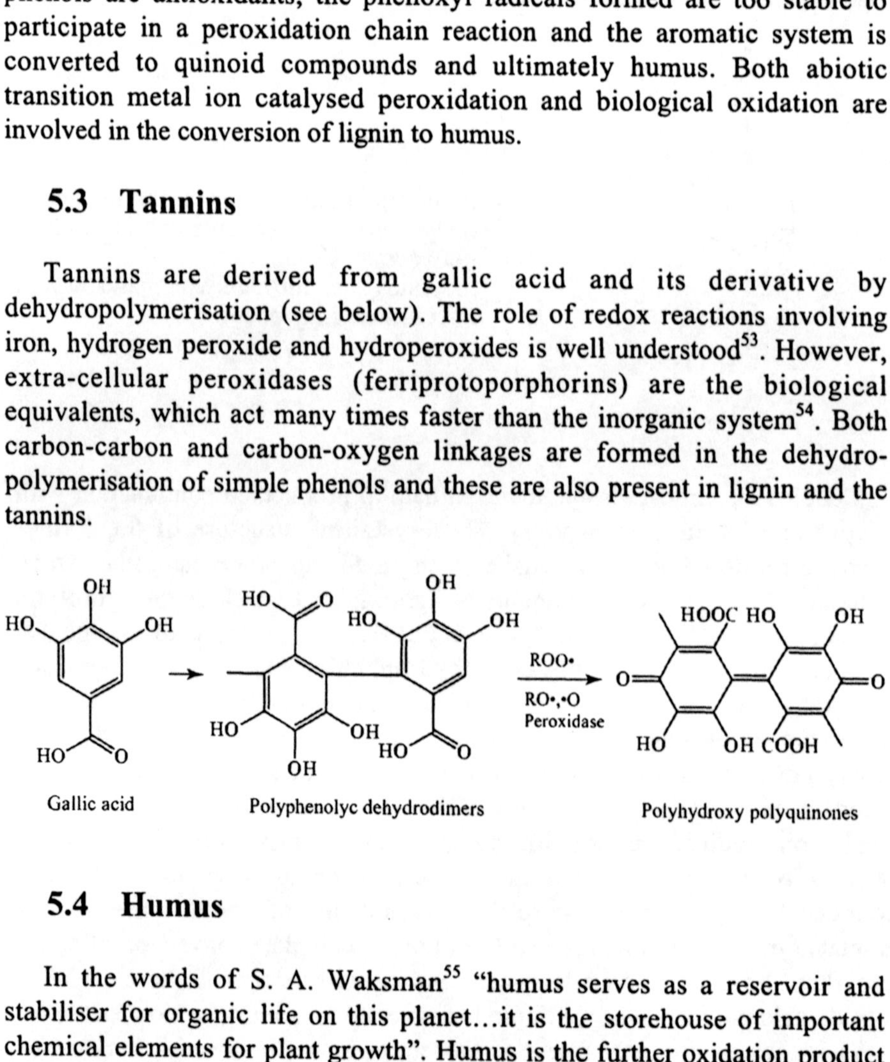

5.4 Humus

In the words of S. A. Waksman[55] "humus serves as a reservoir and stabiliser for organic life on this planet...it is the storehouse of important chemical elements for plant growth". Humus is the further oxidation product of lignocellulose and its brown-black colour is due to the predominance of quinoid polymers that can be partially extracted by organic solvents. It is found in fertile soils to the extent of 1-3%.

It was seen in the previous sections that humus is formed from lignin and the tannins by further oxidation by a variety of peroxidase (polyphenoloxidase) enzymes and during the oxidation process the concentration of carboxyl groups is considerably reduced and the carbon-oxygen ratio increases. However, although lignin and tannic acids are the major source of humus, they are almost certainly not the only source. The peroxidases, which can generate hydroxyl radicals[54], are also able to hydroxylate aromatic rings present in proteins to phenols and hence to humus. Humus also contains substantial amount of nitrogen, chemically attached to the polyphenol-quinone molecules and this is slowly released as fertiliser.

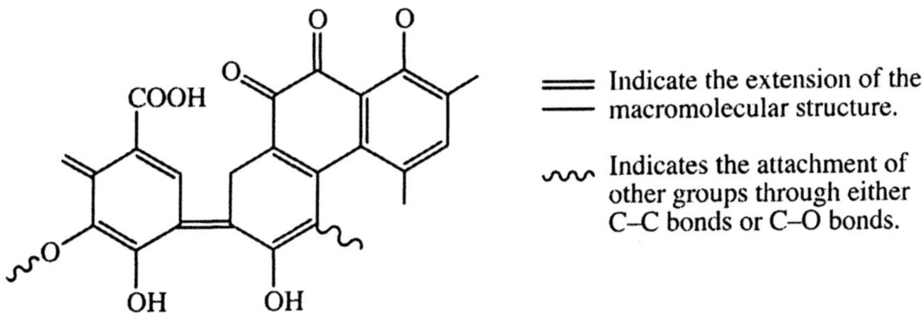

Because of the presence of both carbonyl groups and hydroxyl groups, humus is able to chelate metal ions and is a source of trace elements for plant growth as well as available carbon and nitrogen compounds formed in the breakdown of proteins that act as nutrients for the growing plants. During the digestion of humus there is a slow but steady liberation of carbon dioxide as a consequence of the breakdown of aliphatic carbon sequences but it is important to note that aromatic molecules do not convert rapidly to carbon dioxide. Guillet has noted[56] that when plants are grown on photooxidised polystyrene, approximately 50% of the carbon is absorbed directly by the plant without being liberated to the atmosphere as CO_2. The rate of ultimate mineralisation of both polymers is very similar.

5.5 Polyolefins

Although commercial polyolefins are more oxidatively stable than the polydiene rubbers, they behave very similarly in the environment and like the commercial rubber products, their resistance is due to the antioxidant and photostabiliser packages that have been developed over the past 50 years. As discussed in Section 4.2, polyolefins are peroxidised in the outdoor environment to biodegradable products similar to those formed from the *cis*-PI. It was shown many years ago[57] that S-G degradable polyethylene (see below) may be used after photooxidation as a carbon source by thermophilic fungi at 40 and 50 °C. It was also observed that fragmented polyethylene in conjunction with fertilisers actually had a beneficial effect on the growth of vegetables[57]. The potential for biologically recycling hydrocarbon polymers by accelerated peroxidation after use was even at that time advocated as a way of utilising waste polyolefin packaging materials. However, commercial composting of domestic waste was not then considered to be a 'recycling' process. Consequently biocycling of plastics was initially applied to the manufacture of protective agricultural films with controlled stability that subsequently biodegrade in contact with soils. Even today, environmentalists find it difficult to accept that polymers that are not synthesised biologically can ever biodegrade[36].

Lee and co-workers[58], in a study designed to investigate the use of starch-polyethylene blends (6% starch + mixed transition metal ion pro-degradants selected from Fe, Zn, Ni and Mn) for the disposal of garden waste in compost, examined the effect of a number of lignin-degrading microorganisms. The polymer films were first peroxidised either thermally at 70 °C in an air oven for up to 20 days or by long wave UV irradiation for up to 8 weeks before being exposed to three bacteria (*Streptomyces viridosporus, Streptomyces badius* and *Streptomyces setonii*) and one lignocellulose-degrading fungus (*Phanerochaete chrysosporium*). Using a starch-agar assay, it was found that *S. setonii and P. chrysosporium* were unable to utilise cornstarch but the former did biodegrade polyethylene. Mass-loss measurements were inconclusive due to the difficulty on removing microflora but GPC showed a reduction of polydispersity in the case of the samples incubated with *Streptomyces* spp., indicating the selective removal of lower molar mass species. The authors confirmed previous findings that prior peroxidation is an essential prerequisite to the biodegradation of polyethylene.

More detailed experiments, similar to those described above, have been used to simulate the effect of environmental exposure on the chemical, physical and biological changes occurring in commercial degradable polyethylenes during service and on exposure to the environment[25]. PE films, after peroxidation at composting temperatures or after being subjected to photooxidation, were incubated with bacteria and fungi that had been isolated from soils that were adapted to the presence of partially degraded polyethylene. The peroxidised samples were used as the sole source of carbon for a period of six months. Three different kinds of degradable polymer were used.

a) Photodegradable (photolytic) polymers made by copolymerisation of ethylene with carbon monoxide (E-CO). Union carbide technology[59].
b) Conventional polyethylene containing a transition metal prooxidant blended with starch (E-St). Griffin technology[60].
c) Photo-biodegradable (oxo-biodegradable) polymers based on conventional polyethylene containing a photosensitive transition metal ion complex antioxidant. (S-G). Scott-Gilead technology[61].

Although E/CO initially photodegraded to fragments more rapidly than S-G and E-St, photo-degradation of the transition metal ion catalysed systems continued to a much lower molar mass. After fragmentation, the peroxidised polymers were incubated in the absence of any other source of carbon with three microorganisms isolated from soil in the vicinity of discarded polyethylene. Two were bacteria (*Nocardia asteroides* and *Rhodococcus rhodochrous*) and one was a fungus (*Cladosporium cladosporioides*). It recently has been shown by Delort and co-workers[38] that biofilm formation is very rapid on the surface of peroxidised polyethylene (Fig. 1).

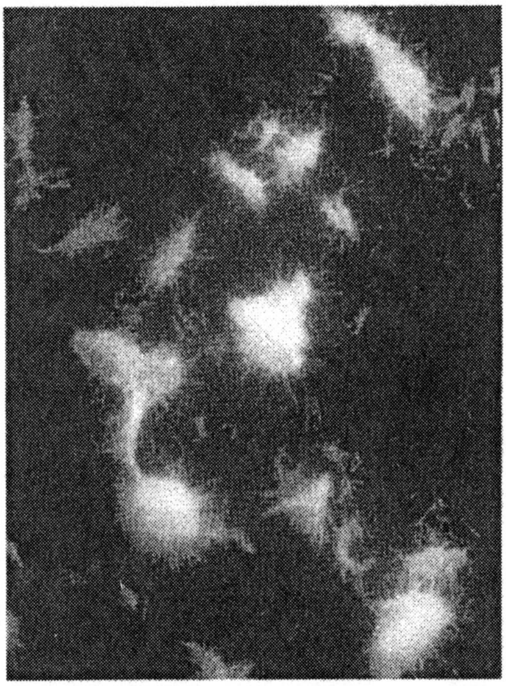

Figure 1. Colonisation of *Nocardia asteroides* (15 min) on peroxidised commercial degradable PE (EPI TDPA™) by epifluorescence spectroscopy (Reproduced with kind permission of A-M.Delort and co-workers, Clermont-Ferrand, University)[38].

Figure 2 shows the next stage in the attack of bacteria on the surface (in this case by *R. rhodochrous*) and shows clearly the separation of the developing bacteria from the craters produced in the polymers surface by "action-at-a-distance' of the micro-organisms noted by other workers.

Nocardia asteroides was particularly effective in bioassimilating all thermally oxidised polymers, whereas *Rhodococcus rhodochrous* bioassimilated photooxidised S-G but had little effect on photooxidised E-CO. The fungus, *Cladosporium* was least effective in reducing the mass of polyethylene samples but it did degrade the photooxidised starch-filled polymer.

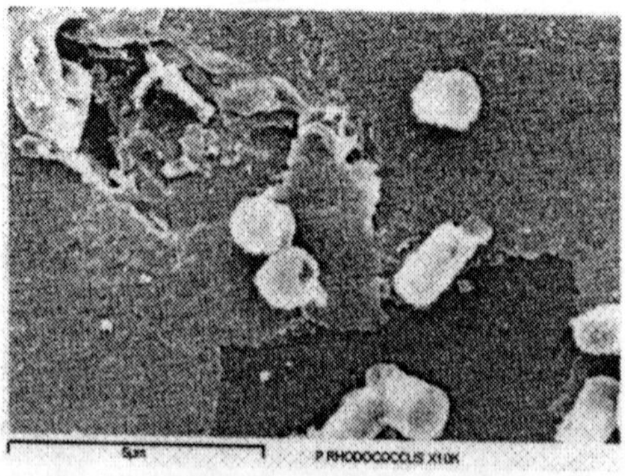

Figure 2. Bioerosion of the surface of peroxidised PE and growth of *Rhodococcus rhodochrous* by SEM after 1 month (Reproduced with kind permission of A-M.Delort and co-workers, Clermont-Ferrand, University)[38].

6. CHARACTERISATION OF BIODEGRADABLE POLYMERS

Biodegradable polymers can be generally classified as hydro-biodegradable (e.g. cellulose, starch and aliphatic polyesters) or oxo-biodegradable (e.g. lignin, lingo-cellulose, humus, rubbers and many related synthetic carbon-chain plastics). In both cases, low molar mass degradation products are formed that are microbially assimilated into the environment. Rapid mineralization is of considerable value when polymers are required to be removed rapidly in water courses or sewage systems, but it is an environmental disadvantage in most agricultural product such as mulching films, tunnels, animal feed bags and baler twines where initial resistance to the environment is essential.

In general the rate of hydrolysis of polymers is determined by physical factors such as hydrophilicity and morphology[62-64], whereas peroxidation is accelerated by transition metal prooxidants and retarded by antioxidants[3,4,18-22]. Oxo-biodegradable polymers such, as polyolefins, polydienes, their copolymers and blends, can be can be made biodegradable in compost by heat or on soil after photooxidation by sunlight. However, rapid degradation under ambient conditions would result in technologically unacceptable materials. Important applications of oxo-biodegradable plastics are discussed in the following sections.

7. APPLICATIONS OF DEGRADABLE PLASTICS IN AGRICULTURE AND HORTICULTURE

There are now many potential applications for biodegradable polymers. Biodegradable packaging has particularly attracted popular attention. In practice, degradable polymers have made much more progress in non-packaging applications where there is a cost benefit, notably in medicine and in agriculture. The former utilizes relatively high cost materials that do not appear in any quantity in the waste stream and standards have little relevance to them. The agricultural industry is much more price-conscious, in spite of the fact that degradable plastics offer a cost-benefit. The applications of biodegradable plastics in agriculture have been extensively reviewed in the technical literature[65-70].

7.1 Mulching Films and Tunnels

Plastics mulching films and tunnels have been used for many years. It has become increasingly evident that their use is crucial to the development of intensive and economically viable agricultural and horticultural systems. The main problem associated with the use of traditional plastics in agriculture is that, unless they are collected from the fields at the end of harvest, they remain on the land substantially intact after use and thus interfere with subsequent crops. Furthermore, in automated commercial horticulture and agriculture, undegraded plastic rapidly clogs the cropping machinery. Hand collection of agricultural plastics debris is expensive to the farmer and the problem of safely disposing of it has in the past proved to be a major problem. Local burning is generally not possible and the recovered plastics are so heavily contaminated by soil and biological waste that mechanical recycling does not give useful products. It is essential then, for economic and technological reasons, that plastic films should disintegrate and subsequently biodegrade in the soil and that no plastics residues or toxic products remain to interfere with the growth of subsequent crops. Degradable protective films have now been in use for twenty years and the experience gained is summarised as follows.

1. Degradable mulching films must have a mechanical performance similar to that of non-degradable films. In particular, toughness and tear resistance must not be compromised during automated lying of the films as the edges of the films are normally 'tucked under' the soil to avoid wind damage.
2. DP films, particularly when used in conjunction with irrigation, must be programmed to remain tough and strong until just prior to the commencement of harvesting in order to maximize crop yields,

minimize the use of water and fertilizers and avoid damage to automatic harvesting equipment. However, premature disintegration leads to loss of yield and is expensive to the farmer so that careful time control is essential.

3. The particle size of disintegrated DP left on the soil must be small enough after harvesting to be ploughed into the ground so that it has no detrimental effect on subsequent crop yields.
4. DP films require different outdoor lifetimes depending on the crop. Disintegration times may vary between several weeks to several months and during this time no disintegration or biodegradation must occur so as to ensure maximum crop yields.
5. There should be no accumulation of particulate residues in the soil.
6. Disintegration and bioassimilation of the polymer should not lead to the formation of soluble toxic products in the soil.
7. Above all, DP mulching films must be cost effective.

Requirements 4, 6 and 7 will be discussed later in the context of standards.

7.2 Agricultural and Horticultural Accessories

The following are the most important uses and potential uses of degradable plastics in farming and garden accessories.

7.2.1 Biodegradable Twines, Nets and Containers

Sisal, originally used to bind hay bales, biodegrades slowly in the environment after discard. For economic reasons, sisal has now been largely replaced by light stabilized polyolefins. Most commercial PP fibres are stabilized by iron oxides that give them an extended durability in the outdoor environment but in the USA these traditional products are being replaced by oxo-biodegradable PP with a service lifetime of about one year. This material loses tensile strength and breaks down into small fragments sharply after one year followed by rapid biodegradation. The use of degradable PP fibre is growing rapidly and like PE it is known to peroxidise to low molecular weight oxidation products (Scheme 1) that are rapidly bioassimilated by microorganisms, particularly in compost[71]. The control of this process is achieved by the use of similar prooxidants to those used in mulching films but more powerful light-sensitive photo-antioxidants are required[12,65,68].

7.2.2 Nets, Fittings and Containers

Degradable PP is currently being evaluated in nets for the protection of fruit crops (including vines) from birds, in degradable PP tapes and clips with programmed lifetimes and in seed trays, pots and other horticultural adjuncts that may be required to last for one or more seasons, generally made from recycled plastics recovered from mixed plastics wastes, reinforced with natural fibre wastes without the use of specific prodegradants. Their lifetime is unpredictable but is generally less important than in other agricultural plastics. PE, PP, PBD, their blends and copolymers can be made to be rapidly peroxidisable to innocuous biodegradable products but PVC is more problematic due to the unknown toxicity of the low molecular weight chlorine-containing degradation products and it is not recommended that this polymer be used in degradable systems.

7.2.3 Silage-Wrap, Hay-Wrap, Animal Feedbags and Fertilizer Sacks

Another developing use of DP is in silage-wrap films that are used to provide an environment for fermentation during storage. They are made almost exclusively from PE but a major problem with silage and hay-wrap films is that they escape from the fields and farmyards where they are used and are carried by the wind to riverbanks, hedges and trees where they are a serious environmental pollutant, particularly in areas of outstanding natural beauty. This costs the rural amenities authorities a considerable amount of money each year to collect the material and dispose of it. Pressure is now being brought on the manufacturers of these materials to make them photo-biodegradable by the same technology that is used in mulching films. Trials have demonstrated that degradable PE with a service lifetime of one year or if required 18 months can solve this problem[72].

It is difficult to draw a sharp distinction from the environmental point of view between animal feed or fertilizer packaging and silage or hay-wrap. The materials used and their environmental lifetimes and ultimate fates are the same. A great deal of agricultural packaging and hay-wrap ends up as durable litter in the countryside in spite of claims from manufacturers that the waste materials are recycled. In practice the detritus is much too contaminated to be recyclable to useful products.

7.2.4 Controlled Fertilizer Release

A recent use of oxo-biodegradable PE is to control the release of nitrogenous fertilizers by encapsulation. The objective is to reduce the eutrophication of rivers and lakes. The hydrocarbon oxidation products remaining after release of the fertilizer have been shown to be biodegradable[73].

8. APPLICATIONS OF DEGRADABLE PLASTICS IN WASTE MANAGEMENT

The increasing use of 'disposable' domestic products such as plastics plates, cutlery and drinking straws are placing an increasing burden on waste disposal authorities, since they are normally incinerated or sent to landfill. The search is on to replace non-biodegradable components by materials that degrade in landfill. There is a fine balance here between durability during use and subsequent environmental disposal. Polymers that hydro-biodegrade may break down during ambient storage and use. Conventional polyolefins on the other hand are much too stable but can be suitably formulated to be stable during use but fragmentable in landfill. Disposable hygiene products such as diapers and tampons are also increasing the cost of waste disposal and there is a need to make the products more biodegradable. Diapers based on degradable PE have recently been developed that have a satisfactory shelf-life but which degrade in landfill, thus increasing the potential capacity of the landfill[74].

Although disposal in landfill is set to decrease over the next ten years, the technology is not yet available for recovery of all the products that at present go into landfill and a major objective is to reduce the volume of active landfills. Landfill is normally composed of alternating layers of mixed domestic wastes, some of which are biodegradable and of soil to protect the surrounding environment from odours due to putrefaction of the waste. The use of soil leads to an undesirable reduction in landfill space and a more environmentally acceptable technique has been developed[74], which replaces the soil by a thin oxo-biodegradable PE membrane. In this process, degradable PE breaks down to fragments within weeks leading to the free circulation of microbes and nutrients within the enclosed landfill volume.

9. OXO-BIODEGRADABLE POLYMERS IN THE SOIL

It will be evident from the applications discussed above that a primary requirement for agricultural plastics is a variable and controllable period of stability in the environment. Oxo-biodegradable synthetic polymers, as was seen earlier, biodegrade in much the same way as natural rubber and lignocellulosic materials such as straw, twigs, wood chips, bark, etc. These natural materials then provide an independent and objective model of biodegradability in the environment.

Polyolefins are biodegraded by similar microorganisms that attack lignin[58]. These are in general oxygenase enzymes, similar to those that attack rubber and it was noted earlier[71] that commercial polyolefins, after extraction to remove antioxidants and exposure to light, lose between 20% and 90% of their mass in compost in five months. The mechanism of biodegradation of peroxidised polyolefins is quite different from that of the hydrocarbon waxes. In the latter case, biodegradation occurs more rapidly with linear chain waxes than with their branched-chain analogues. This indicates that the mechanism of polyolefin biodegradation is essentially a conventional free radical chain reaction as outlined in Section 4, initiated by reactive oxygen species (hydrogen peroxide, superoxide, etc.). This is consistent with the evidence outlined in Section 5 that biodegradation occurs rapidly from the polymer surface in a biotic environment once the stabiliser has been destroyed by heat or light.

A good deal is now known about the kinetics of abiotic peroxidation and stabilisation of carbon-chain polymers and it is possible in principle to extrapolate to the time for ultimate oxidation from laboratory experiments. As already indicated, the key determinant of the time to bioassimilation is the antioxidant and if this is chosen to optimize the service life, bioassimilation can also be achieved as in the case of wood, straw, twigs, etc. It seems that straw is a particularly appropriate model for the biodegradation of the polyolefins since, like the polyolefins, it fully bioassimilated in biologically active soil over a period of about ten years[46]. The most important conclusion from recent work is that nature does not depend on just one degradation mechanism. Abiotically initiated peroxidation is just as important, at least initially as biooxidation.

An interactive model for abio- and biodegradation of hydrocarbon polymers is shown in Figure 3.

Many enzymes but notably the cytochromes, peroxidase and catalase are known to produce reactive oxygen species that can migrate some distance from the cell. Superoxide ($O_2^{\bullet -}$) formed by single electron reduction of oxygen is soluble in polar media but is very unreactive as a free radical[75]. However, its cognate hydroperoxyl radical ($HOO^{\bullet}$) is organo-soluble and can react with carbon-hydrogen bonds in polymers, thus initiating free radical peroxidation and producing hydrogen peroxide in the surface of the polymer. Hydrogen peroxide is also produced by oxidase enzymes in the cell and, like superoxide and hydroperoxyl, it is free to move outside of the cell until it is reduced by transition metal ions, notably Fe^{2+} in the well-known Fenton reaction.

$$H_2O_2 + Fe^{2+} \longrightarrow {}^{\bullet}OH + OH^- + Fe^{3+} \qquad (9)$$

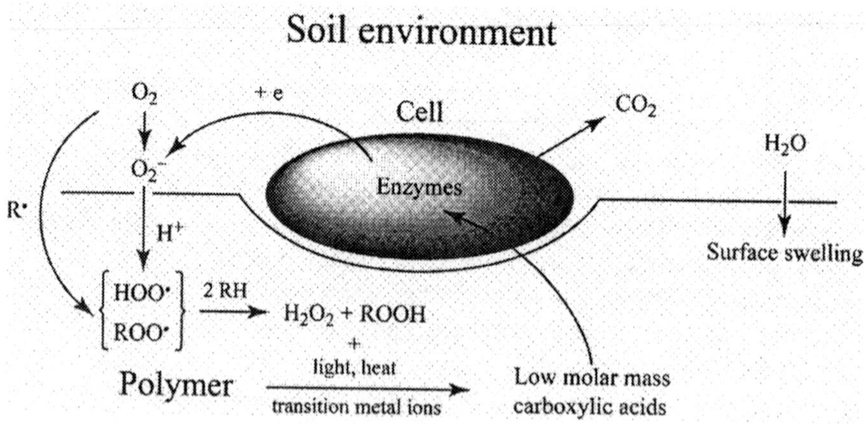

Figure 3. Model for polyolefin bioassimilation after fragmentation[12].

The hydroxyl radical is one of the most reactive free radicals known in chemistry. It is ≈10^7 more reactive than hydroperoxyl[75] and it extracts a hydrogen atom at every encounter with a hydrocarbon. It is thus also one of the most potent initiators of peroxidation known. It is not surprising then that the subsequent polymer degradation reactions are dominated by abiotic peroxidation chemistry.

10. SCIENCE-BASED STANDARDS FOR DEGRADABLE POLYMERS

In the light of the lifetime and post-lifetime requirements of degradable polymers discussed in Sections 7 and 8, it is clear that science-based standards must consider not only potential hazards to the environment involved in the use of synthetic polymers (whether bio-based or fossil-based), but also the longer term benefits in the applications proposed. These may sometimes be in conflict requiring a balanced compromise on advantages and disadvantages, but this must always be based on scientific evidence.

The presently accepted standards for biodegradation of plastics in compost, require that the plastic must be 90% converted to the theoretically obtainable carbon dioxide in six months. This is a quite impractical standard for agricultural plastics and many packaging material, which are required to be stable for up to a year in contact with soil. It may be a convenient way of disposing of domestic packaging but it is not a means of recovering the value of the packaging in compost and consequently does not comply with the EU *Waste Framework Directive* of March 1991, which defines 'recovery' as

"Recycling/reclamation of organic substances........use as fuel to generate energy and spreading on land resulting in benefit to agriculture or ecological improvement including compost and other biological transformation processes."

This directive requires that carbonaceous biomass should be retained in the soil for the benefit of growing plants in the same way as natural lignocellulosic materials[4]. Oxo-biodegradable polymers contribute to soil fertility in the same way as lignocellulosic materials.

10.1 Environmental Effects of Oxo-Biodegradable Polymers

There are two distinct aspects of the environmental effects of degradable polymers. The first is the possibility that partially degraded products may accumulate over a long period. This is a property of the material itself. Figure 4 illustrates in schematic form two types of test procedures (a) and (b) that may be applied.

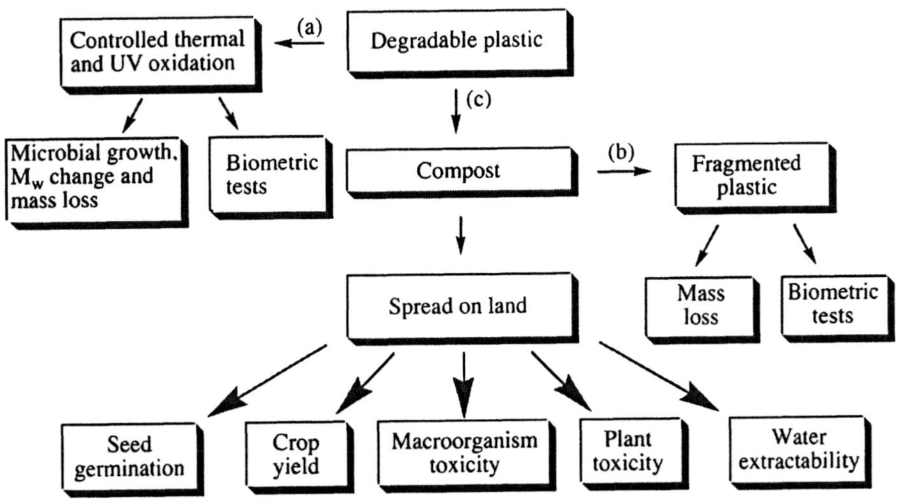

Figure 4. Schematic representation of material tests (a) and (b) and eco-toxicity tests (c) for peroxidised polyolefins.

Routes (a) and (b) are alternative variants to measure the change in physical and chemical behaviour of polyolefins. The length of time a material persists in the environment depends in the first place on user requirements. If a product is intended to last for a year before its purpose is achieved, then biodegradation begins at the end of the peroxidation induction period (IP). At this point, the product contains no antioxidant and the rate at which it breaks down by abiotic peroxidation is well known and is related to the degradation rate at other temperatures by the Arrhenius relationship. In a "standard" soil, a polymer of known structure should biodegrade at a reproducible rate, once the antioxidant has been depleted. Unfortunately, there is at present no such thing as a standard soil and as seen above, the biological activities of soils vary markedly with the nutrients they contain. Furthermore, the rate of biodegradation also depends on the microbial consortia present. In fact, the scavenging of low molecular weight oxidation products by microorganisms has been shown to be very much faster than abiotic peroxidation process[38].

As microorganisms colonise, reactive oxygen species from the cell enzymes interact synergistically with abiotic peroxidation, leading to surface erosion. The evidence for this was shown in Figures 1 and 2 and the proposed model for this process is shown in Figure 3. Abiotically and biotically initiated peroxidations occur together in the later stages of biodegradation, thus reinforcing one another. There is experimental evidence from biometric studies to show that that bio-assimilation is autoaccelerating.

10.2 Eco-toxicity Testing of Fragmented Plastics in Soil

The second requirement of degradable plastics as litter is that they do not interfere with the ecology of the soil and route (c) assesses the biological effects of partially degraded plastics on the growth of plants and on macro-organisms in the soil. Large pieces of plastic in the soil may hinder root growth, affecting seed germination or crop yields. It is not known what is the critical size for physical interference, since no published work has so far shown any significant effects. It may be necessary to establish this critical size for each polymer formulation, but in general terms, the smaller the particle size, the more peroxidised will be the polymer and hence the faster will be the bioassimilation process.

A major concern has been the contamination of the soil by transition metal ions ("heavy metals") that may leach out of degrading polyolefins after repeated use of the same formulation on the same soil. This is to a large extent a function of the low molecular weight additives used in the polymer. However, agricultural soils already vary enormously in transition metal content. Nickel has been studied in some detail because it is used in a number of plastics products and concern has been expressed about its environmental toxicity and possible accumulation in growing plants. Table 3 shows the concentrations of Ni in different types of soils. It is clear that some common soils, notably limestone, contain very considerable concentrations of nickel but no health hazards have been reported in populations farming in limestone areas or even in volcanic regions where the nickel content of the soil is even higher. This has prompted agronomists to investigate the reasons for this and Table 4 shows the concentration of nickel in plants when it is added to the soils at increasing concentrations[67]. No accumulation of Ni in the various parts of the plant could be observed relative to the control soil. It should be noted that the amount of nickel added to the soil was equivalent to the concentrations that would be expected if nickel-containing agricultural films were used on the same soil for up to 180 years. Nevertheless, where potentially toxic transition metal ions are used in degradable plastics formulations, it is necessary to carry out this procedure and as an additional safety step, to study the migration of the metal ions from the plastic into an aqueous environment.

Table 3. Concentration of nickel (ppm) in typical soils[67]

Rock type	Ni	Co
Gabbro	750	100
Gabbro (medium grains)	30	50
Sandstone	90	
Limestone	10-20	
Crystalline (with quartz)	64	

Table 4. Nickel concentrations (ppm) in plants grown on soils that have been doped with nickel sulphate[67]

	Control	60 years*	120 years*	180 years*
Leaves	17.3	15.2	13.5	13.7
Stems	**5.0**	**4.5**	**5.2**	**5.0**
Flesh	2.7	2.0	3.0	3.2
Skin	3.0	3.5	3.2	3.0

*The soil was sprayed with $NiSO_4$ to give nickel concentrations in the topsoil equivalent to the accumulation from S-G mulching films used for the number of years indicated.

11. CONCLUSIONS

It is now recognised that short-term mineralization is not a satisfactory test for the biodegradation of agricultural plastics or for packaging that requires a relatively long but controlled service life. A more relevant standard is required to assess the environmental impact of oxo-biodegradable polymers, irrespective of whether they are fossil-based or bio-based. The presently accepted international standards act more as a hindrance rather than a help in the development of ecologically suitable biodegradable materials in a range of environments and they are counter-productive at a time when the packaging industries are being exhorted by governments to improve their 'green' credentials.

If standards for degradable polymers are to be convincing, they must be based upon a scientific understanding of how nature deals with its own waste. The research discussed in this paper suggests that this is much more complex process than has previously been assumed and that abiotic processes play a major synergistic role with biological processes during the bioassimilation of all polymers. Most polymers require a much longer time for composting if the product is to be acceptable for use in agriculture. Additional standards (based on Scheme 3) are currently under consideration to embrace man-made or man-modified polymers that, like lignocellulose, humus and rubbers, biodegrade by an oxidative mechanism.

ACKNOWLEDGEMENTS

I am grateful to my collaborators, Dr. Anne-Marie Delort and her co-workers at the University of Blaise-Pascal, Clermont-Ferrand for permission to reproduce Figures 1 and 2. I am also indebted to Dr. Martin Patel of the University of Utrecht for helpful discussions and for making available his previously unpublished work and to Mr. Joseph Gho, Dr. David Wiles and their colleagues at EPI Environmental Plastics Inc., Canada for recent information on waste management.

REFERENCES

1. The Green Report; Report of a task force set up by the Attorneys General of the USA to investigate 'Green Marketing', November, 1990.
2. Scott, G., 1999, Management of Polymer Wastes. In *Polymers and the Environment.* Royal Society of Chemistry, Chapter 4.
3. Scott, G., 2000, Green Polymers. *Polym. Degrad. Stabil.* **68**: 1-7.
4. Scott, G., and Wiles, D.M., 2001, Programmed-Life Plastics from Polyolefins: A New Look at Sustainability. *Biomacromolecules* **2**: 615-622.
5. Patel, M., 2002, *Eco-profiles of plastics and related intermediates.* Association of Plastics Manufacturers in Europe, Brussels.
6. Dinkel, F., Pohl, C., Ros, M., and Waldeck, B., (Carbotech), 1996, *Ökobilanz stärkehaltiger Kunststoffe.* BUWAL, Bern, Switzerland.
7. Estermann, R., and Schwarszwälder, W., 1998, for Novamont, Composto. Oltern Switzerland.
8. Conn R.E., 2000, for Cargill-Dow, Presentation at Süddeutches Kuntstoffzentrum, Würzurg, Germany.
9. Patel, M., 1999, PhD Thesis *Closing Carbon Cycles.* University of Utrecht.
10. BIFA, 2001, Interim Report on Starch loose-fill for Flo-Pak. GMBH, Germany
11. Patel, M., 2002, *Proc. Bioplastics Conference.* Europoint, York.
12. Scott, G., 1999, Biodegradable Polymers. In *Polymers and the Environment.* Royal Society of Chemistry, Chapter 5.

13. Lowry, C.D., Egloff, G., Morrell, J.C., and Dryer C.G., 1933, Inhibitors in cracked gasoline. II. Correlation of inhibiting action and oxidation-reduction potential. *Ind. Eng. Chem.* 25: 804-808.
14. Scott, G., 1993, Autoxidation and Antioxidants: Historical Perspective. In *Atmospheric Oxidation and Antioxidants* (G. Scott, ed.), Elsevier, 2nd Ed., Vol. I, Chapter 1.
15. Scott, G., 1965, Peroxides. In *Atmospheric Oxidation and Antioxidants* (G. Scott, ed.), Elsevier, Chapter 2; Scott, G., 1965, Antioxidants: radical chain mechanisms. In *Atmospheric Oxidation and Antioxidants* (G. Scott, ed.), Elsevier, Chapter 4; Scott, G., 1965, Oxidation of Olefinic Oils, Fats and Polymers. In *Atmospheric Oxidation and Antioxidants* (G. Scott, ed.), Elsevier, Chapter 8.
16. Bolland, J.L., 1949, Kinetics of olefin oxidation. *Quart. Rev.* **3**: 1-21.
17. Bateman, L., 1954, Olefin oxidation. *Quart. Rev.* **8**: 147-167.
18. Scott G., 1995, Introduction to the abiotic degradation of carbon chain polymers. In *Degradable Polymers: Principles and Applications* (G. Scott, and D. Gilead, eds.), 1st Ed., Chapman & Hall (Kluwer), Chapter 1; Scott G., 1995, Photo-biodegradable plastics. In *Degradable Polymers: Principles and Applications* (G. Scott, and D. Gilead, eds.), 1st Ed., Chapman & Hall (Kluwer), Chapter 9; see also Scott, G., 2002, Degradation and stabilisation of carbon-chain polymers. In *Degradable Polymers: Principles and Applications*, 2nd Ed., Kluwer Academic Publishers, Chapter 3.
19. Grassie, N., and Scott, G., 1985, Oxidation of polymers. In *Polymer Degradation and Stabilisation.* Cambridge University Press, Chapter 4.
20. Scott G., 1999, Environmental stability of polymers. In *Polymers and the Environment.* Royal Society of Chemistry, Chapter 3.
21. Scott, G., 1993, Oxidation and Stabilisation of polymers during processing. In *Atmospheric Oxidation and Antioxidants* (G. Scott, ed.), 2nd Ed., Vol. II, Elsevier, Chapter 3.
22. Scott, G., 1993, Photo-degradation and photostabilisation. In *Atmospheric Oxidation and Antioxidants* (G. Scott, ed.), 2nd Ed., Vol. II, Elsevier, Chapter 8.
23. R. B. Seymore, and T. Cheng, eds., 1987, *Advances in Polyolefins*. Plenum Press.
24. Osawa, Z., 1993, Metal catalyzed oxidation and its inhibition. In *Atmospheric Oxidation and Antioxidants* (G. Scott, ed.), 2nd Ed., Vol. II, Elsevier, Chapter 6.
25. Arnaud, R., Dabin, P., Lemaire, J., Al-Malaika, S., Chohan, S., Coker, M., Scott, G., Fauve, A., and Maaroufi, A., 1994, Photooxidation and biodegradation of commercial photodegradable polyethylenes. *Polym. Degrad. Stabil.* **46**: 211-224.
26. Albertsson, A-C, Barenstedt, C. Karlsson, S., and Lindberg T., 1995, Degradation product pattern and morphology changes as means to differentiate abiotically and biotically aged degradable polyethylene. *Polymer* **36**: 3075-3083.
27. Shaposnikov, V.N., Rabotnova, L.L., Yarmola, G.A., Kutznetsova, V.M., and Mozokhina-Porshnyakova, 1952, *Microbiologiya* **21**: 146-154.
28. Low, F.C., Tan, A.M., and John, C.K. 1992, Microbial degradation of natural rubber. *J. Nat. Rub. Res.* **7**: 195-205.
29. Tsuchii, A, Suzuki, T., and Takeda, K., 1985, Microbial degradation of natural rubber vulcanizates. *Appl. Environ. Microbiol.* **50**: 965-970.
30. Kajikawa, S., Tsuchii, A., and Takeda, K., 1991, *Nippon Nogeikagaku Kaishi* **65**: 981-986.
31. Ikram, A., Alias, O., and Napi, D., 2000, Effects of added nitrogen and phosphorus on the biodegradation of NR gloves in soil. *J. Rub. Res.* **3**: 104-114.
32. Heisey, R.M., and Papadatos, S., 1995, Isolation of microorganisms able to metabolize purified natural rubber. *Appl. Environ. Microbiol.* **61**: 3092-3097.

33. Ikram, A. Alias, O, Bahri, A.R.S., Fauzi, M.S., and Napi, D., 2001, Effects of added nitrogen and phosphorus on the biodegradation of NR gloves in soil. *J. Rub. Res.* **4**: 102-117.
34. Linos, A., Berekaa, M.M., Reichelt, R., Keller, U., Schmidt, J., Flemming, H-C., Kroppenstedt, R.M., and Steinbüchel, A., 2000, Biodegradation of cis-1,4-polyisoprene rubbers by distinct actinomycetes: microbial strategies and detailed surface analysis. *Appl. Environ. Microbiol.* **66**: 1639-1645.
35. Linos, A., Reichelt, R., Keller, U., and Steinbüchel, A., 2000, A Gram-negative bacterium, identified as Pseudomonas aeruginosa AL98, is a potent degrader of natural rubber and synthetic cis-1,4-polyisoprene. *FEMS Microbiol. Lett.* **182**: 155-161.
36. Sadun, A.G., Webster, T.F., and Commoner, B., 1990, *Breaking down the degradable plastics scam.* Green Peace, Washington D.C.
37. Lechevalier, M.P., Prauser, H.. Labeda, D.P., and Ruan, J-S., 1986, *Int. J. Syst. Bacteriol.* **36**: 29-37.
38. Bonhomme, S., Cuer, A., Delort, A-M, Lemaire, J., Sancelme, M., and. Scott, G., Environmental biodegradation of polyethylene. *Polym. Degrad. Stab.* in press.
39. Scott, G., 1993, Macromolecular and polymer-bound antioxidants. In *Atmospheric Oxidation and Antioxidants.* (G. Scott, ed.), Elsevier, 2nd Ed., Vol. II, Chapter 5.
40. Adler, G., 1977, *Wood Sci. Technol.* **11**: 169-217.
41. Scott, G., 1999, *Polymers and the Environment.* Royal Society of Chemistry, pp. 2-4.
42. Haug, R.T., 1993, *The practical handbook of compost engineering.* Lewis publishers, Boca Raton.
43. Wehmer, C., 1915, Degradation of woody matter ("Holzsubstanz") by fungi. *Ber.*, **48**: 130-134.
44. Hammouda, G.H.H., and Adams, W.A., 1989, in *Compost: Production, quality and use* (M. De Bertoldi, M.P. Ferranti, P. L'Hermite, and F. Zucconi, eds.), Elsevier App. Sci., pp 245-253.
45. Horwath, W.R., Elliot, L.F., and Churchill, D.B., 1995, *Compost Sci. Util.* **3**: 22-30.
46. Jansson, S.L., 1963, in *The use of isotopes in soil organic matter studies.* Report of the FAO/IAEA
47. Adhi, T.P., Korus, R.A,, and Crawford, D.L., 1989, Production of major extracellular enzymes during lignocellulose degradation by two streptomycetes in agitated submerged culture. *Appl. Environ. Microbiol*, **55**: 1165-1168.
48. Adhi, T.P., Korus, R.A., Pometto, A.L., and Crawford, D.L., 1988, Lignin degradation and production of microbially modified lignin polymers by Streptomyces viridosporus in slurry reactors. *Appl. Biochem. Biotechnol.* **18**: 291-301.
49. Ramachandra, M., Crawford, D.L., and Hertel, G., 1988, Characterization of an extracellular lignin peroxidase of the lignocellulolytic actinomycete Streptomyces viridosporus. *Appl. Environ. Microbiol.* **54**: 3057-3063.
50. Hirai, H., Kondo, R., and Sakai, K., 1994, Screening of lignin-degrading fungi and their ligninolytic enzyme activities during biological bleaching of kraft pulp. *Mokzua Gakkaishii* **40**: 980-986.
51. Katagiri, N., Tsutsumi, Y., and Nishida, T., 1995, Correlation of brightening with cumulative enzyme activity related to lignin biodegradation during biobleaching of kraft pulp by white rot fungi in the solid-state fermentation system. *Appl. Environ Microbiol.* **61**: 617-622.
52. Paice, M.G., Reid, I.D., Bourbonnais, R., Archibald, F.S., and Jurasek, L., 1993, Manganese peroxidase, produced by Trametes versicolor during pulp bleaching, demethylates and delignifies kraft pulp. *Appl. Environ. Microbiol.* **59**: 260-265.

53. Scott, G., 1965, *Atmospheric Oxidation and Antioxidants* (G. Scott, ed.), Elsevier, pp. 125-132.
54. Metodiewa, D., and Dunford, B., 1993, Medical aspects and techniques for peroxidases and catalases. In *Atmospheric Oxidation and Antioxidants* (G, Scott, ed.), Elsevier, Vol. III, Chapter 11.
55. Waksman, S.A. 1938, *Humus*, Baillièri, Tindal and Cox, p. xii.
56. Guillet, J., 1995, Plastics and the environment. In *Degradable Polymers: Principles and Applications* (G. Scott, and D. Gilead, eds.), Chapman and Hall, London, 1st Ed., p. 216-246.
57. Eggins, H.O.W., Mills, J., Holt, A., and Scott, G., 1971, Biodeterioration and biodegradation of synthetic polymers. In *Microbial Aspects of Pollution* (G. Sykes, and F.A. Skinner, eds.), Academic Press, pp. 267-279.
58. Lee, B., Pometti, A.L., Fratzke, A., and Bailey, T.B., 1991, Biodegradation of degradable plastic polyethylene by Phanerochaete and Streptomyces species. *Am. Soc. Microbiol.* **57**: 678-685.
59. Harlan, G., and Kmiec, C., 1995, Ethylene-carbon monoxide copolymers. In *Degradable Polymers: Principles and Applications* (G. Scott, and D. Gilead, eds.), Chapman & Hall, 1st Ed., Chapter 8.
60. Griffin, G.J.L., 1994, Particulate starch based products. In *Chemistry and Technology of Biodegradable Polymers* (G.J.L. Griffin, ed.), Blackie Academic & Professional, Chapter 3.
61. Scott, G., 1995, Photo-biodegradable plastics. In *Degradable Polymers: Principles and Applications* (G. Scott, and D. Gilead, eds.), Chapman and Hall, 1st Ed., Chapter 9.
62. Pitt, C.G., 1992, Non-microbial degradation of polyesters: mechanisms and modifications. In *Biodegradable Polymers and Plastics* (M. Vert, J. Feijen, A-C. Albertsson, G. Scott, and E. Chiellini, eds.), Royal Society of Chemistry, pp. 7-19.
63. Li. S., and Vert, M. 1995, Biodegradation of aliphatic polyesters. In *Degradable Polymers: Principles and Applications* (G. Scott, and D. Gilead, eds.), Chapman & Hall (Kluwer), 1st Ed., Chapter 4.
64. Hammond, T., and Liggatt, J.J., 1995, Properties and applications of bacterially derived polyhydroxyalkanoates. In *Degradable Polymers: Principles and Applications* (G. Scott, and D. Gilead, eds.), Chapman & Hall (Kluwer), 1st Ed., Chapter 5.
65. Gilead, D., and Scott, G., 1982, Time-controlled stabilization of polymers. In *Development in Polymer Stabilisation-5* (G. Scott, and D. Gilead, eds.), App. Sci. Pub., Chapter 4.
66. Gilead, D., 1995, Photodegradable plastics in agriculture. In *Degradable Polymers: Principles and Applications* (G. Scott, and D. Gilead, eds.), Chapman & Hall, 1st Ed., Chapter 10; see also Scott, G., and Wiles, D.M., 2002, Degradable hydrocarbon polymers in waste and litter control. In *Degradable Polymers, Principles and Applications.* Editor G. Scott, Kluwer Academic Publishers, 2nd Ed., Chapter 13.
67. Fabbri, A., 1995, The role of degradable polymers in agricultural systems. In *Degradable Polymers: Principles and Applications* (G. Scott, and D. Gilead, eds.), Chapman & Hall, 1st Ed., Chapter 10.
68. Scott, G., 1997, Abiotic control of polymer biodegradation. *Trends Polym. Sci.* **5**: 361-368.
69. Scott, G., 2000, Biodegradable plastics in agriculture – Working with nature. In Proceedings of the 6th Arab Conference on Materials Science, (S. Kandil, ed.), J. Appl. Polym Sci. in press.
70. Scott, G., 1990, Photobiodegradable plastics: their role in the protection of the environment. *Polym. Degrad. Stabil.* **29**: 135-154.

71. Pandey. J.K, and Singh, R.P., 2001, UV-Irradiated Biodegradability of Ethylene-Propylene Copolymers, LDPE, and I-PP in Composting and Culture Environments. *Biomacromolecules* **2**: 880-885.
72. Scott, G., unpublished work.
73. Kawai, F., Shibata, M., Yokoyama, S., Maeda, S., Tada, K., and Hayashi, S., 1999, Biodegradability of Scott-Gelead photodegradable polyethylene and polyethylene wax by microorganisms. In *Degradability, Renewability and Recycling.* 5th International Scientific Workshop on biodegradable Plastics and Polymers, Macromolecular Symposia, (A-C. Albertsson, E. Chiellini, J. Feijen, G. Scott, and M. Vert, eds.) Wiley-VCH, pp. 73-84.
74. Gho, J.G. EPI, 2000, Personal communication.
75. Scott, G., 1997, *Antioxidants in science, technology, medicine and nutrition.* Albion Chemical Sciences Series, p. 10.
76. Scott, G., 1999, Antioxidant control of polymer biodegradation. In *Degradability, Renewability and Recycling.* 5th Int. Sci. Workshop on Biodegradable Plastics and Polymers, Macromolecular Symposia, (A-C. Albertsson, E. Chiellini, J. Feijen, G. Scott, and M. Vert, eds.) Wiley-VCH, pp. 113-125.

Biodegradability and Compostability

The International Norms

FRANCESCO DEGLI INNOCENTI
NOVAMONT S.P.A. via Fauser 8 I-28100 Novara (Italy)

1. EVERYTHING IS BIODEGRADABLE. CAN EVERYTHING BE BIO-RECYCLED?

In principle, any organic compound is biodegradable. The meaning of this statement is that the organic carbon of any substance, after a series of degradation processes, more or less extreme and prolonged, can be oxidized into carbon dioxide through microbial respiration. In the absence of oxygen, methane and CO_2 are produced, as known. This conversion process from organic carbon to inorganic carbon is defined *mineralization*. The terms *biodegradation* and *mineralization* are frequently used as synonyms. In reality, not always a biodegradation process leads to a complete mineralization of the substrate. On the other hand, other environmental factors, besides biodegradation, such as temperature, water, irradiation etc., can have a part in the mineralization process. But, whatever is the degradation pathway, the essential result of mineralization is the closing of the carbon bio-geo-chemical loop.

The statement that everything is biodegradable collides with the common observation that plastics, that is to say the modern synthetic polymeric materials, seem to persist in the environment for infinite time. The puzzle lies in the temporal scale. The traditional plastics have degradation times longer than the capacity of human beings to perceive a substantial change. But, with appropriate means, by using for instance radio-labelled polymers[1],

Biodegradable Polymers and Plastics, Edited by Chiellini and Solaro
Kluwer Academic/Plenum Publishers, New York, 2003

it is possible to determine the mineralization even if, under our eyes, nothing is apparently happening.

Obviously the impact on our current society of polymers with degradation times of decades and decades is different from the impact of fast degrading polymers. The key factor in waste management is the rate of waste production, which, nowadays, is very high. The great diffusion of plastics in the human society can only be solved by a disposal rate similar to the production rate. As a matter of fact, incineration with energy recovery is a very fast *mineralization process*, and is considered the proper way of disposal of traditional non-biodegradable plastics. This also applies to the biodegradable plastics whose utility in waste management is linked to a fast degradation rate. Together with disintegrability, biodegradability is necessary to avoid accumulation in the soil. Otherwise, composting would merely be the process of transformation of *visible* solid waste into *invisible* waste, which is then spread into the soil. This would just be like sweeping the dust under the carpet. Besides, in the long term the effect of these xenobiotics on the soil ecology is unknown.

So, the statement everything is biodegradable is true, but this does not necessarily mean that everything can be profitably recycled by means of biological treatments.

This preface is to exemplify how the term biodegradable can be misleading. In spite of its very large use (and abuse), the term biodegradable is not helpful because it is not informative. The term does not give any information about the specific environment where the biodegradation is supposed to happen, the rate which will regulate the process (fast, slow, geological), and the extent of biodegradation (partial or total conversion into CO_2).

2. ROLE OF STANDARDIZATION

In order to avoid misunderstandings and endless disputes, it is important to distinguish at an early stage between an academic and a practical approach. The two domains have different objectives and different methodologies. It can be frustrating to apply the scheme of one domain to the other one and vice-versa. The first approach aims at giving an accurate description of reality, finding new phenomena and new correlations, regardless the time scale of the phenomena and the effective relevance for the everyday life. The second approach aims at finding technical solutions to specific practical problems (i.e. waste management) with the highest social, political and scientific consensus and acceptability. The first domain is Science; the second domain is Standardization.

The example of the work performed by the European Committee for Standardization (CEN) to define biodegradability and compostability is illustrative of the role of standardization in this field.

When in the mid nineties CEN was appointed by the European Commission with the Mandate M200 to produce standards in the field of compostability of packaging, as a follow-up of the European Directive on packaging waste (94/62/EC), the approach of the working group 2 led by Dr.-Ing. Karl-Friedrich Ziegahn of the Fraunhofer Institut für Chemische Technologie of Pfinztal was very pragmatic. What are the requirements and the expectations of the stakeholders, namely the composting managers, the farmers, the packaging producers, and the public administrators?

From the composting manager viewpoint, the "compostable" plastic material, which can be accepted as a feedstock in his own composting plant is the one that does not impair his business. The material shall disintegrate in a composting cycle, before the final screening. In case the plastic material due to the poor degradability is screened out at this stage, then an alternative disposal, such as dumping in a landfill or incineration, has to be found. This implies extra costs.[2] On the other hand, if the partly destroyed material passes through the screen and goes into the refined compost, but it is still visible, then the risk is not to sell the compost or to sell it as a second choice compost because "dirty" with plastics. Needless to say, compost must also comply with the national quality requirements, which, in particular require low content of heavy metals. Therefore, for the composting manager the most important feature compostable plastics must have is: disintegrability within a composting cycle and low content of heavy metals.

The farmer wants good quality compost, free from substances, which can interfere with the agricultural production either in the short or in the long term. Therefore, the compost must be free not only from visible contaminants, such as plastic residues, but also from recalcitrant xenobiotics (*xenobiotics* are foreign compounds in biological systems) produced during degradation. The farmer wants to be sure that the continuous application of compost will not alter the productivity of his field. If this is not guaranteed, the farmer will not take any risk and will rather use other compost (produced using only pure green waste, without any man-made substance) or no compost at all. Therefore, in order to satisfy the requirement of the farmer, the biodegradation of plastics during the composting phase shall be fast and possibly complete. The farmer does not want his field to become the location where the second phase of a municipal organic waste bio-treatment takes place. It must be remarked that the presence of wood residues and lignin more or less converted into humus is not considered by the farmer as a risk for his field, no matter how long it takes to lignin to fully mineralize into CO_2. Lignin and humus are natural fractions of soil and do not represent a problem for it (they are "GRAS" = General Recognized As Safe). Thus absence of phyto- and ecotoxicity and total biodegradability are the most important properties for the farmer.

The packaging producers and, eventually, the companies selling goods to the final consumers have the liability of the packaging. Needless to say, the interest of this part is not to suffer from any consequence coming from a negative environmental impact of compostable packaging. No negative publicity or claims coming from the public opinion. Therefore, the packaging producers and users are very conservative.

What about the public administrators, the legislators, the politicians? Since public health and food production could be affected by a massive application of biodegradable plastics in compost, the public administrators indubitably prefer any option increasing safety rather than decreasing it. This is particularly true after the recent case of the Mad Cow disease and the presence of dioxins in chickens. Summarizing: very low heavy metals content, no ecotoxic effects, no plastic residues, visible or invisible, brought in the field together with compost.

The CEN has developed his work with the declared objective of offering to the stakeholders, that is the composting managers, the farmers, the laboratory analysts, the packaging producers, and the politicians, a set of standard test methods and criteria capable of reassuring them all about the applicability of biodegradable plastics. The full acceptance of this approach is demonstrated by the norm EN 13432:2000 "Requirements for packaging recoverable through composting and biodegradation- Test scheme and evaluation criteria for the final acceptance of packaging", unanimously approved by the CEN national members and then accepted by the European Commission as a harmonized European norm.[3] The use of an harmonized norm gives presumption of conformity with the European Directive.

In the following paragraph the norm will be illustrated and explained.

3. COMPOSTABILITY OF PACKAGING: THE EN 13432

3.1 Packaging Materials

Compostability is a neologism indicating an additional property of plastic materials, never taken into account before: the overall compatibility of a material with composting, the biological system of treatment of organic solid waste. The definition of compostability given by CEN (European Committee of Standardisation) is:

- *Compostability is a property of a packaging to be biodegraded in a composting process.*
- *To claim compostability it must have been demonstrated that a packaging can be biodegraded and disintegrated in a composting system (as can be shown by standard test methods) and completes its biodegradation during the end-use of the compost*
- *The compost must meet the relevant quality criteria. Quality criteria are e.g. heavy metal content, no ecotoxicity, no obviously distinguishable residues.*

According to the EN13432 a packaging is compostable if it is formed by components, which have been, each individually qualified as compostable. In this way the analysis of compostability of a packaging is simplified and traced back to the analysis of compostability of the single constitutive materials. Therefore, it is sufficient to use compostable materials in order to obtain a final compostable packaging.

3.2 Compostability Is a Set of Properties

The European norm defines the specific properties of the compostable packaging materials and the test methods needed to verify their conformity. The compostable packaging shall be endowed with four main features.

- Biodegradability, that is the metabolic conversion of the packaging material into carbon dioxide (absence of chemical pollution).
- Disintegrability, that is fragmentation and loss of visibility in the final compost (absence of visual pollution).
- Absence of negative effects on the process of composting.
- Absence of negative effects on the final compost (i.e. reduction of the agronomic value and presence of ecotoxicological effects on the plant growth).

Each of these points is required for the definition of compostability but is not sufficient, alone (Figure 1). A biodegradable material is not necessarily compostable, because it must also disintegrate during the composting cycle and it must cause no problem either to the process or to the final product (the compost).

The evaluation procedures of the packaging materials are reported in the following sections.

3.2.1 Characterization

In a preliminary phase, information on the packaging material is gathered. The constituents, namely the ingredients used for the production of the material, are identified and the presence of toxic substances, heavy metals in particular, is verified. The heavy metals concentration in the organic waste entering into the composting plants must be low because the heavy metals concentration in the final compost is regulated by law. As a consequence of the composting process, a 50% mass reduction of the organic waste is usually attained. Thus, the heavy metals introduced together with the feedstock can even concentrate in the final compost.

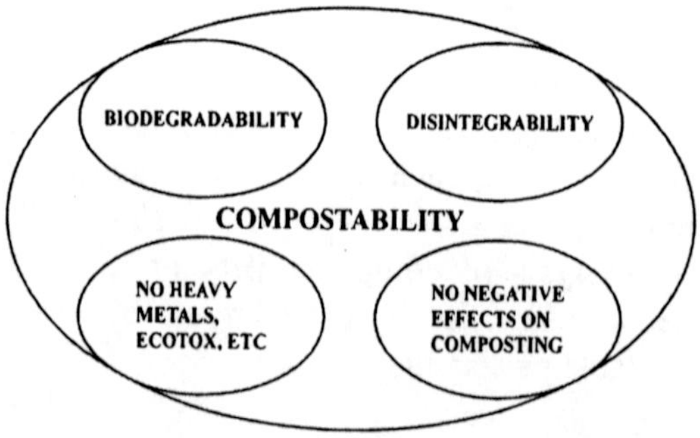

Figure 1. Compostability is a set of properties. Each of them is necessary but not sufficient

The maximum concentrations of heavy metals allowed in the compostable packaging are specified by the EN13432. In this case, the limits are lower than those required by the Directive 94/62/EC, because the release of heavy metals in the final compost is highly detrimental for the quality of compost. These limits have been derived by the heavy metal limits allowed in final compost, according to the European Eco-label for soil improvers[4] and considering a mass reduction of 2.

Furthermore, the test material is analysed to determine the total organic carbon, the dry weight, and the volatile solids, all information needed in the test of biodegradability.

3.2.2 Biodegradability

The meaning of two terms frequently used as synonyms (even by the experts) must be clarified: biodegradability and biodegradation.

Biodegradability refers to a potentiality (i.e. the ability of a polymer to be degraded by a bio-logical agent). It is the general knowledge that there is at least one enzyme in the biosphere able to break the chemical bonds of a given polymer chain.

Biodegradation refers to a process, occurring under certain conditions, in a given time, with measurable results.

How can we determine the biodegradability of a polymer? In theory, if we knew the catalytic sites of all enzymes and all possible biochemical mechanisms we could deduce the biodegradability from the chemical structure of the polymer. This will be probably possible in the future. Today we must apply a laboratory approach: in order to verify the biodegradability of a polymer we must follow a biodegradation process and measure what happens. **The biodegradability of a polymer is inferred by studying biodegradation processes.** The biodegradation of a polymer is observed under specific laboratory conditions and, from its behaviour, the conclusion that the polymer is biodegradable (namely, *it can be biodegraded*) can be drawn.

The biodegradability of compostable polymers is assessed by measuring in the laboratory the mineralization under conditions as similar as possible to the composting conditions (solid state, high temperature, proper oxygen and water contents, compost used as an inoculum). The specific test method is the controlled aerobic composting test prEN 14046, which is technically identical to the ISO 14855:1999 "*Determination of the ultimate aerobic biodegradability and disintegration of plastic materials under controlled composting conditions - Method by analysis of evolved carbon dioxide*". The method simulates the environmental and microbiological conditions of a composting process. The test material is generally reduced to powder and mixed with a certain amount of mature compost, which is used as a source of microorganisms and nutrients, brought to the correct degree of moisture and maintained at a constant temperature of 58 °C. From the measurement of the CO_2 produced under these conditions, the degree of conversion (mineralization) of the organic carbon of the bio-based material is determined. In parallel the biodegradation of the reference material, microcrystalline cellulose, is measured. According to the EN 13432, the biodegradation of the test material, measured through the controlled composting test, must be at least 90% (conversion percentage of the organic carbon into CO_2) or 90% of the level reached by the cellulose at the same time (relative biodegradation), for a maximum time of six months. Cellulose must reach at the end of the test a biodegradation of at least 70%; otherwise the test is invalid. The cellulose is used as an internal control, to assure that the test environment is active. Being a fully biodegradable material, the mineralization level of cellulose is considered as the maximum mineralization achievable under the test conditions.

In alternative to the ISO 14855, two ISO methods (14851 and 14852) applied to measure the biodegradability in aqueous environment can be used, for those cases in which the composting method is not appropriate (ink, additives, colorants, etc.).

Test duration and biodegradation threshold are arbitrary values, which derive from long discussions inside and outside the CEN working group. Criticisms have been voiced about the test duration that is considered too long. The test of biodegradation under composting conditions is an optimized simulation of the composting environment. The temperature is set at 58 °C and kept constant, humidity is controlled at optimal levels and so it is the aeration. On the other hand, the thermophilic phase of real composting processes lasts normally no more than 2-3 months. Furthermore, even during the thermophilic phase, temperature is not constantly high: after a first hot period, temperature slowly decreases towards ambient temperature. Not always water availability and aeration will be optimal. Therefore, the application of the laboratory scale biodegradation data to the real composting process could be very misleading. Biodegradation of the polymer could be rather limited under real conditions and hence final compost full of residues. According to these reviewers, the test duration should be 2-3 months maximum, similarly to the real composting processes.

For others the test duration is too short. The biodegradation test should be considered just as a means to verify the inherent biodegradability. The duration of a real composting process is not a point. The pilot-scale composting test with the determination of the disintegration degree (see next point) is considered as an adequate proof that the material under real conditions is sufficiently degraded. Biodegradation can then be completed during the application phase of compost, when spread in the soil.

The test duration set by the EN13432 is the result of combining these different viewpoints. It takes into account the duration of the real composting phase which can effectively be of 6 months, considering also the curing phase. On the other hand, the test conditions are optimized in order to avoid false negative results caused by poor environmental conditions and thus to verify the inherent biodegradability of the materials. A 90% biodegradation in 6 months is considered as a good indication that the material is inherently biodegradable and with a rate compatible with the composting cycle.

3.2.3 Disintegration under Composting Conditions and Verification of the Effects on the Process

In order to verify that the test material, in its final physical form, is disintegrated during a composting cycle without leaving residues (a visual pollution is not acceptable in a commercial compost), a real composting process must be performed. Samples of the test material are mixed together with fresh organic waste and co-composted at pilot-scale level in a 200 litres bin at a concentration of 1%. At the end of the process, a screening is performed on the final compost using a 2 mm sieve. The >2mm fraction is assumed to be the non-disintegrated fraction and used to determine the threshold for disintegration degree. The method is described in the standard prEN 14045 (equivalent to the ISO 16929). Obviously, in this case it is not possible to use powdered material (namely already mechanically disintegrated). The basic material must be converted into the final packaging or in the semi-manufactured product. Therefore, in this test method the sample can be a film, a foil, a sheet, foam, or the packaging itself. The thickness of the specimens used in the disintegration test is important because it fixes the maximum thickness at which the packaging material under study can be applied to the market. The disintegration rate generally decreases with the increase of the thickness. Therefore, a positive result obtained in the disintegration test allows the use of the material at the tested thickness or at lower thickness but is does not guarantee the compostability of the material if used at higher thickness. Having to use higher thickness, it is necessary to repeat the test verifying the disintegrability of thicker specimens. At the end of the 3 months cycle the disintegration is verified by sieving. The pilot scale composting is also used to verify possible negative effects of the test material on the composting process and to produce the compost needed for the ensuing quality analysis and ecotoxicity testing. As an alternative, full-scale test can be performed to assess disintegrability. The pilot scale composting test is in any case much more convenient, in fact it is a real composting process even if run at a reduced scale with a good control of the main parameters (humidity, aeration, temperature). The amount of specimens needed to run a full-scale test with an initial 1% concentration is generally huge and unmanageable, depending on the composting plant size.[5,6] With lower concentrations and because of the impossibility of screening the whole compost, the risk is to perform a not significant sampling and do not find residues not because of full disintegration but just because of dilution. This problem can be solved using a test method developed by the Knoten Weimar University (Germany). The test material specimens are mixed with biowaste and closed in net bags. The bags are then added to the composting mass at the beginning of the process. At the end of the composting cycle the bags are retrieved form the compost and the content screened to verify the disintegrability of the specimens. This test procedure assures the traceability of the test material.

3.2.4 Compost Quality: Chemical and Ecotoxicological Analysis

The test material can be considered as a "raw material" of a composting process. Therefore, it must not decrease the characteristics of the final product, the compost. Samples of compost, obtained mixing the test material with organic waste, are compared with samples of reference compost, produced only with organic waste, without the test material. The results must not differ significantly. The required analyses are: volumetric weight (density), total dry solids, volatile solids, salt content, pH, and levels of nutrients (total nitrogen, ammonium nitrogen, phosphorus, magnesium, and potassium). Furthermore, the effect of the compost samples on the plant growth are assessed, using the method described in the same norm, to show that the test material, during degradation, does not release into the compost substances toxic for the plants and the environment. A very high initial test material concentration (10%) is required for ecotoxicity testing.

3.2.5 Natural Materials

Chemically unmodified natural constituents, such as wood, wood fibre, cotton fibre, starch, paper pulp or jute are considered as biodegradable, without the need of testing their biodegradability. Still, all the other characteristics that concur to show compostability are required (absence of ecotoxicity, low heavy metals concentration, disintegrability, no negative effects of compost quality, etc.). This exception is due to the fact that some natural products (most notably lignin) do not comply with the biodegradability criteria (90% biodegradation in 6 months). The critics of the EN 13432 consider this result as a proof that the criteria are not satisfactory. Lignin is a very complex natural material, slowly biodegradable. Compost is the product of the initial rapid stage of decomposition. During the composting process carbohydrates, lipids, and proteins are fast degraded while lignin remains. As a consequence, compost is rich in lignin under humification. In several cases the slow biodegradation of lignin causes an accumulation of this substances in soil as humus. The accumulation of lignin in the environment is a natural event, which is beneficial for the soil fertility. On the other hand, the accumulation of xenobiotics cannot be encouraged because, while it is well known that lignin is ultimately degradable and beneficial for environment and soil structure, this cannot be claimed for synthetic products, whose behaviour in natural environment is not known. Therefore, the compostability criteria have been devised to refuse materials, which do not degrade during composting and then may build up in the soil. Unavoidably, the system "recognises" lignin as a material potentially causing accumulation. This is a proof that the system works properly; lignin in fact does cause build up in the soil. However, in this case, the accumulation is beneficial and accepted by the stakeholders.

4. OTHER NOTABLE STANDARDS ON COMPOSTABILITY

Other standards on compostability originated at national level are known also at international level. Most notably the DIN V 54900 "Prüfung der

Kompostierbarkeit von polymeren Werkstoffen" and the ASTM D6400-99 "Standard Specification for Compostable Plastics". These standards are very similar to the EN 13432. As a matter of fact, the three standards were developed simultaneously as the result of an intense discussion at the ISR-ASTM[7] in the USA and at the Organic Reclamation and Composting Association.[8]

A main point of differentiation among the EN 13432 on one side, and the ASTM D6400-99 and the DIN 54900 on the other side, is the minimum percentage of biodegradation. According to the DIN and ASTM norms, a 60% biodegradation in 6 months is a sufficient proof of biodegradability for homopolymers and copolymers with random distribution of monomers. All the other polymers must reach the 90% biodegradation level. On the other hand, the EN 13432 fixes the biodegradability requirement to 90%, for any material. The reasoning behind the DIN and ASTM is that if a copolymer or, even better, a homopolymer reaches a biodegradation of 60% this implies that the whole structure is biodegradable. On the other hand, the CEN considers that if a copolymer is formed by two monomers, one biodegradable and the other not, a 60% biodegradation level can be reached, depending on their mass ratio, as the result of 100% biodegradation of one moiety and 0% biodegradation of the other moiety. It would be a sort of "average" biodegradation, which is unacceptable because it would have the consequence of causing the accumulation in the soil of the recalcitrant monomer.

A further differentiation between the DIN 54900 and the ASTM is that the test duration can be up to one year if a radio-labelled product is tested.

5. NEW FRONTIERS IN STANDARDISATION

Several products made with biodegradable polymers are not supposed to be disposed of by composting at the end of their life. Other applications of biodegradable plastics require the biodegradation of the products in soil. Strictly speaking, the environmental fate of most biodegradable materials is to be applied and finally integrated into the soil directly (i.e. mulch films) or indirectly after a composting treatment. The biodegradable plastics used in agriculture or littered in forests are supposed to biodegrade in the soil, while the compostable materials collected together with the biowaste must pass through a controlled biological treatment (i.e. the composting process) before being added to the soil. Since the agricultural soil is the medium for the production of food for humans and farm animals, the absence of negative effects linked to the *in situ* disposal of the plastics and the absence of a build-up of residues must be verified. Therefore, the definition of standard test methods and specific criteria to verify biodegradability in soil and absence of eco-toxic effects are issues as important as the definition of compostability and deserve to be addressed and characterized carefully. This issue is presently scrutinized by the CEN TC 249 (Plastics) with a specific working group (WG9).

The issue of biodegradability in soil seems more complicated than the compostability. The plastic items used in agriculture can have different environmental exposures. The mulch film is ploughed under by tillage, while strings, clips, and other small plastic items are dropped after their use and left on the ground, not necessarily buried. Nursery pots are usually buried after a period in the greenhouse, under protected environment. Generally speaking, two phases can be distinguished in the life of a biodegradable object located in soil: a first phase on the surface, under the action of sun and other climatic factors, and a second phase, buried in the soil in contact with the active microorganisms. Usually the first phase is the functional phase: the object must satisfy some functional requirement: for example the mulch film must control the growth of weeds. If biodegradation occurs during this phase it may be a negative factor. The second phase corresponds with the phase of disposal, when the object must disappear and be recycled through natural processes; in this phase biodegradation is a positive factor. In both phases no ecotoxic molecules have to be generated.

The soil is not a homogeneous environment. It is affected by several uncontrolled parameters. The temperature, controlled by the regional climate and the seasonal fluctuations; the soil water content, dependent on one side on rainfalls (also a climatic factor) and irrigation (if and when applied) and, on the other side, influenced by the soil water retaining capacity; the chemical composition (mineral compounds and organic matter), a geographical and climatic factor; and finally the pH. All these factors joint together in different combinations create different environments and strongly affect the ecology of soil. The first dilemma encountered when addressing the issue of predicting the biodegradation in soil is the definition of the environmental parameters the polymer will be exposed to and which one should be considered when planning a test system. This problem was less critical when defining the scheme for biodegradability under composting conditions, being the variability of the composting environment low. The composting environment is a rather homogeneous ecological niche and can be considered as a consistent micro-cosmos. This is due to the fact that compost is the result of an *industrial process*. Any composting manager, in any latitude, will impose similar conditions to his composting plant, in spite of the different engineering, in order to reach the same purpose: a fast conversion of the acidic, fermenting waste into a stabilised, earth-smelling, marketable compost. On the other hand, the environmental factors present in the soil environment can be very different and consequently different can be the rate of degradation.

These are some of the problems, which need to be addressed to define a standard for utilization of plastic biodegradable materials in agriculture.

ACKNOWLEDGEMENTS

Many thanks to Sara Guerrini for reading the manuscript.

REFERENCES

1. Albertsson A.C., 1978, Biodegradation of Synthetic Polymers. II. A Limited Microbial Conversion of ^{14}C in Polyethylene to $^{14}CO_2$ by some Soil Fungi. *J. Appl. Polym. Sci.* **22**: 3419-3433.
2. Bianchi D., Orlando M. Degli Innocenti F., and Versari M., 2000, Effects of different Organic Waste Collection Options on the Economics of Composting. Presented at the Int. Conference on "*Biodegradable Polymers : production, marketing, utilization and residues management*" Wolfsburg (Germany) 4-5 Sept. 2000.
3. *Official Journal of the European Communities* L 190 12/07/2001 p. 0021-0023.
4. *Official Journal of the European Communities* 7. 8. 98 L 219/39 Commission Decision of 7 April 1998 establishing the ecological criteria for the award of the Community eco-label to soil improvers.
5. Degli Innocenti F., Piccinini S., Rossi L., and Bastioli C., 1995, Degradation of Mater-Bi ZI01U Sheets in a Composting Pilot. *Proc. ORCA Congress.* 18-19 January 1995, Brussels Belgium.
6. Piccinini S., Rossi L., Degli Innocenti F., and Bastioli C., 1996, Behaviour of biodegradable Mater-Bi ZI01U Plastic Layers in a Composting Pilot Plant. In *The Science of Composting* (M. De Bertoldi, P. Sequi, B. Lemmes, and T. Papi, eds.). Blackie Academic & Professional, p. 1271.
7. ASTM/ISR, 1996, Reports on the Compostability Testing of Degradable Polymeric Materials.
8. ORCA Compostability criteria - Guidelines for the Evaluation of Feedstock for Source Separated Biowaste Composting and Biogasification, 1994, Mechelse Drukkerijen n.v. Ijzerenleen 26 Mechelen.

Study of the Aerobic Biodegradability of Plastic Materials under Controlled Compost

Development of the Screening Test Method for Biodegradation by Analysis of Evolved Carbon Dioxide

AKIRA HOSHINO[a], MASAO TSUJI[a], MICHIO ITO[a], MASANORI MOMOCHI[a], AKIKO MIZUTANI[a], KYOHEI TAKAKUWA[a], SACHIKO HIGO[a], HIDEO SAWADA[a], and SHOGO UEMATSU[b]

a: Biodegradable Plastics Society, Grande Bldg., 26-9, Hatchobori 2 Chome, Chuo-ku Tokyo 104-0032, Japan b: School of Pharmaceutical Sciences, University of Shizuoka, 52-1, Yada, Shizuoka-shi 422-8526, Japan

1. INTRODUCTION

Management of municipal solid waste is a serious problem in the current world. Synthetic plastics have been used for various purposes, and annual production of the plastics in 2001 reached about 15 million tons in Japan, and 177 million tons in the world. However, synthetic plastics with high performance and stability cause serious problems in waste management. To solve the problem of plastic waste, biodegradable plastics have attracted the special attention of the public as the plastics of the 21st century, since they are biologically degradable and environmentally friendly. However, before the use of biodegradable plastics spreads, it is crucial to establish the test methods for their biodegradation in the environment. The Biodegradable Plastics Society[1,2] has conducted two field experiments for estimating the degradation of biodegradable plastics in soil at more than 20 locations (including one location in the U.S.) over a period of 3 years. The results showed that all of the tested plastics were degraded in soil. Hoshino[3] *et al.* reported on the influence of weather conditions, and soil properties for the degradation of biodegradable polymers. Nishide[4] *et al.* reported that the degradation of biodegradable polymers occurred under aerobic conditions as compared to anaerobic conditions. Many studies on the degradation behaviour in aquatic and terrestrial environments were reported and several standard test methods were established in 2000 in the world, such as JIS K6950[5] (ISO 14851) and JIS K6951[6] (ISO 14852), relevant to aquatic environments and JIS K6953[7] (ISO 14855) for controlled composting condition, respectively. Especially JIS K6953 based on ISO 14855 is an excellent test-method, which automatically calibrates the evolved carbon dioxide by gas chromatography although it is yet presenting some difficulties. First, the reactor is suitable for large-scale tests and the apparatus is expensive. Second, it costs a great deal to examine specimens to develop biodegradable polymers, compounds and manufactured goods. The focus of this study is to develop the evaluating and screening test method for the biodegradation of plastic materials with a simplified apparatus on a laboratory scale.

2. MATERIALS AND METHODS

2.1 Apparatus

MODA-apparatus is shown in Fig. 1. The principle of this method is to measure the increasing weight of the absorption column charged with soda lime due to evolved carbon dioxide.

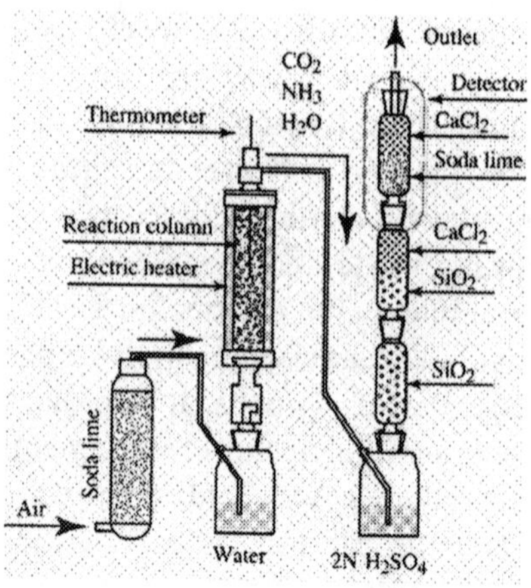

Figure 1. Schematic Representation of the Microbial Oxidative Degradation Analyzer.

It consists of 4 reactors (volume: 500 ml x 4), 4 absorption columns charged with soda lime, an absorption column to remove carbon dioxide from air, a bottle of saturation with water vapour and an air-supplying pump. All reactors are covered with controlled temperature band-heaters. Each reactor is equipped with an absorbent bottle of 2N H_2SO_4, a column charged with silica gel and $CaCl_2$ to absorb ammonia gas and prevent water from passing into the absorption column charged with soda lime.

2.2 Materials

The investigated plastic materials and reference samples are shown in Table 1. The forms of specimens are powder (< 250 μm) and film (50 μm thickness); the films were used as 10x10 mm^2 squares. Cellulose, which was used both as powder and filter paper, was selected as reference material. The reactor mass consisted of mature compost, the investigated specimen, and sea-sand. The YK-1 and YK-2 compost samples used as inoculum originated from poultry-manure, and their chemical properties are shown in Table 2. YK-1 and YK-2 correspond to the manufacturer lot number.

Table 1. Specimens of biodegradable plastics and reference materials employed in the degradation experiments

Sample	Component	Molecular formula	Manufacturer
Celgreen	PCL	$(C_6H_{10}O_2)_n$	Daicel Chemical Industries Co., Ltd.
Bionolle	PBS	$(C_8H_{12}O_2)_n$	Showa Highpolymer Co., Ltd.
LACEA	PLA	$(C_3H_4O_2)_n$	Mitsui Chemicals, Inc.
Avicel	Cellulose (powder)	$(C_6H_{10}O_5)_n$	Merck Co., Ltd.
No. 51A	Cellulose (filter paper)	$(C_6H_{10}O_5)_n$	ADVANTEC Co., Ltd.

Table 2. Characteristics of YK-1 and YK-2 compost samples[a]

Parameters	YK-1	YK-2
Total dry solid (g/kg) [b]	420	573
Volatile solid (g/kg) [c]	700	688
pH [d]	7.0	7.4
Total Organic Carbon (g/kg) [e]	120	181
Total Nitrogen g/kg [f]	17.0	25.2
C/N ratio	6.8	7.2

[a] YK-1 and YK-2 are the lot number of compost samples manufactured by Yahata Bussan Co., Ltd. [b] Compost was dried in a drying oven at 105 °C for 10 hrs. [c] Total dry solid was burnt to ashes at 550 °C in an electric furnace. [d] pH was determined on 1:5 compost:water suspension with a pH meter manufactured by TOA. [e] Carbon content was measured by the Tyulin-method. [f] Total nitrogen was measured by the Devarda's alloy-sulphuric acid method.

Reagent grade sea-sand (16 mesh-pass) was used with the purpose of holding moisture and to support the microorganism activity in the inoculum and the test material as well as to maintain the soil texture. All components were mixed together gently and then put into the reactor.

2.3 Procedure

Under the ring testing conducted by BPS, experiments were performed by using three MODA apparatuses. The four reactors of each apparatus were set with blank, reference sample, and two test specimens, respectively. The components added to each reactor are reported in Table 3.

Table 3. Components, water content, and aeration rate used in the experiments[a]

Specimen	Reactor content	Shape	Polymer (g)	Water content (%)	Aeration rate (ml/min)
PCL and PLA	Blank	-	-	35	35
	Control	powder & sheet	10	35	35
	Sample	powder & film	10	35	35
	Sample	powder & film	10	35	35
PBS	Blank	-	-	40	40
	Control	powder & sheet	10	40	40
	Sample	powder & film	10	40	40
	Sample	powder & film	10	40	40

[a] Avicel and No.51A filter paper were used as control for powder and sheet samples, respectively; the reported weights refer to dry samples. 60 g compost and 320 g sea sand were placed in each reactor.

The biodegradation of the selected polymer samples was investigated under controlled conditions depending on a series of preliminary experiments. The following conditions were set according to JIS K6953 (ISO 14855) standard: pH 7.5-9.5, 35-40 ml/min aeration rate, 35-40% moisture content, and 58±2 °C reactor-temperature. The mixture in each reactor was turned after 14, 30, 41 days; the water content in the reactor was adjusted to 35-40% at the same time intervals.

2.4 Analysis

The degradation extent was periodically evaluated by measuring the increasing weight of the evolved carbon dioxide as absorbed by the column charged with soda lime. An electronic balance with ±0.1 mg precision was used. The measurements were carried out at the end of the experiment within the prescribed time intervals during 45 days.

The percentage of biodegradation Dt (%) for the test samples was calculated from the amount of carbon dioxide evolved for each measurement interval using the following formula:

$$Dt = 100\,(\Sigma CT - \Sigma CB) / TC$$

where $\sum CT$ is the weight of carbon dioxide evolved in the test material reactor between the start of the test and time t, $\sum CB$ is the weight of carbon dioxide evolved in the blank reactor between the start of the test and time t, and TC is the theoretical amount of carbon dioxide calculated for the complete oxidation of the test materials.

The biodegradation of the reference material (cellulose) used to check the compost activity was evaluated in the same way.

2.5 Statistical analysis

Single-correlation multivariate analysis of all experimental data was performed by Excel 97 statistics (SSRI Inc.) to determine the statistical significance of the treatment and degradation levels using $p<0.05$ (t-test, Table 4).

Table 4. Correlation among the degradation levels of the reference material and plastic samples in reactors

Specimens and forms	Avicel	PCL	PBS	PLA	Filter paper	PCL	PBS	PLA
Powder	**	**	**	**	-	-	-	-
Films	-	-	-	-	-	**	**	**
Sheets	-	-	-	-	**	-	-	-
Repeat number (n)	3	6	4	6	3	6	2	6

**: significant at 1% level.

3. RESULTS AND DISCUSSION

YK-1 and YK-2 compost samples were used as the inoculum for powders and films, respectively; however, no major degradability difference was observed in the two sets of experiments. This result can be attributed to the large nitrogen content of the two compost samples that maintained excellent activity levels in the reported experiments.

Figure 2 shows the kinetics of PCL samples. After 45 days, powder and film specimens attained approximately 83.0% and 81.8% degradation, respectively. The reported data were averaged over 6 experiments (n = 6). On the contrary, Avicel and No.51A filter paper reference controls reached an average degradation level (n = 3) of 77.2% and 56.5%, respectively. This result can be related to the different molecular structure of Avicel (polysaccharide) as compared to the tested polymer samples (polyesters). In general, the microorganisms of the inoculum are not always susceptible to the polymeric materials under the conditions employed in the biodegradation. However, judging from the degradation-extent recorded for Avicel, a validation of the tests can be stated.

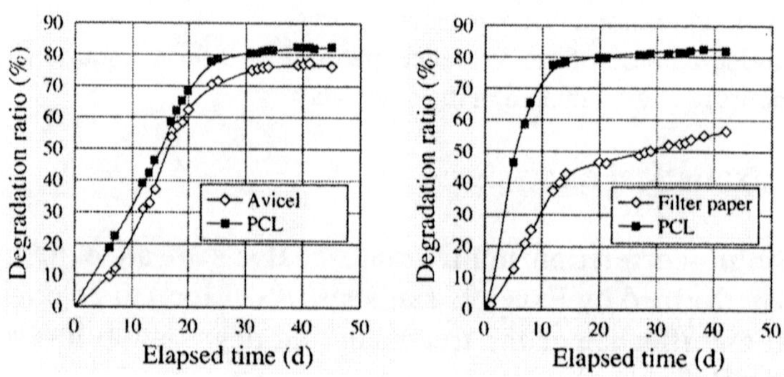

Figure 2. Degradation of PCL powder (left) and film (right).

Figure 3 shows the degradation kinetics of PBS. After 45 days, powder and film samples attained 32.8% (n = 4) and 17.1% (n = 2) average degradation, respectively. On the other hand, Avicel and No.51A filter paper as reference material reached 74.3% and 70.3% average degradation (n = 3), respectively. PBS degradation extent was lower than that recorded for any other specimen. Hoshino[8] et al. reported on the changes of molecular structure of poly(3-hydroxybutyrate-*co*-3-hydroxyvalerate) (PHB/V), PCL, PBS, PLA, and poly(butylene succinate adipate) (PBSA) by the estimation of FT-IR and NMR on the degradation of these biodegradable polymers in soil burial tests. PBS molecular structure consists of a random copolymer of 1,4-butanediol and succinic acid, whereas the other specimens are homopolymers consisting of caprolactone and lactic acid monomeric units. The microorganisms of the inoculum must then act on the two monomeric components, that is 1,4-butanediol and succinic acid under the conditions employed in the biodegradation. The degradation extent of PBS was found to practically coincide with the results of ISO 14855.

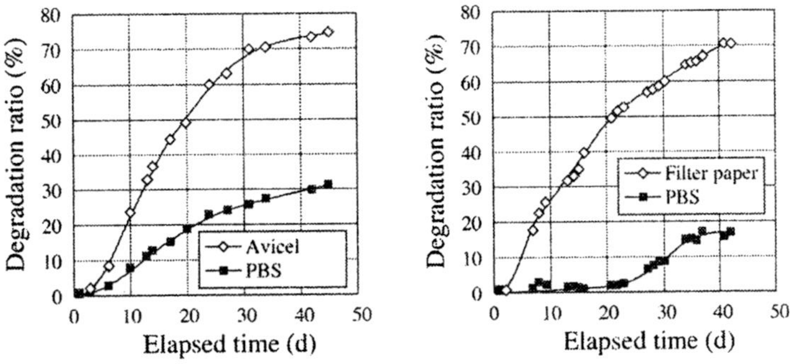

Figure 3. Degradation of PBS powder (left) and film (right).

Figure 4 shows the degradation profiles recorded for PLA. The degradation extent of powder and film reached 61.2% and 73.5% as the mean (n=6) of the experiment during 45 days, respectively. On the contrary, Avicel and filter paper as reference materials reached 49.2% and 70.5% as the mean (n = 3), respectively. On the other hand, judging from the degradation of PLA, it was faster compared to the field test performed by BPS from 1999 to 2000 in Japan. Therefore, we have confirmed that this method is valid for the evaluation of the biodegradability of plastic materials in the presence of mature compost.

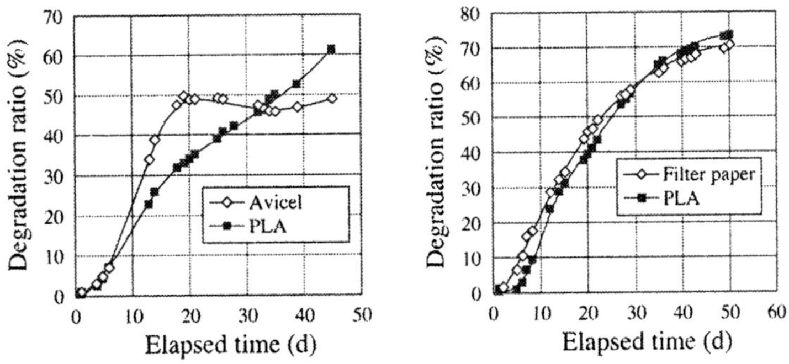

Figure 4. Degradation of PLA powder (left) and film (right).

4. CONCLUSIONS

In a series of experiments, the results of this method using MODA-apparatus approximately correlated with JIS K6953 (ISO14855). It was suggested that the aeration-ratio and the water content played important roles in evolving carbon dioxide for the biodegradation of plastic materials. Degradations studies showed that other factors, such as the turning out intervals of the mass in a reactor and the water-holding capacity also have important roles in the degradation of them. In conclusion, under the controlled composting conditions during the experiment, this method demonstrates an effective screening test method for the compostability of plastic materials in a laboratory scale.

ACKNOWLEDGEMENTS

The authors would like to thank Saida Ironworks Co., LTD. for supplying the MODA-apparatus. The experiments were carried out in cooperation with the technology committee members in BPS. We also thank Daicel Chemical Industries, LTD. Showa High Polymer Co., LTD. and Mitsui Chemicals, INC. for supplying the plastics. We wish to thank Dr. K. Ohshima (BPS) for his scientific advice and Prof. M. Kimura (Nagoya University) for his soil biology advice.

REFERENCES

1. Biodegradable Plastics Society, 1995, *Field Testing of Biodegradable Plastics (I)*. Biodegradable Plastics Society, Tokyo (in Japanese)
2. Biodegradable Plastics Society, 1998, *Field Testing of Biodegradable Plastics (II)*. Biodegradable Plastics Society, Tokyo (in Japanese)
3. Hoshino, A., Sawada, H., Yokota, M., Tsuji, M., Fukuda, K., and Kimura, M., 2001, Influence of Weather Conditions and Soil Properties on Degradation of Biodegradable Plastics in Soil. *Soil Sci. Plant Nutr.* **47**: 35-43
4. Nishide, H., Toyota, K., and Kimura, M., 1999, Effects of Soil Temperature and Anaerobiosis on Degradation of Biodegradable Plastics in Soil and Their Degrading Microorganisms. *Soil Sci. Plant Nutr.* **45**: 963-972
5. Japanese Industrial Standards Committee, 2000, Determination of the ultimate aerobic biodegradability of plastic materials in an aqueous medium - Method by measuring the oxygen demand in a closed respirometer (JIS K6950 (ISO 14851))
6. Japanese Industrial Standards Committee, 2000, Determination of the ultimate aerobic biodegradability of plastic materials in an aqueous medium - Method by analysis of evolved carbon dioxide (JIS K6951 (ISO 14852))
7. Japanese Industrial Standards Committee, 2000, Determination of the ultimate aerobic biodegradability and disintegration of plastic materials under controlled composting conditions - Method by analysis of evolved carbon dioxide (JIS K6953 (ISO 14855))
8. Hoshino, A., Tsuji, M., Fukuda, K., Nonagase, M., Sawada, H., and Kimura, M., 2002, Changes in Molecular Structure of Biodegradable Plastics during Degradation in Soils Estimated by FT-IR and NMR. *Soil Sci. Plant Nutr.* **48**: 469-473.

Environmentally Degradable Plastics and ICS-UNIDO Global Program

STANISLAV MIERTUS and XIN REN
International Centre for Science and High Technology of the United Nations Industrial Development Organization (ICS-UNIDO), AREA Science Park, 34012 Trieste, Italy

1. INTRODUCTION

The production and consumption of polymeric materials in the last decades have generated plastic wastes, which are increasingly putting stress on the environment all over the world. In the 21st century, the demand for polymeric materials are in favour of a 2 to 3 fold production increase as a consequence of the increased consumption in developing countries and countries in transition.

However, the rapid increase in production and consumption of plastics has contributed to serious plastic waste problems and landfill depletion, due to their high volume to weight ratio and resistance to degradation. Moreover, as over 99% of plastics are of fossil fuel origin, the limited non-renewable resources of our planet are increasingly pressed by this rapid growth. Environmentally degradable polymers and plastics (EDPs) from renewable sources will be neutral with respect to the carbon dioxide cycle, thus helping to reduce greenhouse gas emission and global warming.

In summary, EDPs may serve as a promising solution to: a) litters of disposable plastic products and related problems; b) over-loaded landfills by diverting part of bulky plastic packaging to other methods of waste management; c) organic waste management by eliminating the cost involved in removing the collection bags before entering compost facilities; d) preserve the valuable non-renewable resources and make the global economy more sustainable. Besides environmental concerns, the advantageous functionalities of EDPs increasingly find their application in medical and pharmaceutical fields.

2. EDPS AND WASTE MANAGEMENT

With the introduction of these new plastics, the need for organic recycling such as composting and bio-gasification will increase. This may lead to competition with other waste management technologies such as landfill, incineration and mechanical or chemical recycling. More integrated and comprehensive approaches are thus required. These aspects are of particular importance for developing countries and emerging economies, where the concept of waste management and rational production of plastic items need to be matched.

Environmentally sound solid waste management (or generally shortened as SWM), as defined by United Nations Environment Programme, means taking all practical steps to ensure that wastes are managed in a manner that will protect human health and the environment against the adverse effects, which may result from such wastes.

The logical starting point for solid waste management is to reduce the amounts of waste at the source. For the wastes that nevertheless are generated, a strong control should be involved over the life cycle of the product from design, production and use to after-use stages. Waste management should follow the hierarchy of *Reduce, Reuse, Recycle* (or recovery of materials) if cannot be reused, and *Recovery the energy* content if not recyclable, before final disposal usually landfill.

2.1 EDPs and Recycling

Introduction of EDPs is mainly driven by the need of handling the non-recyclable in an economically viable way. Inherently, EDPs are not created for material recycling. A typical plastic recycling process involves re-heating, during which EDPs will usually decompose and make further processing impossible. Mixing of EDPs in the feedstock of recycling will damage the process and the quality of recycled products. Therefore, an effective sorting becomes more important after EDPs are widely adopted. Policy makers as well as industry will need to prepare in advance.

2.2 EDPs and Incineration

Incineration is not the desired destination of EDPs, not much research is yet available regarding the possible effects of EDPs if combusted in incinerators for wastes. The major environmental and health concern over the incineration of waste plastics are heavy metals and hazardous by-products such as dioxin, which is applicable to EDPs.

2.3 EDPs and Organic Recycling (Composting)

Organic recycling (e.g. composting) is the most relevant waste treatment technology for EDPs, particularly for those applications other than medical and pharmaceutical. So far, the international accepted definitions and standards for EDPs are all based on their compostability. The success of EDPs will depend on the availability of composting facilities. On the other hand, demand for EDPs will stimulate development of organic recycling/composting.

The success and failures in many countries demonstrate that the quality of finished product (compost) with acceptably low contaminant level is essential for municipal waste composting as a successful recycling for organic wastes. Mixed wastes composting should be replaced by composting of well-sorted organic waste with minimum contaminants.

2.4 EDPs and Landfill

In many less developed countries, the landfills are usually substandard. Entry of EDPs will further worsen the existing biodegradation and subsequently the contamination of ground and surface water and of the environment.

It is clear that application of EDPs calls for source separation and a more integrated approach towards the whole waste management system. Effective source separation need co-operation of a well-motivated public by educational programs, environmental awareness raising campaigns and mass media. More important is the integration in policies, regulations, economic instruments so as to create concerted pressure as well as economic incentives to promote waste source reduction and separation. Obstacles imposed by improper or fragmented policies and regulations should be identified and modified. Furthermore, a holistic view and life cycle approach should be extended to production and consumption systems.

3. EDPS AND RENEWABLE RESOURCES

To produce EDPs from renewable resources is a natural and promising approach to tackle the plastic waste problem, depletion of fossil fuel resources and global warming.

There are three basic routes to produce polymers from renewable resources feedstock. Direct extraction yields polymer materials such as cellulose, starch, fibres, oils and proteins from which plastic materials can be developed. The second pathway is to convert raw materials first into bio-monomers by hydrolysis, and then to polymers by chemical synthesis. A good example is PLA, the most commercialised so far. The third route is to obtain polymeric materials directly by microbial way from carbon sources through biosynthesis (fermentation). A typical example is the production of PHAs by bacteria.

There is a long history to extract polymeric materials from renewable resources. Now focus has been given to the study of structure–function-biodegradability relationships, preparation of composites of natural with synthetic materials for improved properties and various applications, and development of processing technologies, such as foaming technology. Study and application of other abundant non-starch part of plants, natural fibers (jute, kenaf), oil, fats and proteins are active too.

The key factors that influence microbial production of EDPs are the fermentation productivity, yield from carbon sources, and the ease of recovery and purification of the product. Therefore, R&D in metabolism and utilization of locally abundant resources are of great importance. Genetic technology is being explored for the improvement of yield and functionality. Modifying bacterial genes or transplanting other genes into microbes might allow previously less cost-competitive carbon sources to become competitive with starch and sugars.

To produce polymeric materials directly from plants is regarded as the most efficient and elegant way. However, the quantity and quality of the plastics (e.g. PHB/V) accumulated in plants using transgenic technology need to be raised significantly if the product is to reach the marketplace. Further metabolic investigation will be required. Nevertheless, the work have opened up new possibility for R&D in biotechnology and polymer science.

4. LIFE CYCLE CONSIDERATION

So far only limited life cycle assessment (LCA) have been carried out for biodegradable polymer products, resulting in conclusions favourable to EDPs, though in some other cases the opposite is true. Introducing life cycle consideration into design of novel polymers and polymeric products is a new challenge facing polymer scientists and producers. The so-called design for the environment (DfE) or Eco-design should go beyond the traditional logic and procedure of product design to take into account the after-use stage of the product. The adverse impacts on human health and the environment should be minimised by proper design, not only during production and use phases, but also in final disposal after the product is discarded. For example, one principle of Eco-design is to minimise the use of toxic chemicals and heavy metals unwanted by all disposal methods. Keep this in mind while developing new polymers and plastics will facilitate waste management through decreasing the potential of contamination. All crucial stages of a product's life cycle should be born in mind while developing EDPs and planning for possible applications.

5. SITUATION AND NEEDS IN DEVELOPING COUNTRIES

In many developing countries, as reported during annual expert group meetings organized by the International Centre for Science and High technology of the United nations Industrial Development Organization (ICS-UNIDO) over the last few years, legislation and regulations regarding EDPs are still not existing. Government and industry are either unaware of or have little interest in EDPs, not alone R&D and production of EDPs. On the other hand, due to increased plastic demand, environmental awareness and worsening of environmental problems related to plastic wastes in almost all developing countries, EDPs become more attractive. Government and industry in many countries begin to look into R&D and technology transfer of EDPs.

5.1 Sound Waste Management Implementation and EDPs

Many developing countries have waste management laws and regulations in place. What they lack is enforcement of law. Many countries as more important identify how to promote their implementation. Standards and operational guidelines are required to facilitate implementation of laws. Many countries including developed and developing countries call for international co-operation and harmonisation of biodegradation standards and certification, speed-up of the development of standards and guidelines for composting and compost since composting is the most common and proved disposal of EDPs after use. Developing countries show a stronger interest in enhancing solid waste management since all analysis leads to the conclusion that EDPs is not a single solution, but an option for waste problem thus cannot replace a sound waste management. It is agreed that

EDPs would be useless without effective organic recycling (composting/digestion) facilities. EDPs development and application should be incorporated as a part of solid waste management. Integration of various policies, regulations and methods should be given close consideration.

It is hoped that awareness on EDPs and sound waste management would be raised through international initiatives such as workshops and training organised jointly by ICS-UNIDO with developing countries, particularly in the absence of a similar legislative initiative like EU pushing environmentally sound waste management. External expertise is needed in preparing a national policy and action plan on plastic waste management. A guideline on EDPs, or on a broader sense, sustainable polymers and plastics is expected.

Another major problem regarding plastic in many developing countries is littering habit and social and environmental implications of plastic litters. Policy initiatives and government/industry interventions in this aspect are becoming strong in several countries (e.g. China, India). But the main obstacle is still the enforcement of laws and regulations.

5.2 Other Barriers in Developing Countries

Although law implementation is regarded as the major barrier and driver for EDPs in developing countries, comparable performance and cost of EDPs with conventional plastics are other main factors for the wide acceptance of EDPs.

Consideration of cost is one driver for the utilisation of locally abundant renewable resources for the production of EDPs in developing countries. Much work underway in Europe and North American is mainly focusing on agro-wastes and by-products typically available in these regions. On the other hand, R&D and commercialisation of products in developing countries, for example cassava starch in South East Asia, bagasse in India and South America, sweet potato starch, rice and wheat straw in China, chitin sources in coastal regions, molasses in sugar producing countries etc., are still in their infancy. Therefore, capacity–knowledge building in EDPs and waste management, training for the managers, technology transfer for EDPs, international co-operation in regulations, technology and waste management are urgently needed.

However, when choosing which technology to transfer, one has to be aware that the situation and driving force for EDPs in different countries vary. For example in India it is how to better use natural resources, and to a less extent, concern over waste management. On the contrary, in China the concern over litter is the most urgent. Therefore, in China, a sounder plastic waste management is viewed as the most appropriate short-term solution, while EDPs and polymers made from renewable resources are regarded as an important long-term strategy.

5.3 EDPs R&D Activities in Developing Countries

In a couple of developing countries where there is an active interest in and support to EDPs, blending starch and polyolefins (HDPE or LDPE etc.) is currently the main stream. However, internationally this is not anymore the case, due to their questionable biodegradability. Blending can be a starting point, but effort should be focused on utilising locally rich renewable resources to develop polyester type of EDPs. Developing countries should carefully select their way in developing EDPs and avoid the mistakes and repetition of what the developed countries have made in EDPs and plastic waste management. This also indicates that information support is very much needed in many developing countries. This can be achieved by workshops, training courses, creation of databases, or regional networks with focal points either through already existing regional associations/network or building up new consortia on EDPs and waste management.

There is concern about the competition from the demand for food on bio-based materials in some countries. For example in India, edible starch cannot be used other than as food. This should be kept in mind while selecting cleaner technology for developing countries. Life cycle assessment (LCA) could be a useful tool to assist decision-making but not universally applicable. More understanding of and subsequent training in LCA is desired by many people working in the field of waste management and EDPs.

6. ICS-UNIDO ACTIVITIES ON EDPS

The International Centre for Science and High Technology (ICS), an institution within the legal framework of UNIDO located in Trieste, Italy, focuses on the transfer of know-how and technology from industrialised to developing countries to promote sustainable development. According to the United Nations World Commission on Environment and Development, sustainable development is "the development that meets the needs of the present without compromising the ability of future generations to meet their own needs". It requires harmonisation of economic growth with environmental conservation and protection. In this way, the three main parts – the economy, the environment, and the health of society – can be sustained into future. At present, ICS-UNIDO activities focus on specific sectors within the areas of:

- Pure and applied chemistry
- Earth, environmental and marine sciences and technologies
- High technology and new materials
- Technology management and transfer.

Within the chemistry area, there are four subprograms:

- ➢ Environmentally degradable plastics (EDPs), as one major branch of ICS-UNIDO activity aims at bringing the updated knowledge directly to developing countries, and to stimulate the diffusion of harmonic decisions on the global issue of plastic waste to the benefit of these countries;
- ➢ Remediation technologies, dealing with decontamination of soil and waters that were polluted through domestic and industrial activities;
- ➢ Combinatorial Chemistry and Technologies, for the development of new chemicals, not only to help industry competitiveness but also for environment protection by optimising industrial processes;
- ➢ Catalysis and sustainable chemistry, for cleaner industrial production as process optimisation depend to a great extent upon the improvement of catalyst performance in bulk and fine chemical production.

6.1 ICS-UNIDO's EDPs Subprogram

All activities under the EDPs subprogram are designed to fulfil the gap and needs identified in developing countries. Major projects and initiatives in the field of EDPs, recently completed or being developed at or jointly with ICS-UNIDO are:

1. *Management of Innovation in Environmentally Degradable Plastics,* a project funded by the European Commission within the framework of the Leonardo da Vinci Programme. Major outcomes are: an information package, a training package and database of EDPs technologies, companies, relevant regulations, standards and waste management issues.
2. *Plastic Waste Management and EDPs in Egypt and Turkey,* aims to produce and update an infopack and database on the situation of plastic waste and its management in countries of the Mediterranean region (Egypt and Turkey) and disseminating the best practices in the relevant institutions.

3. *Eco-compatible Bioplastic Packaging in China based on Polyesters from Renewable Resources,* aims at accelerating in China the production of PHAs obtained from renewable resources to develop foamed rice-bowls and kitchenware to alleviate the problem of "white pollution" in China and also in other Asian countries.
4. *Development of a Project Proposal on Industrial Promotion of EDPs Concept in Korea,* will involve both academic institutions and companies. The demonstration project is expected to result in the transfer of technology for the production of EDPs from renewable resources and manufacturing of packaging items.
5. *Sustainable Plastic Waste Management and EDPs from Renewable Resources in Indonesia,* a research project proposed jointly with Agency for Assessment and Application of Technology in Indonesia with the aim of identifying plastic waste problems and treatment technology for the region and EDPs technology transfer using locally rich renewable resources.
6. *Participation under European Commission program* in new project proposals in the development and integration of nanotechnology and biomaterials.
7. *Participation in various proposals for European Network of Excellence* with specific aim to link this network activity to ICS-UNIDO initiative in EDPs. Example of such proposals is chemistry for sustainability through the use of renewable materials and biotechnology.

In addition, the following actions were identified as follow-up initiatives to enhance the dissemination of information, knowledge and technology to developing countries:

- Technology transfer from companies already in the production of EDPs through licensing or joint ventures.
- Promotion of feasibility studies on local resources for production of EDPs. Local needs and potential market should be investigated with some seedling support by ICS-UNIDO.
- Preparing and finalising action plans by governmental institutions or companies interested in EDPs production and utilisation.
- Capacity building through training activities, pilot projects and so on.

International or national workshops continue to be the major mean for training and capacity building. Focus of year 2002-2003 is on key decision makers (government, industry, local officials, academia), to bring awareness and support by ICS-UNIDO and international experts. To achieve this goal, a half day special briefing sessions with ICS-UNIDO documentation on the selected topic of the country was included in the following international workshops on EDPs and plastic waste management held in 2002:

- China: 25-27 October 2002, title: *Sustainable Development and EDPs.* 102 participants – representing companies (50%), academics, and government officials.
- Thailand: 21-23 October 2002; title: *Plastics Recycling and Development of EDPs.* 138 participants representing agricultural producers and plastics industry, government officials, and academia. Chile: 25-27 November 2002, title: *EDPs, Plastics Recycling and Polymer Waste Management,* 28 participants representing Latin American countries.

For the year 2003 the following international workshops are planned within the ICS programme on EDPs

- Uganda: September 2003, proposed title: *Plastics recycling and EDPs.* The workshop will be addressed to African countries
- Iran: September/October 2003, title: *Plastics recycling and EDPs*, The workshop will be addressed to Asian countries.

6.2 ICS-UNIDO's Initiatives in Promoting Tools for EDPs and Plastics Waste Management

In order to efficiently promote sustainable polymers and sound plastic waste management worldwide, in particular in developing countries, barriers mentioned above should be overcome and gaps should be filled. Various means and tools have been proposed and reviewed for this purpose, resulting in the following ones that are considered of higher priority and relevance: information networking focal points, decision support tools (DST), guidelines and database.

- Focal point network and database

Establishment of focal point for the exchange of information on EDPs in Central Eastern European countries is undergoing. Similar action will be undertaken in Mediterranean and Middle East countries, Latin America, Africa and Asia so as to form a network of information exchange. Based on this, a database on EDPs will be able to be constructed and linked in future. ICS-UNIDO will continue to take advantage of its co-ordinating and catalytic role as an UN organisation to support this initiative as well as the derived co-operation and pilot projects. Such a global network needs co-operation from all those involved and full use of possibility provided by the advance of computer and intertnet technology.

- EDPs Guidelines

Brief guidelines on EDPs are under preparation. They cover sustainable polymers and plastics, their relation with waste management, degradation mechanism, key issues of global importance such as standardisation and certification, and major applications of EDPs. These guidelines can be used in awareness raising and training targeted at government officials, decision makers, managers, professionals in other fields and general public.

- Decision Support Tools

To make more quantitative comparison and evaluation of the environmental performance and economic feasibility of different waste management options so as to choose the most suitable technology requires decision support tools (DST). Developing countries showed a strong interest in such tools. Several models on solid waste management and life cycle assessment are being reviewed by ICS-UNIDO for their reliability and usability under international context, particularly for their applicability in various developing countries and economies in transition. It is envisaged that such kind of tools would evolve into a decision support system for waste management with EDPs as an element.

7. CONCLUSIONS

This paper has given a general introduction about the international efforts in the field of environmentally degradable plastics (EDPs), exemplified by activities at ICS-UNIDO. These activities and projects reveal some major problems that concern both industry and the society, in particular those in developing countries. EDPs provide new chances and yet new challenges to plastic industry, waste management and sustainable development as a whole. To what extent these new products will benefit both the environment and the society will depend much on how effectively we can tackle the challenges facing us. A holistic view, a life cycle strategy and an integrated approach proved to be necessary for the success of large-scale application of EDPs.

REFERENCES

1. ICS-UNIDO, September 2000, Proc. Int. Workshop: *Environmentally degradable plastics: industrial development and application*. Seoul, South Korea.
2. S. Miertus, and X. Ren, 2002, Environmental degradable plastics and waste management. *Polymery,* **47,** 28-33.
3. ICS-UNIDO, 2002, *Report on expert group meeting on environmentally degradable plastics*. Dec., 2001, Trieste, Italy.

Biodegradable Plastics

Views of APME (Association of Plastics Manufacturers in Europe)

FREDDY MARÉCHAL
Director, APME Technical & Environmental Centre

1. INTRODUCTION

1.1 About APME

APME is the voice of the plastics manufacturing industry, representing over 90 per cent of Western Europe's polymer production capacity - with a turnover of more than 29 billion €. Combined with the European polymer converting industry and the machinery manufacturers, the plastics industry represents a major contributor to Europe's economic strength employing well over one million people, generating sales in excess of 135 billion € and representing an important sector of the European chemical industry, Europe's second largest industry

1.2 Why an APME position on biodegradable plastics?

The demand for such position came from customers and converters somewhat puzzled by the various, sometimes questionable and contradictory messages and ideas launched in public debates. The relation between biodegradability and starting raw materials, litter, or environmental impact, as well as the comparative merits of plastics based on crude oil or on biomass, are amongst the aspects that deserve some basic clarification.

Biodegradable Polymers and Plastics, Edited by Chiellini and Solaro
Kluwer Academic/Plenum Publishers, New York, 2003

2. APME POSITION

- Biodegradable plastics are polymer specialities. They provide special properties in selected applications and can offer certain additional advantages during use and/or waste management.

- Technical quality and functionality of plastics products are essential in order to meet the product requirements of the respective application.

- Biodegradable polymers can be made from both renewable and fossil resources.

- For biodegradable plastics waste, as for any other plastics waste, all recovery options should be open in order to achieve an ecologically sensible and economically feasible, i.e. eco-efficient, use of the biodegradable plastics waste fraction.

- Biodegradable plastics, which are compostable can be treated biologically together with other bio-waste.

- The compostability is independent of the resources used as raw materials. Therefore, the market should decide which raw material is best for the respective biodegradable plastics application.

- The compostability of biodegradable plastics should not encourage irresponsible behaviour leading to litter. Instead, improved awareness and education on proper use and disposal should be promoted.

3. BACKGROUND

3.1 Definitions

A "*degradable plastic*"[1] is a material designed to undergo a significant change in its chemical structure under specific environmental conditions. The resulting loss of material properties can be measured by standard test methods. This position paper refers to biodegradable plastics only and not to photodegradable plastics.

[1] Definition ISO/CD 15315

When degradation is caused by biological activity, especially by enzymatic action, it is called *"biodegradation"*. If the biodegradation process is sufficient to mineralise organic matter into carbon dioxide or methane respectively, water and biomass the material is termed *"biodegradable"*[2]. A material is *compostable* when it is biodegradable under composting conditions.

The *"biodegradability"* of plastics is dependent on the chemical structure of the material and on the constitution of the final product, but not on the resources used for its production. This fact is proven both scientifically and technically. Therefore, no distinction should be made based on the source of the raw material.

3.2 Production, Application and Usage

Most commercial biodegradable plastics are currently based on combinations of polymers of vegetable and/or fossil origin, using the synergy potential of both raw materials, renewable and fossil resources. Today, production of vegetable-based polymers requires a significant use of fossil fuel, agricultural land and water. For most biodegradable plastic applications, the partial use of fossil-based raw materials is necessary in order to fulfil the requirements of quality and functionality of the final product.

Biodegradable plastics are specialities developed for selected applications, which offer certain additional advantages during use and/or recovery. This can be demonstrated by some typical examples:

- In agriculture, they can be used as e.g. flowerpots, which completely biodegrade in the soil while functioning as a soil conditioner, leaving biomass. Mulch-films assist in the growth of plants and also have a positive effect on weed control. The usage of biodegradable foils eliminates the need for mechanical removal and thus damage to plants is avoided. After their use, biodegradable mulch films can be ploughed in as they biodegrade in the soil.

- In packaging, biodegradable plastics can be used as biodegradable kitchen-waste bags, which can be composted together with their biodegradable content, enhancing the functioning of green waste collection systems while providing a hygienically safe solution and simplified handling.

[2] Definition ISO/CD 16929

3.3 Recovery and Disposal

In addition to the conventional possibilities of waste management – mechanical recycling, feedstock recycling and energy recovery – biodegradable plastics waste can also be treated by composting (aerobic degradation) or digestion (anaerobic degradation). In principle, all recovery options should be open in order to achieve an ecologically sensible and economically feasible, i.e. eco-efficient, use of the biodegradable plastics waste stream.

Biodegradable plastics scrap can be recycled in a "closed loop" manufacturing operation or reworked at the converter. This is common practice when the biodegradable plastics waste is clean and of a pure grade. Since post-consumer plastics waste does not normally fulfil such quality criteria, other recovery routes must be considered. Here, biological treatment together with biowaste is a sensible option when a consistent use of biodegradable plastics is possible such as in the example of compostable kitchen-waste bags explained above.

Education of the consumer on the benefits of plastics products is an important part of the plastics industry's message. It is essential that the availability of biodegradable or compostable plastics is not a "licence-to-litter" and that information is given on the need to separate the compostable waste fractions at source.

3.4 Standardisation and Certification

Biological treatment, e.g. composting or digestion, is an important method for the management of biodegradable plastics waste. International normalisation institutes have developed, or are in the process of developing, standard test methods to confirm the biodegradability or the compostability of materials and of products (e.g. EN 13432 for compostable packaging). In this context, concepts for characterisation, labelling and identification are being developed.

Biodegradable plastics meet stringent norms with regards to their complete biodegradability, compost quality and product safety. Conformity with a standard can be declared by self-assessment or by third party certification.

The European plastics manufacturing industry is of the opinion that both biodegradable materials and the resulting compost product should be standardised.

4. CONCLUSIONS

Biodegradability can bring benefits in specific applications, particularly at the end of life of some plastics products; but it is not an end by itself and it should be highlighted that end-of-life is generally not the main source of environmental impact in a product life. Fitness for use is the main driver for plastic type selection for any application and, in the perspective of an optimised use of existing resources, eco-efficiency over the whole life cycle should be the main criterion against which the combined environmental and economic impacts of any product have to be assessed.

Plastics are very versatile and innovative materials. New developments, including in the field of biodegradable polymers, deserve financial support and enabling legislation, but legal measures should never create undue market distortions in favour or at the expense of any material type.

Molecular design, compounding, inter-plastics or multi-material combination as well as product designers' creativity continuously extend the field of plastics' applications. And biodegradable plastics have their own role to play in such developments.

Plastics are really the material of choice for the 21st century.

For more information on plastics, see www.apme.org

Market Introduction of Compostable Packaging

Consumers' Acceptance and Disposal Habits in the Kassel Project

JÖRAN RESKE
INTERSEROH GmbH, Stollwerckstr. 9A, D-51149 Köln, Germany

1. INTRODUCTION

After about a decade of technical development, the application of biodegradable polymers (BDPs) is technically feasible for a wide range of products including packaging. The packaging sector is one of the most promising, because high amounts of BDPs could be used. Especially food packaging usually only needs to protect the product for a very limited time span. Furthermore, when it comes to the disposal of the packaging, contamination by food does not represent a problem when the packaging is treated in a biological recovery system as for example a composting plant. This perspective has lead to several activities with the aim of developing the market for compostable (food) packaging. Among these, the Kassel project – named after the place where it took place from spring 2001 until autumn 2002 – has attracted very much interest both at national German level and worldwide. This attempt was the first to place a spectrum of compostable packaging in the market and to use the existing routes of distribution and waste management.

The acceptance of consumers for the new type of packaging and their disposal behaviour were examined with scientific methods. A summary of the most important findings is given here, indicating very high consumer acceptance and good results for the source-separated disposal after use. From these findings, general conclusions might be drawn also for other places with regard to the recovery of compostable packaging via the source-separated collection of biowaste.

Biodegradable Polymers and Plastics, Edited by Chiellini and Solaro
Kluwer Academic/Plenum Publishers, New York, 2003

2. BACKGROUND: THE SITUATION BEFORE THE KASSEL PROJECT

Over the last years, several manufacturers have achieved a tremendous technical progress providing compostable plastics and products. At the same time, the technical framework and legislation have been developed in many countries, making possible the routine application of BDPs e.g. for packaging. Especially in Germany, the federal government supported this development by funding not only R&D projects, but also projects to establish the needed technical infrastructure. As a result of these efforts, two important preconditions for the safe application of BDP packaging could be fulfilled:

1. A Certification system was established to ensure safe compostability. This was a precondition for the acceptance of BDPs in the composting plants. The system has been set up by the International Biodegradable Materials Association IBAW (Berlin, Germany) in cooperation with the Associations of the German farmers, municipalities, waste management companies and others. The system is run since 1998 by DIN CERTCO (Berlin, Germany) and has lead to the implementation of a global network on certification for compostable products, as can be seen in Fig. 1.

Figure 1. Participants in the International Compostability Certification Network.

2. A logo was chosen by the stakeholders involved in the development of the certification system. The logo (see Fig. 2) is a clear indication for consumers in order to provide the information that the marked packaging is compostable and shall be disposed of

via the biowaste collection system. In order to be as individual as possible and thus to differ obviously from other labels, a hexagon was added to the original seedling logo held by IBAW.

kompostierbar

Figure 2. Logo for certified compostable products.

3. THE PROJECT: ISSUES AND PARTICIPANTS

With the preconditions for the application in place, the last crucial question for the proposed waste management of compostable packaging via the biobin was, whether it would be possible to educate consumers about the logo and the right way of disposal, in order not to risk additional misplacement into the biobin. The last objection compost plant operators would raise was, that they did not trust consumers to differentiate between compostable plastic packaging and traditional – non-compostable – packaging. In this case, as a consequence, misplacement of traditional plastic packaging into the biobin could rise.

The second important question in the model project was, whether consumers would accept compostable packaging for their food and whether they would accept higher prices, which as a rule still occur with compostable packaging. The two main questions are visualized in Fig. 3.

The Kassel project being the first attempt worldwide to provide the complete infrastructure and full cooperation of the federal, state and local institutions and some of the leading retail chains, several companies from the BDP industry were interested in this approach and participated in the project. Fig. 4 gives an overview of the participants. Detailed information about the companies and about the entire project may be obtained from the Internet at www.modellprojekt-kassel.de.

In Kassel, big amounts of compostable packaging were distributed and disposed of by consumers via the biobin for the first time ever. Together with the biowaste of the households, the packaging was collected by the local waste hauler (Stadtreiniger Kassel) and treated in the composting plant of the City Göttingen. To guarantee for the safe treatment, only certified packaging was allowed to be distributed in the model project. The whole waste management part was organized by Interseroh, which included the acknowledgement of the system by the Hessian Ministry of Environment, Agriculture and Forestry. This organizational model was chosen to fulfil the obligations of the relevant German ordinances (packaging ordinance, biowaste ordinance).

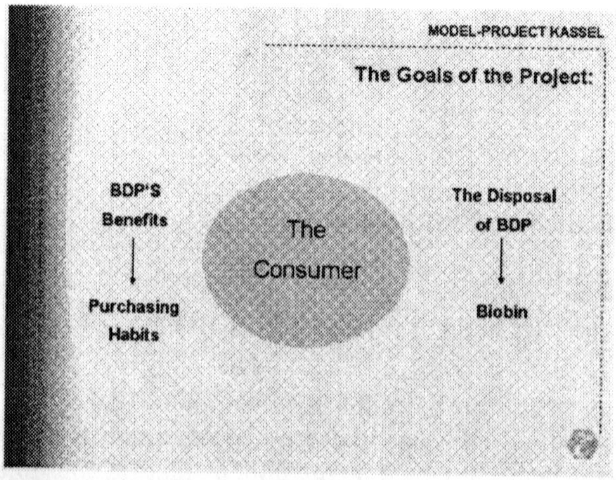

Figure 3. Main questions in the Kassel project.

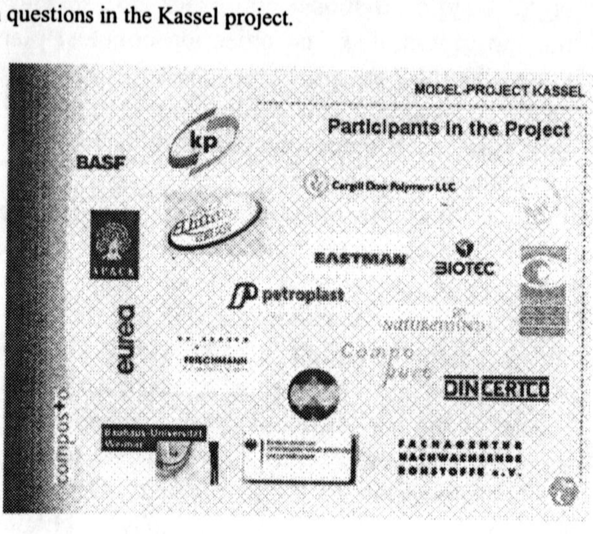

Figure 4. Participants in the project.

A broad spectrum of BDP packaging was tested in Kassel, the majority of which were shopping bags and small fruit bags. All kinds of biodegradable films (starch based, Copolyester-type, PLA), rigid starch and PLA trays and dishes, as well as butter wrappers made from extrusion coated paper and plant pots made from starch and cellulose were applied. Figure 5 displays some of the distributed packaging types.

Figure 5. Examples for Packaging distributed during the Kassel Project.

An important task in the project was the distribution of information. Some of the most important partners were therefore the media. Especially local and regional newspapers featured news and stories about the project and the next packaging types to be placed in the markets. Also TV-stations at the regional and federal level were reporting. Additionally, information material was spread to the households and several events informed the consumers about the project.

4. RESULTS

For the success of compostable packaging in the markets, consumers play the decisive role. Their acceptance will influence the decisions of producers to pack the goods in the new type of packaging. Also retailers were very interested in the opinion of the consumers, since in the application of packaging made from renewable resources they see a marketing opportunity. For all these reasons, the focus of the Kassel project was on the consumer acceptance of compostable packaging. They were asked generally, what they thought about the idea to replace conventional by compostable packaging. Another important question was, what consumers would be willing to pay additionally for the innovative packaging and how satisfied they were, in case they had bought BDP-packed goods. Figures 6 to 8 show some of the results.

Besides the consumer acceptance, a similar important aspect was their disposal behaviour. They were asked where they had put the packaging after use - to find out whether they knew the logo, identified the respective packaging and disposed it in the right way. The results of the survey were compared to the examinations of Bauhaus-University of Weimar, which allowed telling how much of the distributed packaging was collected in the biobin. In general, the answers of the consumers related very well to the findings of Bauhaus-University. Most of the compostable packaging ended up in the biobins, a remarkable amount was composted by the consumers in their gardens and only minor amounts were disposed in the other waste collection systems (see Fig. 9). The most important result of the model project with regard to the waste management was, that no rise in the misplacement of conventional plastic packaging in the biobins could be detected.

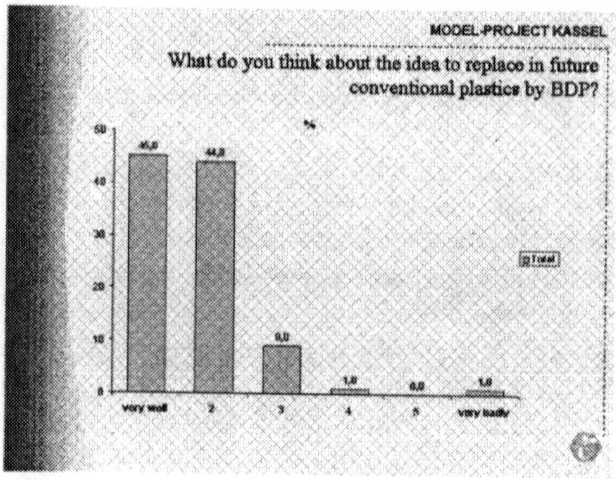

Figure 6. General opinion of consumers.

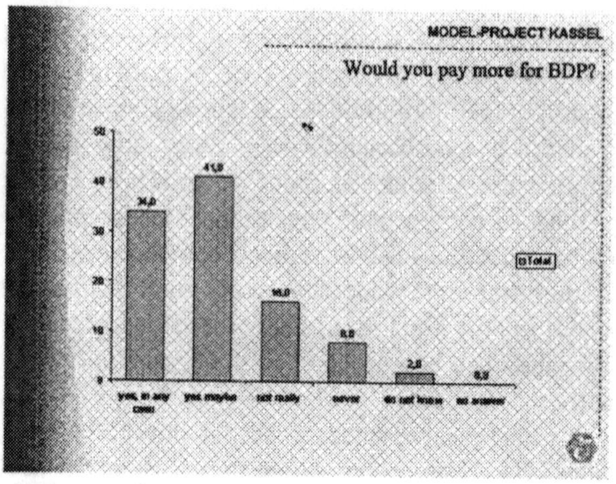

Figure 7. Willingness of consumers to pay more.

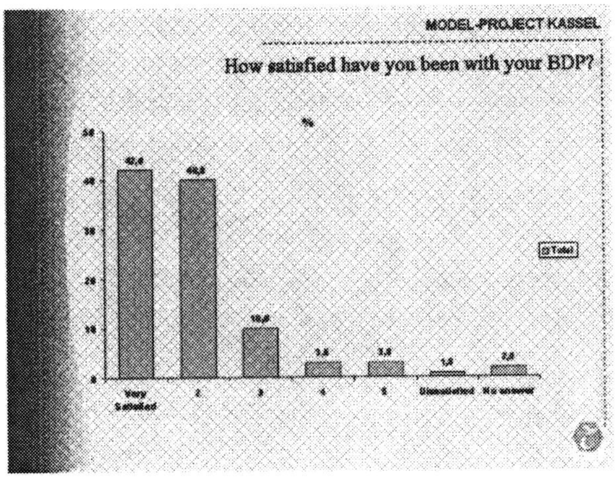

Figure 8. Satisfaction of consumers with BDP packaging.

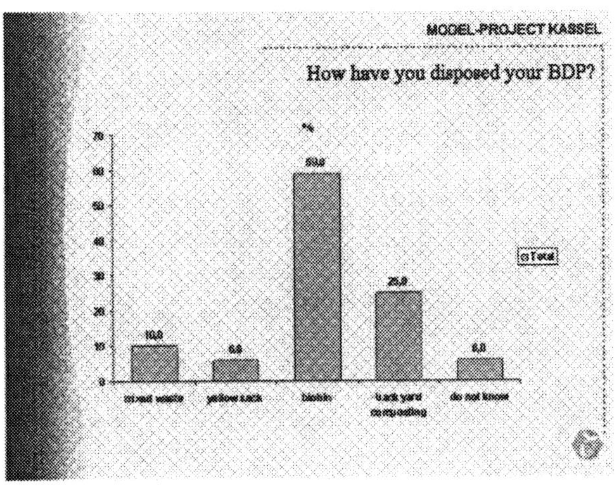

Figure 9. Disposal routes chosen by consumers for compostable packaging.

In order to close the material cycle from renewable resources via packaging made thereof to the disposal of the packaging into the biobin, the composting and the use of the resulting compost as a fertilizer in agriculture, a very important aspect is the quality of BDP compost and the acceptance of the farmers to apply such compost. To address these issues, a field trial was performed with compost evolving from the Kassel project. With only certified packaging distributed in Kassel, the experts expected no differences between BDP compost and compost without BDP input. So were the results, proving the good quality of compost made from the joint collection of biowaste from the household and certified compostable packaging.

In conclusion, the demonstration project in Kassel showed in a very convincing way that the proposed material cycle with certified compostable packaging works and is highly appreciated by consumers and retailers.

NOTES

Detailed reports are available either from IBAW (www.ibaw.org) or from Bauhaus-University Weimar (www.uni-weimar.de).

ACKNOWLEDGEMENTS

The project was supported by grants from the German Ministry of Consumer Protection, Nutrition and Agriculture. The project coordinator Mr. Martin Lichtl of LICHTL Consultancy (Frankfurt, Germany) and Mr. Markus Weber of DIN CERTCO (Berlin, Germany) provided figures.

PART 2

BIOBASED SYSTEMS

Do Biopolymers Fulfill Our Expectations Concerning Environmental Benefits?

MARTIN PATEL
Department of Science, Technology and Society, Utrecht University, Padualaan 14, 3584 CH Utrecht, Netherlands

1. BIOPOLYMERS – A RELEVANT TOPIC?

This book chapter discusses results from Life Cycle Assessment (LCA) studies for the commercially most important biopolymers. Biopolymers are defined here as polymers that are fully or partially produced from renewable raw materials, covering both biodegradable and non-degradable polymers.

Looking back at the development in the last two decades one can conclude that the renaissance of biopolymers in Europe began with the commercialisation of *biodegradable* polymers. These were originally developed and introduced to the markets for two main reasons. Firstly, the limited volume of landfill capacity became more and more a threat and secondly, the bad general image of plastics in public called for more environmentally friendly products. While the first issue has largely disappeared from the top of the agendas due to the introduction of plastics recycling schemes and due to newly built incineration plants, the environmental performance is currently the main argument for biodegradable polymers. The environmental motive also played an important role for the development and manufacture of biopolymers that are not biodegradable.

Biopolymers have been benefiting from the progress made in biotechnology in the recent past. Apart from biotechnology also emerging nanotechnology is about to offer new opportunities for bio-based polymers.

Biodegradable Polymers and Plastics, Edited by Chiellini and Solaro
Kluwer Academic/Plenum Publishers, New York, 2003

In 1998, the production of bio-based polymers in the European Union amounted to about 25,000 tons[1]. The total market is projected to grow autonomously to about 500,000 t until 2010 (28% yearly) and to about 1 million tons if supportive policies and measures are implemented (36% yearly[2]). So far, there are no policies and measures at the EU level while national legislation provides advantages for bio-based polymers in certain areas (e.g. due to the packaging ordinance in Germany).

The fact that environmental considerations have been and will continue to be an important motivation to develop and introduce biopolymers calls for a comparison of their environmental performance with their petrochemical counterparts. To this end, life cycle assessment (LCA) can be applied, which is a standardised method to quantify environmental impacts[2]. LCA studies, however, do not address environmental risks (e.g. related to outcrossing of genetically modified species) and they neither cover ethical, social, and economic aspects.

The biopolymers covered in this book chapter are: Starch polymers, polyhydroxyalkanoates (PHA), polylactides (PLA), lignin-epoxy resins, epoxidised linseed oil and composites reinforced with natural fibres such as flax, hemp, and china reed (miscanthus). The first three materials are biodegradable while this is not the case for the remaining studied materials.

The types of end products covered are primary plastic materials (mainly pellets, i.e. granules; not to be confused with transportation pallets), loose-fill packaging material (packaging chips), films, bags, mulch films, printed wiring boards (for electronics), thickener for lacquer, two different panels for passenger cars, and transport pallets. These products are compared with equivalent products made from petrochemical polymers – in many cases polyethylene, polypropylene or polystyrene.

Some of the studies reviewed are rather limited in scope by assessing only energy use and CO_2 emissions. They are nevertheless included because they contribute to a better understanding of the environmental aspects by addressing additional types of materials and by providing an indication of the uncertainty of the results.

2. ENVIRONMENTAL SUPERIORITY? – HAVING A CLOSER LOOK AT STARCH POLYMERS

This section discusses the group of polymers that has by far the greatest commercial importance today: starch polymers. We will first study *pure* starch polymers and will then move on to those starch polymer types that also contain petrochemical co-polymers.

Table 1 shows the (non-renewable) energy requirements for pure starch polymers. All energy data are expressed in primary energy terms. They include also the requirements for extraction, transportation, and pre-processing of fuels (also referred to as energy requirements for energy or precombustion energy). According to CARBOTECH[3], which is one of the most prominent studies in the field, a total of about 25.5 Gigajoules (GJ) of primary, non-renewable energy are required for one ton of starch polymer. This amount also referred to as "cumulated energy demand", excludes the energy content of the biomass input. Expressed in percentages, about 12% of the total energy use is required for cultivation, 42% for starch and sorbite production, 44% for processing (destructurisation of starch by raising the temperature beyond the melting point, followed by the manufacture of thermoplastic granules by means of extrusion) and 1% for transportation and waste management[3].

To assess the uncertainties, the results according to CARBOTECH were compared to other sources. The comparison was made in two steps, firstly for cultivation and starch production and secondly for polymer production:

- As the first step, Table 1 compares data for cultivation and starch production according to CARBOTECH[3] with results according to TFA[9] and to our own calculations based on a combination of three sources (Kaltschmitt/Reinhardt[4], BIFA/IFEU/Flo-Pak[5], and Geier *et al.*[6]). In some cases the outcome is quite close (e.g. for potato cultivation according to CARBOTECH versus TFA), but it can also deviate by a factor of 2 to 4 (e.g. for starch extraction from wheat). The different allocation procedures are the main reason for this wide variation (see footnotes of Table 1). The values for the total process chain (last column of Table 1) deviate by a maximum of roughly 50% for potato starch and maize starch and by a maximum factor of about 2 for wheat starch.
- Within the second step, the energy for processing (mixing and heating) was studied in more detail. Data were provided by Novamont (Novara, Italy) for two types of starch polymers and by Biotec (Emmerich, Germany) for one product (survey by N. Kopf[7]). The values range from 0.8 to 2.5 GJ electrical energy per ton of polymer. Combining these data with a wide range of efficiencies for power generation (23-40%, including energy requirements for mining, transportation and pre-processing) results in a minimum value of 2.0 GJ primary non-renewable energy while the upper value equals 11.2 GJ.

By combining the data of the two steps, the overall uncertainty can be estimated: As shown in Table 2, the energy requirements of the total system range between 14.7 and 29.7 GJ/t at most. Compared to the value originally established by CARBOTECH[3], i.e. 25.5 GJ/t, this represents a maximum range of –42 to +16%.

Table 1. Energy requirements for the production of pure thermoplastic starch polymers (without petrochemical copolymers)

Reference	Crop	Cumulated energy demand (in GJ/t starch) Cultivation	Starch/sorbite extraction	Total
CARBOTECH[3]	Potatoes [a]	3.8	7.2	10.9
	Maize (sorbite) [a]	1.7	17.9	19.7
TFA[9]	Potatoes [b]	4.1	4.4	8.5 (-12.9)[c]
	Maize [b]	5.9	3.5	9.4 (-10.1)[c]
	Wheat [b]	4.9	3.3	8.2 (-9.9)[c]
Own calc. based on Kalt-Schmitt[4], BIFA[5], Geier *et al.*[6]	Potatoes	5.8-10.9 [d]	< 3.6 [g,j]	< 9.1 - 14.5
	Maize	2.5 - 3.1 [e]	< 10.4 [h,j]	< 12.9 - 13.5
	Wheat	3.1 - 3.6 [f]	< 15.3 [i,j]	< 18.4 - 18.9

[a] By-products of starch production are fodder and raw materials for other syntheses. The energy content was used as the basis for allocating energy use and the environmental impacts to the main product and the by-products. For potatoes, this gives an allocation ratio of main products to by-products of 2:1; for sorbite production, the ratio of main products to by-products is 0.52.

[b] It is not obvious whether the allocation of energy use and environmental impacts was based on prices, the energy content, or the mass of the various products.

[c] The value in brackets includes the share of energy for the production of all co-products and hence represents the absolute maximum.

[d] According to BIFA/IFEU/Flo-Pak[5] about 5.0 tons of potatoes are required for 1 ton of starch and substantial amounts of co-products. Allocation according to prices results in an allocation fraction of 87% for starch, while allocation according to mass result in a fraction of 46%; both approaches are used, resulting in the range. The primary energy requirements for potato cultivation amount to 2.5 GJ/t (fresh matter) according to Kaltschmitt/Reinhardt.[4]

[e] According to BIFA/IFEU/Flo-Pak[5] 1.52 tons of maize is required for 1 ton of starch and substantial amounts of co-products. Allocation according to prices results in an allocation fraction of 81% for starch, while allocation according to mass result in a fraction of 66%; both approaches are used, resulting in the range. The primary energy requirements for maize cultivation amount to 2.5 GJ/t (fresh matter) according to Geier *et al.*[6]

[f] According to BIFA/IFEU/Flo-Pak[5] 1.28 tons of wheat is required for 1 ton of starch and the co-products. Allocation according to prices results in an allocation fraction of 92% for starch, while allocation according to mass result in a fraction of 78%; both approaches are used, resulting in the range. The primary energy requirements for wheat cultivation amount to 2.5 GJ/t (fresh matter) according to Kaltschmitt and Reinhardt[4].

[g] According to Meuser and German[8], quoted in BIFA/IFEU/Flo-Pak[5]: 2.09 GJ primary energy equivalents for electricity per ton of starch and 1.19 GJ natural gas per ton of starch.

[h] According to Meuser and German[8], quoted in BIFA/IFEU/Flo-Pak[5]: 4.366 GJ/t steam (produced at 71.4%), 0.46 GJ natural gas, 1.109 GJ electricity (produced at 37.2%).

[i] According to Meuser and German[8], quoted in BIFA/IFEU/Flo-Pak[5]: 7.322 GJ/t steam (produced at 71.4%), 1.40 GJ electricity (produced at 37.2%).

[j] The value includes the share of energy for the production of all co-products and is hence overestimated. An average of 9% has been added to account for the extraction of primary energy, its transportation, and preprocessing (precombustion).

Table 2. Uncertainty analysis - Energy requirements for the production of pure thermoplastic starch polymers (without petrochemical copolymers)

Cumulated energy demand (CED)	Unit	CARBOTECH[3]		Sensitivity analysis, MAX		Sensitivity analysis, MIN	
		Potatoes	Maize (sorbite)	Potatoes	Maize (sorbite)	Potatoes	Maize (sorbite)
Starch production	GJ/t starch	10.9	19.7	14.5	19.7	8.5	19.7
		67% ↘	↙ 33%	67% ↘	↙ 33%	67% ↘	↙ 33%
Raw materials	GJ/t starch polymer	13.8		16.2		12.2	
Processing	GJ/t starch polymer	11.2		11.2		2.0	
Transp. & waste management	GJ/t starch polymer	0.5		0.5		0.5	
Total system	GJ/t starch polymer	**25.5**		**27.9**		**14.7**	

It is important to note that the availability of further, independent datasets could lead to different outcomes. Especially the non-availability of independent data of sorbite production from maize needs to be pointed out.

Apart from *pure* starch polymers (as discussed so far), starch polymers containing petrochemical copolymers are also commercially available. The more of these copolymers is added, the higher the overall energy requirements are (Table 3). Nevertheless, the values are still clearly lower than those for polyethylene (PE), which belongs to the petrochemical polymers with the lowest energy requirements for production.

Table 3. Life cycle energy requirements and CO_2 emissions for pure thermoplastic starch and for starch copolymers (various sources; compare Patel *et al.* [10])

Type of plastic	Share of petrochemical compounds (% wt)	Energy use throughout life-cycle (production and waste incineration) [a] (GJ/t product)	Fossil CO_2 emissions throughout life-cycle (production and waste incineration) (kg CO_2/t product)
TPS [b]	0%	25.5	1140
TPS/poly(vinyl alcohol) [c]	15%	25.0	1730
TPS/polycaprolacton [c]	52.5%	48.4	3360
TPS/polycaprolacton [c]	60%	52.4	3600
LDPE [d]	100%	80.7	4840

[a] Non-renewable energy (fossil and nuclear).

[b] Source of data in this row: CARBOTECH[3], p.51. The CO_2 to energy ratio according to this dataset is very low (45 kg CO_2/GJ). The reason might be co-firing of biomass waste.

[c] Fraunhofer ISI[11].

[d] APME[12]. Embodied carbon: 3140 kg CO_2/t PE. For comparison, the values for HDPE are 79.9 GJ/t and 4.84 t CO_2/t.

3. ENVIRONMENTAL COMPARISON – A BIRD'S VIEW

Table 4 provides data for polycaprolactone (PCL) and polyvinyl alcohol (PVOH), which are both used as co-polymers for starch plastics. Life-cycle practitioners consider these data to be subject to major uncertainities[13]. This is supported by the considerable range of values for energy use in the case of polyvinyl alcohol and for CO_2 emissions for both PCL and PVOH (Table 4).

In the case of starch polymer pellets, energy requirements are mostly 25-75% below those for polyethylene (PE) and greenhouse gas emissions are 20-80% lower. These ranges originate from the comparison of different starch/copolymer blends, different waste treatments, and different polyolefin materials used as reference. Regarding the latter, APME data for LLDPE (72.3 MJ/kg) and LDPE (80.6 MJ/kg) were assumed, which are lower than the value according to Carbotech (91.7 MJ/kg, see Table 4). The lower APME values serve also as reference for the comparison with the other biopolymers (see below).

Starch polymers score better than PE also for all other indicators listed in Table 4, eutrophication being the sole exception. The lower the share of petrochemical copolymers, the smaller is generally the environmental impact of starch polymers. However, the application areas for pure starch polymers and blends with small amounts of copolymers are limited due to inferior material properties. Hence, blending can extend the applicability of starch polymers and thus lower the overall environmental impact at the macroeconomic level.

The cradle-to-factory gate energy requirements for PLA are 20-30% below those for polyethylene, while GHG emissions are about 15-25% lower. The results for PHA vary greatly (only energy data are available). Cradle-to-factory gate energy requirements in the best case (66.1 GJ/t) are 10-20% lower than those for polyethylene. PHA does not compare well with petrochemical polymers for more energy intensive production processes.

Since all data in Table 4 refer to the current state-of-the-art, technological progress, improved process integration and various other possibilities for optimisation are likely to result in more favourable results for biopolymers in the future.

The results for starch polymer loose fills (Table 5) differ decisively depending on the source. Much of these differences can be explained by different assumptions regarding the bulk density of the loose fills (see second column in Table 5) and different approaches for the quantification of the ozone depletion potential (inclusion versus exclusion of NO_x). It seems therefore more useful to compare the results of each study separately.

Table 4. Summary of key indicators from the LCA studies reviewed (state-of-the-art technologies only)

Type of plastic	Cradle-to-gate non non-renewable energy use [a] (MJ/functional unit)	Type of waste treatment assumed for calculations	GHG emissions (kg CO_2eq./ functional unit)	Ozone precursors (g ethylene eq.)	Acidification (g SO_2 eq.)	Eutrophication (g PO_4 eq.)	Ref.
Petrochemical polymers							
HDPE	79.9	Incineration	4.84 [b]	n/a	n/a	n/a	12
LLDPE	72.3	Incineration	4.54 [b]	n/a	n/a	n/a	12
LDPE	80.6	Incineration	5.04 [b]	n/a	n/a	n/a	12
LDPE	91.7	80% incineration + 20% landfilling	5.20 [b]	13.0	17.4	1.1	3
Nylon 6	120	Incineration	7.64 [b]	n/a	n/a	n/a	12
PET (bottle grade)	77	Incineration	4.93 [b]	n/a	n/a	n/a	12.
PS (general purpose)	87	Incineration	5.98 [b]	n/a	n/a	n/a	12
EPS	84	Incineration	5.88 [b]	n/a	n/a	n/a	12
EPS	88	None	2.80	43.0	170.0	5.8	13
EPS (PS + 2%SBR + Pentan + Butan)	87	None	2.72	1.2	18.5	1.5	5
Petrochemical co-polymers							
Polycaprolactone (PCL)	83	Incineration	3.1 [b]	6.1	5.5	0.5	14
Polycaprolactone (PCL)	77	Incineration	5.0-5.7 [b]	n/a	n/a	n/a	7
Polyvinyl alcohol (PVOH)	102	Incineration	2.7 [b]	8.9	8.0	0.9	13
Polyvinyl alcohol (PVOH)	58	Incineration	4.1-4.3 [b]	n/a	n/a	n/a	7
Bio-based plastics (pellets)							
TPS	25.4	Incineration	1.14	n/a	n/a	n/a	3
TPS	25.5	80% incineration + 20% compost.	1.20	4.7	10.9	4.7	3
TPS	25.4	100% composting	1.14	5.0	10.6	4.7	3
TPS (maize starch+5.4%maize grit +12.7%PVOH)	18.9	None [c]	1.10 [c]	0.2	4.6	0.5	5
TPS + 15% PVOH	24.9	Incineration	1.73	n/a	n/a	n/a	3
TPS + 52.5% PCL	48.3	Incineration	3.36	n/a	n/a	n/a	3
TPS + 60% PCL	52.3	Incineration	3.60	n/a	n/a	n/a	3
Mater-Bi foam grade	32.4	Composting	0.89	5.5	20.8	2.8	13
Mater-Bi foam grade	36.5	Waste water treatment plant	1.43	5.8	20.7	3.1	13
Mater-Bi film grade	53.5	Composting	1.21	5.3	10.4	1.1	14
PLA	57	Incineration	3.84	n/a	n/a	n/a	15
PHA by fermentation	81	n/a	n/a	n/a	n/a	n/a	16
PHB, various processes	66 - 573	n/a	n/a	n/a	n/a	n/a	17

[a] Total process and feedstock energy. Non-renewable energy only, i.e. total fossil and nuclear energy. In the "cradle-to factory gate" concept the downstream system boundary coincides with the output of the polymer or the end product, no credits are ascribed to valuable by-products from waste management (steam, electricity, secondary materials).

[b] Only CO_2. Embodied carbon: 3.14 kg CO_2/kg PE. 2.34 kg CO_2/kg nylon 6. 2.29 kg CO_2/t PET. 3.38 kg CO_2/t PS. 2.32 kg CO_2/t PCL, 2.00 kg CO_2/t PVOH.

[c] No credit for carbon uptake by plants.

One can conclude from both the Composto and the BIFA/IFEU/Flo-Pak studies[5] that starch polymer loose fills generally score better than their equivalents made of virgin EPS. GHG emissions represent an exception where the release of CH_4 emissions from biodegradable compounds in landfills results in a disadvantage for starch polymers (only according to BIFA/IFEU/Flo-Pak[5]). Loose fills produced from recycled polystyrene may represent an attractive option in environmental terms compared to starch polymers[5].

Table 5. Summary of LCA key indicators for end products (some of the products listed are commercialised, others are not; see text)

Type of plastic	Functional unit	Cradle-to-gate non-renewable energy use [a] (MJ/functional unit)	Type of waste treatment assumed for calculations	GHG emissions (kg CO_2 eq./ functional unit)	Ozone precursors (g ethylene eq.)	Acidification (g SO_2 eq.)	Eutrophication (g PO_4 eq.)	Ref.
Loose fills								
Mater-Bi starch loose fills	1 m^3 (10 kg)	492	Waste water treatment plant	21.0	115	276	39.0	13
FloPak starch loose fill	1 m^3 (12 kg)	277	30% incin., 70% landfilling	33.5	10	83	9.9	5
EPS loose fill	1 m^3 (4.5 kg)	680	Incineration	56.0	1200	325	42.0	13
EPS loose fill	1 m^3 (4 kg)	453	30% incin., 70% landfilling	22.5	57	85	8.0	5
EPS loose fill (by recycling of PS waste)	1 m^3 (4 kg)	361	30% incin., 70% landfilling	18.6	55	107	9.9	5
Films and bags								
TPS film	100 m^2, 150 µm [b]	649	80% incin.+ 20% landfilling	25.30	100	239	103.0	3
Mater-Bi starch film	100 m^2, 20 µm [b]	133	Composting	2.98	14.0	26.5	2.8	14
PE film	100 m^2, 150 µm [b]	1340	80% incin.+ 20% landfilling	66.70	180	238	15.0	14
Printed wiring boards								
Conventional epoxy [c]	1 kg	~17 [d]	Disposal as MSW	100%				18
Kraft lignin/epoxy	1 kg	~12 [d]	Disposal as MSW					18
Organosolve lignin/epoxy	1 kg	~10 [d]	Disposal as MSW					18
Lacquer								
Conventional petrochemical thickener	1 kg lacquer	220	None	9.6 [e]		26 [f]		19
Epoxidised linseed oil as thickener	1 kg lacquer	25	None	1.3 [e]		3 [f]		19
Under-floor panel for passenger car								
Glass-fibre reinforced polypropylene	1 panel	155						20
Flax reinforced polypropylene	1 panel	134						20
Glass fibre mat	1 kg fibre mat	54.7						20
Flax fibre mat	1 kg fibre mat	9.6						20
Interior side panel for passenger car								
ABS copolymer	1 panel	132	None	5.4				21
Hemp fibre/epoxy composite	1 panel	73	None	4.7				21
Transport pallet								
Glass-fibre reinforced polypropylene (GF)	1 pallet	1400	None	75.3	208	653	68.2	21
China reed reinforced polypropylene (CF)	1 pallet	717	None	40.4	133	432	62.8	21
Glass-fibre reinforced polypropylene (GF) [g]	1 kg pallet	93	None	5.0	14	44	4.5	21
China reed reinforced polypropylene (CF) [g]	1 kg pallet	61	None	3.4	11	37	5.3	21

[a] Total process and feedstock energy. Non-renewable energy only, i.e. total of fossil and nuclear energy. In the "cradle-to factory gate" concept the downstream system boundary coincides with the output of the polymer or the end product, no credits are ascribed to valuable by-products from waste management (steam, electricity, secondary materials).
[b] An important explanation for the large difference between the values reported in the following columns is that Carbotech assumes a film thickness of 150 µm while it is only 20 µm in the case of Composto.
[c] Epoxy resin (FR4) cured with dicyandiamide (DICY)
[d] Reference 18 provides only energy data for two cradle-to-grave systems. These are firstly 100% incineration for metal reclamation and secondly a combination of 50% disposal as municipal solid waste and 50% incineration for metal reclamation. Based on the results for these two cases we have roughly estimated the cradle-to-gate energy use.
[e] Only CO_2 emissions.
[f] Calculated only on the basis of NO_X emissions since no information was available for SO_2 and other acidifying emissions; assumed conversion factor: 1 kg NO_X = 0.7 kg SO_2 (see ref. 22).
[g] The data in these two rows have been calculated from the preceding two rows by using the pallet weight (1 GF pallet = 15 kg; 1 CF pallet = 11.8 kg).

By analogy to loose fills, the range of results for starch polymer films and bags is to a large extent understandable from the differences in film thickness. Taking this factor into account, the environmental impacts of the starch films/bags are lower with regard to energy, GHG emissions and ozone precursors. The situation is less clear for acidification. For eutrophication, PE films tend to score better.

For Printed Wiring Boards (PWBs), about 30-40% energy can be saved and the GHG mitigation potential is estimated to lie in a similar range. Exceptionally high savings (around 90% for energy and GHG emissions) have been established for epoxidised linseed oil as thickener for lacquers. Similar savings (above 80% for energy) are reached only by substituting flax fibre mats for fibreglass mats (Table 5). For complete components (end products), the use of natural fibres is reported to save between 14% (underfloor panels) and 4-50% energy (interior side panels and transport pallets).

Leaving PHA aside as the only exception, it can be summarised that biopolymers and natural fibres typically enable savings of around 20% (energy and CO_2). Substantially higher savings up to 50% and beyond are considered feasible for certain starch polymers, printed wiring boards, certain lacquers, and natural fibre composites.

Apart from assessing bio-based materials in terms of the relative decrease of environmental impacts (in percent, as just discussed), comparison of savings per kg of bio-based polymer (Table 6) can provide additional insight. These results show that printed wiring boards offer relatively low saving potentials and that GHG emissions for fibre composites can be disadvantageous. Otherwise, Table 4 and 5 confirm the finding that very attractive potentials for energy saving and GHG emission reduction exist for bio-based plastics (pellets), non-plastics (lacquers), and fibre composites.

Table 6. Energy and GHG savings by biopolymers relative to their petrochemical counterparts

	Energy savings[a] (MJ/kg bio-based polymer)	GHG savings[a] (kg CO_2 eq./kg bio-based polymer)
Bio-based plastics (pellets)		
TPS	51	3.7
TPS + 15% PVOH	52	3.1
TPS + 52.5% PCL	28	1.4
TPS + 60% PCL	24	1.2
Mater-Bi foam grade	42	3.6
Mater-Bi film grade	23	3.6
PLA	19	1.0
Printed wiring boards	5	n/a
Lacquer	195	8.3
Flax fibre mat	45	n/a
Interior side panel for pass. car	28	-0.9
Transport pallet	33	1.6

[a] Max. ±15% depending on whether LDPE or LLDPE is chosen as reference according to APME.

4. ARE WE CRITICAL ENOUGH?

The comparison of the main assumptions made in the various studies and the comparison with the current state of the art reveals a number of uncertainties and caveats, which are discussed in this chapter.

- As mentioned above, LCA data for polycaprolactone (PCL) and polyvinyl alcohol (PVOH) are subject to considerable uncertainties. In view of the widespread use of these compounds in biodegradable materials and given the strong impact on the final results especially for some starch polymers, reliable LCA data need to be generated.
- A literature survey revealed that the process energy requirements for propylene oxide differ substantially depending on the source. The value used by ITUC[19] is in the upper range. Since the production of propylene oxide contributes more than 55% to the total energy requirements of the petrochemical thickener for lacquers[19], the energy input for the petrochemical thickener as a whole might be overestimated. This might explain to a large extent why the saving potential related to the replacement of petrochemical by bio-based lacquer is exceptionally high compared to the other products (Table 6). In-depth analysis of the key assumptions would be required to understand the reasons and, if required, to reduce the concomitant uncertainties.
- The data used for composting are subject to major uncertainties. This is partly explicitly stated by the authors (Ref. 13, p. 16) partly it becomes obvious by comparing the assumptions made in the various studies (wherever these are described in detail). According to COMPOSTO[14], 40-60% of the carbon absorbed in the vegetable material is released to the atmosphere during composting. To avoid the underestimation of GHG emissions, the COMPOSTO Studies[13, 14] assume that 60% of the absorbed carbon is released. The assumption can be considered as safe if compared to Schleiss and Chardonnens[23] according to whom the average carbon dissipation in the form of CO_2 amounts to 40% (average of all composting plants in Switzerland). While these data refer to the average of all inputs and outputs of a composting plant, the question arises whether it also holds true for biodegradable starch polymers: since the quality of biodegradable polymers is that they decompose to a large extent within a short period of time, the question arises whether the approach chosen by BIFA/IFEU/Flo-Pak[5] might be more accurate where it was assumed that the build up of organic matter and hence, the effect of carbon sequestration is negligible. According to biodegradation tests conducted by several institutes, the degradation of starch polymers during composting (59 °C, 45 days) amounts to about 80 to 90% (test refers to mixture of 15% starch polymers and 85% pure cellulose). Since

biodegradation in the subsequent maturation phase is negligible, Novamont draws the conclusion that an average conversion rate of 80% is realistic (personal communication, L. Marini, 2001). The specific characteristics of the considered starch polymer and the type of composting technology applied may influence the biodegradation fraction.

- The various studies differ in the accounting method for waste incineration of biopolymers. Even though the detailed assumptions are hardly ever spelled out, it is quite obvious that the chosen approaches are not comparable. For example, the BIFA/IFEU/Flo-Pak study[5] assumes that incineration takes place in waste-to-energy facilities, resulting in a net output of electricity and/or heat. Credits are assigned to these useful products. In contrast, the COMPOSTO studies do not account for co-produced electricity/steam. It is unlikely that this reflects the differences in the share of energy recovery (waste-to-energy facilities versus simple incineration without energy recovery) among the countries studied; it rather represents different choices of system boundaries.
- The environmental assessment of the incineration of mulch films with adhering organic waste (soil) raises particular questions. In one of the sensitivity analyses, the COMPOSTO study[14] introduces a CO_2 penalty in order to account for the emissions resulting from the incineration of adhering organic waste. This may be justified if the moisture of the organic waste is so high that the vaporisation of the water contained requires more energy than the calorific value of the organic waste. In this case the incineration of the adhering waste represents a net energy sink. In practice, this is typically compensated by co-firing of fossil fuels or of other high-calorific combustible waste leading to CO_2 and other environmental impacts. On the other hand, it is also possible that the moisture content of the adhering waste is low, resulting in a net energy yield in the incineration process. Moreover, if the organic waste is of biogenic origin, its incineration is neutral in CO_2 terms (due to extraction of CO_2 from the atmosphere during plant growth). These considerations show that the specific circumstances determine whether the co-combustion of adhering organic waste – be it soil, organic kitchen waste, or any other type of biogeneous waste – results in net environmental benefits or disadvantages.
- Biopolymers generally have lower heating values than most petrochemical bulk polymers (Table 7). In some cases the difference is negligible (e.g., PH3HB versus PET), while in other cases it is substantial (starch polymers versus PE). In practice, the difference in recoverable heat may be even larger than indicated by Table 7 due to the feature of most biopolymers to absorb water rather easily. The choice of the waste management system may therefore have a considerable impact on the

overall conclusions. Regarding energy use, cradle-to-factory gate analyses, landfilling, and waste incineration without energy recovery are in favour of biopolymers; on the other hand, incineration in waste-to-energy facilities, especially with high energy-recovery yields, is in favour of petrochemical polymers (in energy terms). This calls for studying the material options by types of waste management technologies individually. Moreover, the actual situation in the studied country and region should be analysed. It should be taken into account here that energy recovery yields from waste-to-energy facilities are generally low at present. It is estimated that one quarter of the heating value of the waste is converted to final energy in the form of power and useable heat. The generation of the same amount of final energy from regular fuels in power plants and district heating plants requires only half of the energy input. As a consequence, the credit for energy recovery is only one half of the heating value of the combusted plastics. The advantage of petrochemical materials over biopolymers is therefore half the difference of their heating values. Depending on the studied petrochemical and bio-based polymer, this difference can still be substantial but it may also be negligible.

Table 7. Heating value of bio-based and petrochemical polymers (heating values calculated according to Boie, compare Reimann and Hämmerli[27])

Polymer	Lower heating value GJ/ton (dry matter)
Starch polymers	13.6
Polyhydroxybutyrate (P3HB)	22.0
Polyhydroxyvalerate (P3HV)	25.0
Poly(lactic acid) (PLA)	17.9
Lignin (picea abies)	24.2
China reed	18.0
Flax	16.3
Hemp	17.4
Kenaf	16.5
PE	43.3
PS	39.4
PET	22.1
PVC	17.9

- In the case of landfilling, some studies account for methane emissions due to anaerobic fermentation while others do not take this into consideration. This can have a considerable impact on the results due to the relatively strong greenhouse gas effect of methane ($GWP_{100} = 23$). As a consequence, the overall GHG emissions from biodegradable polymers manufactured from renewable raw materials may be higher than for petrochemical plastics depending on the waste management system chosen for the latters[5]. On the other hand, none of the reviewed studies analyses the effect of collecting landfill gas, which may mitigate or even overcompensate the effects of release if used for electricity/heat production.

- Being another important greenhouse gas, the release of N_2O from fertilizer use in agriculture may contribute substantially to the overall global warming impact of bio-based products.
- Environmental comparisons including recycling as a waste management option are rarely made. Moreover, most of the biopolymers except starch can be processed by mechanical or even feedstock recycling (back to monomer). Mechanical recycling is in principle possible even for thermoplastic polymers reinforced with natural fibres[24]. More attention must be paid to these options in future studies.
- The characterisation factors for global warming used in most of the reviewed studies are outdated (GWP100 for methane and nitrous oxide. The GWP equivalence factors used in the various studies are 11 or 21 for CH_4 and 270 or 310 for N_2O, while - according to the current state of research - more accurate figures are 23 and 296 for CH_4 and N_2O, respectively[25]). Since the contribution of CO_2 dominates the overall GHG effect, this uncertainty is considered to be less important.
- When making comparisons with conventional fossil fuel-based polymers it must be borne in mind that LCA data for these products are also uncertain and continue to be corrected. This is in spite of the fact that petrochemical polymers are manufactured by use of mature technologies that are applied globally with only limited variations. For example, energy data for polyethylene (PE) production range between ca. 65 and 85 GJ/t according to a comparison of various sources[26] while the CARBOTECH study assumes about 92 GJ/t. If, for example, compared to the TPS data determined by CARBOTECH, the wide range of the values for PE does not change anything about the conclusion that TPS is most beneficial in terms of energy use and CO_2 emissions. However, it is unclear whether the final conclusions for the other environmental parameters covered by the CARBOTECH study (e.g., emissions) are also insensitive to larger variations in the PE data.

The problem related to these uncertainties can be resolved to some extent by taking into account the significance of the difference in values for the compared systems (thresholds for the categories "significantly higher", "higher", "comparable" etc., see COMPOSTO[13]). In addition, it is an important goal of future research to reduce further the existing uncertainties. At the same time, it should be remembered that according to the simple uncertainty analysis presented in Section 2 for pure starch polymers, the original value for cumulated energy demand (25.5 GJ/t) determined by CARBOTECH[3] is quite reliable, avoiding underestimation in most cases.

In all reviewed studies, ecological ranking was assessed by determining for how many indicators the environmental impact is lower for biopolymers as compared to petrochemical polymers. The disadvantage of this approach is that the selection of the compared indicators can have an influence on the final conclusions. Together with the fact that the relative difference in the results for the various impact categories (a few per cent versus a few hundred per cent) is hardly ever accounted for this shows the urgent need for the further development of the LCA methodology (e.g. by introduction of significance thresholds). When interpreting the results, it must finally be taken into account that the reviewed studies partly differ in regional scope. Since the results are to some extent subject to country specific circumstances (e.g., GHG emissions from national power production) care must be taken when drawing more general conclusions. On the other hand, the uncertainties related to conclusions can be reduced if several independent analyses for different countries get similar conclusions.

5. WHAT CAN WE CONCLUDE?

The number of published LCAs for biopolymers and natural fibres is quite limited. This seems to be in contrast to the general public interest for this issue and to the more recent interest by policy makers. For example, no comprehensive LCAs have been published so far for PLA (plant-based), cellulose polymers (plant-based), and some fossil fuel-based biodegradable polymers, such as BASF's product Ecoflex.

The existing LCAs contain uncertainties, which should be addressed by future research and analysis. A prominent example is the environmental assessment of the composting process for biodegradable polymers. In some studies further sensitivity analyses would be required to ensure that the final findings are well underpinned (e.g. for smaller PE bags in COMPOSTO[14]). Moreover, many of the environmental analyses choose a cradle-to-factory gate perspective (i.e., the analysis ends with the product under consideration). While this approach provides valuable results, additional analyses taking a cradle-to-grave perspective by inclusion of the waste management stage should also be conducted. Due to their strong impact on the final results, all major waste management options should be studied (landfilling, composting, MSWI plants, waste-to-energy facilities, digestion, and recycling).

To assist life cycle practitioners in making use of the lessons learnt from this review, a checklist has been prepared which will be published by

RAPRA[28]. In addition to this checklist, any LCA study must comply with the requirements specified in the ISO standards 14 040 to 14 043 [2].

Apart from some methodological shortcomings, three LCA studies evaluate products for which considerable technical problems related to production and product properties still need to be overcome. This is the case for the two china reed-based products studied by FAT/CARBOTECH[29] and for the transport pallet studied by the Swiss Federal Institute of Technology[30]. This calls firstly for further R&D and secondly for caution when making use of the LCA results for these specific applications.

In spite of these restrictions and the uncertainties and the information gaps mentioned above, the body of analysed work overwhelmingly indicates that biopolymers offer important environmental benefits today and for the future. This applies to all three studied groups, i.e. to bio-based plastics, bio-based non-plastics (with one type of lacquers as the only representative), and composites based on natural fibres.

Of all studied bio-based plastics (pellets), starch polymers are considered to perform best in environmental terms under the current state of the art – with some differences among the various types of starch polymers. Compared to starch polymers, the environmental benefits seem to be smaller for PLA (LCA results only available for energy and CO_2). For PHA, the environmental advantage currently seems to be very small compared to conventional polymers (LCA results only available for energy use). For both PLA and PHA, the production method, the scale of production, and the type of waste management treatment can influence decisively the ultimate conclusion about the overall environmental balance.

The only available analysis for non-plastics (lacquer thickener based on linseed oil) revealed an exceptionally high saving potential, which calls for further analysis.

For natural fibres, the extent to which these can replace fibreglass (which is heavy and energy intensive to produce) determines mainly the net environmental benefits. The advantages according to cradle-to-factory gate analyses were rather limited in one case (-14% for underfloor panel) and very attractive in the other two cases (-45% to –50% for interior side panel and transport pallet).

The case studies for reinforced products demonstrate that savings in the use phase – sometimes also referred to as "secondary savings" – are as high or even clearly higher than the "primary savings" (typically related to cradle-to-factory gate systems) and that they may even be the main driver for substitution. In the case of the china reed pallet and the hemp fibre-based interior side panel, secondary energy savings are up to three times as high. Exceptionally high secondary savings are reported to be obtainable with tyres containing starch-based fillers where secondary energy savings exceed primary savings by a factor of more than 20. The conclusion of the ECCP

Working Group on Renewable Raw materials[1], according to which total secondary savings may exceed the primary savings by about one order of magnitude, is hence clearly confirmed by the case study for tyres while it seems somewhat too optimistic if compared to the results for natural fibres.

Starch polymers are currently the only type of bio-based polymer for which several comprehensive LCA studies are available. According to these assessments, starch polymers do not perform better than their fossil fuel-based counterparts in *all* environmental categories, including biodiversity and soil quality, which are generally outside the scope of LCAs. However, most studies come to the conclusion that starch polymers (pellets and end products) are more beneficial in environmental terms than their petrochemical counterparts; this conclusion is drawn without weighing and in most cases without significance thresholds. The preferences among the environmental targets determine whether biopolymers are considered to be environmentally attractive. Full-sized LCA studies for further bio-based materials are indispensable to allow deriving conclusions and recommendations that are better underpinned and more focussed.

For the time being, it is not possible to make a concluding *general* judgement whether bio-based plastics should be preferred to petrochemical polymers from an environmental point of view. This has partly to do with the limited availability of comprehensive LCAs. But even if more LCA studies were available, one would be left with considerable uncertainties, e.g. because it will never be feasible to cover all possible products and all possible impact categories[31]. In spite of these limitations, one can conclude that already today the results for the use of fossil energy resources and GHG emissions are more favourable for most biopolymers. As an exception, landfilling of biodegradable polymers can result in methane emissions (unless landfill gas is captured), which may make the system unattractive in terms of reducing greenhouse gas emissions. As a potential source of N_2O emissions, fertilizers also require special attention.

By comparing the use of biomass for the manufacture of materials (polymers and fibres) on the one hand and for energy purposes (bioenergy) on the other hand, insight can be gained about the most effective options for land use and cultivation. Important findings of the CARBOTECH study[3] and the LCA prepared by the Swiss Federal Institute of Technology[30] are hence that materials based on starch, kenaf, and china reed offer larger opportunities for energy saving and GHG mitigation per unit of land than bioenergy (ref. 3, p. 12 and 92; partly based on ref. 29). In contrast, Kurdikar *et al.*[32] argue that bioenergy contributes more to GHG emission reduction than biomass-derived feedstocks. The main reason for this contrasting finding seems to be that the product and process Kurdikar *et al.*[32] studied – i.e., the production of polyhydroxyalkanoates in plants – currently cannot compete with conventional products in energy terms. Wherever the opposite applies – and this is the case for most of the other analysed products– the available results indicate that biomaterials offer higher environmental gains than bioenergy. This issue will systematically have to be studied for biopolymers by comparing the benefits per km^2 of cultivated land, thereby also accounting for options of multiple land use and making use of agricultural by-products. In other words, comparative assessments will continue to be needed in order to keep track of the aspects of competition and complementarity between bioenergy and biomaterials. This is also necessary in order to account for innovations in both areas. It would ease such comparisons and the usefulness for decision-makers if future studies dealing with bioenergy and biomaterials always also studied the land use requirements of the various options.

To maximise the environmental benefits from biopolymers, further R&D will be necessary in order to optimise the production by increasing the efficiencies of the various unit processes involved (e.g. separation processes) and by process integration. Substantial scope for improvement can be expected here considering scale economies and given the fact that all biopolymers are still in their infancy while the manufacture of petrochemical polymers has been optimised for decades. Some of the LCA discussed above were already outdated when these conclusions were drawn since substantial progress had been made in manufacturing and processing biopolymers (e.g., for films). This means that the real environmental impacts caused by biopolymers tend to be lower than established in the reviewed LCA studies.

As a guide for future R&D, good practice targets for environmentally advantageous bio-based products could be very useful. Based on the results presented in the preceding sections, a first attempt is made here to specify such targets. It is recommended that, relative to their petrochemical counterparts, biopolymers should

- save at least 20 MJ (non-renewable) energy per kg polymer,
- avoid at least 1 kg CO_2 per kg polymer,
- reduce most other environmental impacts by at least 20%.

A good practice target will also have to be specified for land use, i.e. in terms of GJ energy saved per ha land cultivated. In parallel to these environmental targets, cost reduction must continue to be a priority.

A promising line for R&D in the longer term could be the development of biomass-derived polymers that can be recycled mechanically and/or back to feedstock/monomers. Preferably, this should be possible also in combination with petrochemical polymers. Such recyclable polymers made from renewable raw materials have good chances to be unrivalled in environmental terms provided that their manufacture is not too resource-intensive in the first place. This may offer longer-term prospects to PHA, PLA, and some other biopolymers.

To summarize, the available LCA studies and environmental assessments strongly support the further development of biopolymers. Careful monitoring of the various environmental impacts continues to be necessary both for decision makers in companies and in policy. If combined with good-practice targets, this may accelerate and focus the ongoing product and process innovation. For some materials, the environmental benefits achieved are substantial already today. In many other cases the potentials are very promising and need to be exploited.

ACKNOWLEDGEMENTS

This chapter is based on earlier documents which were only possible due to the large support from various experts in the field, especially to Mrs. Catia Bastioli and Mr. Luigi Marini, both from Novamont (Italy) and to Mr. Eduard Würdinger, Bayrisches Institut für angewandte Umweltforschung und –technik (BIFA, Germany). These three experts are co-authors of an extensive review of LCAs for biopolymers[10]. I would also like to thank Mr. Gerald Scott (United Kingdom) for his critical comments and his interest in this type of analysis and Prof. Emo Chiellini, University of Pisa (Italy) for bringing these topics to a wider audience.

REFERENCES

1. ECCP (European Climate Change Programme), 2001, *Long report.* http://europa.eu.int/comm/environment/climat/eccp_longreport_0106.pdf, Brussels, Belgium.
2. ISO (1997-1999), International Organisation for Standardisation:
   - ISO 14 040 (DIN EN ISO 14040), 1997, *Environmental Management – Life Cycle Assessment – Principles and Framework.* Berlin: DIN (Deutsches Institut für Normung).
   - EN ISO 14 041, 1998, *Environmental Management – Life Cycle Assessment – Goal and scope definition and inventory analysis.* Berlin: DIN (Deutsches Institut für Normung).
   - EN ISO 14 042, 1998, *Environmental Management – Life Cycle Assessment – Life Cycle Impact Assessment.* Draft. Berlin: DIN (Deutsches Institut für Normung).
   - EN ISO 14 043. 1999, *Environmental Management – Life Cycle Assessment – Life Cycle Interpretation.* Draft. Berlin: DIN (Deutsches Institut für Normung).
3. CARBOTECH (1996): Dinkel, F., Pohl, C., Ros, M., Waldeck, B., *Ökobilanz stärkehaltiger Kunststoffe* (Nr. 271), 2 volumes. Commissioned by the Bundesamt für Umwelt und Landschaft (BUWAL), Bern, Switzerland.
4. Kaltschmitt, M., and Reinhardt, G. A., 1997, *Nachwachsende Energieträger – Grundlagen, Verfahren, Ökologische Bilanzierung.* Vieweg Publishers, Braunschweig/Wiesbaden, Germany

5. BIFA/IFEU/Flo-Pak, 2001, *Kunststoffe aus nachwachsenden Rohstoffen - Vergleichende Ökobilanz für Loose-fill-Packmittel aus Stärke bzw. aus Polystyrol* (confidential interim report). Bayrisches Institut für angewandte Umweltforschung und –technik, Augsburg (BIFA, project leader), Institut für Energie- und Umweltforschung Heidelberg (IFEU), Flo-Pak GmbH, Germany
6. Geier, U., Frieben, B., Haas, G., Molkenthin, V., and Köpke, U., 1998, *Ökobilanz Hamburger Landwirtschaft – Umweltrelevanz verschiedener Produktionsweisen, Handlungsfelder Hamburger Umweltpolitik.* Prepared by the Institute of Organic Agriculture of the Rheinische Friedrich-Wilhelm-Universität Bonn, Part 1. Schriftenreihe für organischen Landbau, Verlag Dr. Köster, Berlin
7. Kopf, N., 1999, *Kunststoffe aus nachwachsenden Rohstoffen – Polyhydroxybutyrat und Stärkekunststoffe – Abschätzungen zum kumulierten Energieaufwand und zu CO_2-Emissionen.* Student's report, prepared at the Fraunhofer Institute for Systems and Innovation Research (FhG-ISI), Karlsruhe for Engler-Bunte-Institut at Karlsruhe University. Karlsruhe, Germany
8. Meuser, F., German, H., 1981, Enrgiebedarf und Energiefluss einer Kartoffelstärkefabrik. *Starch/Stärke* **33** (10): 332-337
9. Winter, D., Fürniß, B., Klein-Vielhauer, S., Leible, L., Nieke, E., Rösch, Ch., and Taugen, H., 1993, *TFA (Technologiefolgenabschätzung zum Thema Nachwachsende Rohstoffe)*, Landwirschaftsverlag, Münster, Germany
10. Patel, M., Bastioli, C., Marini, L., and Würdinger, E., 2003, Life-cycle assessment of bio-based polymers and natural fibres. In *Biopolymers*, Vol. 10, Wiley-VCH, forthcoming in 2003, see also downloadable documents on http://www.chem.uu.nl/nws/www/nws.html
11. Fraunhofer ISI, 1999, C-STREAMS - Estimation of material, energy and CO_2 flows for model systems in the context of non-energy use, from a life cycle perspective – Status and scenarios. Volume I: Estimates for the total system (in German, English abstract). M. Patel, E. Jochem, F. Marscheider-Weidemann, P. Radgen, N. von Thienen, Karlsruhe, Germany
12. APME (1999, 2000) *Eco-profiles of plastics and related intermediates* (about 55 products). Association of Plastics Manufacturers in Europe (APME). Downloadable from the internet (http://lca.apme.org), Brussels, Belgium
13. COMPOSTO (July 2000): *Life cycle assessment of Mater-Bi and EPS loose fills.* Study prepared by by R. Estermann, B. Schwarzwälder and B. Gysin, Composto, for Novamont, Olten, Switzerland for Novamont, Novara, Italy
14. COMPOSTO (September 1998): *Life cycle assessment of Mater-Bi bags for the collection of compostable waste.* Study prepared by R. Estermann and B. Schwarzwälder, Composto, Olten, Switzerland for Novamont, Novara, Italy
15. CARGILL DOW, 2001, *NatureWorks – A new generation of biopolymers.* Presentation by E. Vink on 29 March 2001, Birmingham, United Kingdom
16. Gerngross, T. U., and Slater, S., 2000, How Green are Green Plastics? *Scientific American*, August, 2000, pp. 37-41
17. Heyde, M., 1998, Ecological considerations on the use and production of biosynthetic and synthetic biodegradable polymers. *Polym. Degrad. Stab.* **59** (1-3): 3-6.
18. Kosbar, L. L., Gelorme, J. D., Japp, R. M., and Fotorny, W. T., 2001, Introducing biobased materials into the electronics industry. *J. Ind. Ecol.* **4** (3): 93-105.
19. Diehlmann, A., Kreisel, G., 2000, *Ökologische Bilanzierung ausgewählter Lackrohstoffe: Vergleich von Bindemitteln auf nativer und petrochemischer Basis.* ITUC (Institute for Technical Chemistry and Environmental Chemistry), Jena University, Germany, pp. 95
20. Diener, J., and Siehler, U., 1999, Ökologischer Vergleich von NMT- und GMT-Bauteilen. *Angew. Makromol. Chem.* **272**: 1-4.

21. Wötzel, K., Wirth, R., and Flake, R., 1999, Life cycle studies on hemp fibre reinforced components and ABS for automotive parts. *Angew. Makromol. Chem.* **272**: 121-127.
22. Heijungs, R., Guinee, J. B., Huppes, G., Lankreijer, R. M., Udo de Haes, H. A., Wegener, and Sleeswijk, A., 1992, CML (Centre for Environmental Science Leiden), Netherlands Organisation for Applied Scientific Research (TNO), Fuels and Raw Materials Bureau (B&G): Environmental Life-Cycle Assessment of Products – Guide and Background. Leiden, Netherlands
23. Schleiss, K., and Chardonnens, M., 1994, Stand und Entwicklung der Kompostierung in der Schweiz 1993. BUWAL Umwelt-Materialien Nr. 21. Bundesamt für Umwelt, Wald und Landschaft (BUWAL), Bern, Switzerland
24. Hauspurg, Ch., Reußmann, T., and Mieck, K.-P., 2000, Recycling von Naturfaserverbundwerkstoffen im Plastifizier-Pressverfahren. *GAK* 8/2000, Jahrgang 53, 523-527.
25. *Climate Change 2001 – The scientific basis*, 2001 (Houghton, J. T., Ding, Y., Griggs, D. J., Noguer, M., van der Linden, P. J., Da, X., Maskell, K., and Johnson, C. A., eds.), Contribution of Working Group I to the Third Assessment Report of the Intergovernmental Panel on Climate Change (IPCC). Full text downloadable from www.ipcc.ch.
26. Patel, M., 1999, PhD thesis *Closing Carbon Cycles: Carbon use for materials in the context of resource efficiency and climate change.* (ISBN 90-73958-51-2, http://www.library.uu.nl/digiarchief/dip/diss/1894529/inhoud.htm), Utrecht University, Utrecht, Netherlands.
27. Reimann, D. O., and Hämmerli, H., 1995, *Verbrennungstechnik für Abfälle in Theorie und Praxis.* Schriftenreihe Umweltschutz, Bamberg, p. 246.
28. Patel, M., 2003, Environmental life cycle comparisons of biodegradable plastics. In *Handbook on Biodegradable Materials* (C. Bastioli, ed.), RAPRA Technology Ltd., forthcoming in 2003.
29. FAT/CARBOTECH, 1997, Beurteilung nachwachsender Rohstoffe in der Schweiz in den Jahren 1993-1996 – Vergleichende Betrachtung von Produkten aus ausgewählten nachwachsenden Rohstoffen und entsprechenden konventionellen Produkten bezüglich Umweltwirkungen und Wirtschaftlichkeit, (U. Wolfensberger and F. Dinkel, eds.). Prepared by FAT (Eidgenössische Forschungsanstalt für Agrarwirtschaft und Landtechnik) and CARBOTECH for the Federal Office of Agriculture, Bern, Switzerland.
30. Corbière-Nicollier, T., Gfeller Laban, B., Lundquist, L., Leterrier, Y. Manson, J.-A.E., and Jolliet. O., 2001, Life cycle assessment of biofibres replacing glass fibres as reinforcement in plastics. Swiss Federal Institute of Technology, Lausanne, Switzerland. *Resour. Conserv. Recy.* **33**: 267-287.
31. Finnveden, G., 2000, On the limitations of life cycle assessment and environmental system analysis tools in general. *Int. J. Life Cycle Assess.* **5** (4): 229-238.
32. Kurdikar, D., Paster, M., Gruys, K. J., Fournet, L., Gerngross, T. U., Slater, S. C., and Coulon, R., 2001, Greenhouse gas profile of a plastic derived from a genetically modified plant. *J. Ind. Ecology.* **4** (3): 107-122.

Biobased Polymeric Materials

HYOE HATAKEYAMA*, YASUHIRO ASANO*, and TATSUKO HATAKEYAMA[#]
** Department of Applied Physics and Chemistry, Fukui University of Technology, Fukui, Japan; [#]Department of Textile Science, Otsuma Women's University, Tokyo, Japan*

1. INTRODUCTION

For the development of environmentally compatible polymers, it is essential to understand that nature constructs a variety of materials, which can be used in human life. Plant materials such as cellulose, hemicellulose and lignin are the largest organic resources. However, they are not very well used except for cellulose. Hemicellulose has not yet been utilized. Lignin, which is obtained as a by-product of the pulping industry, is mostly burnt as fuel and only increases the amount of carbon dioxide in the environment, although lignin is one of the most useful natural resources.

Biomaterials span the range from elastic solids to viscous liquids. The complexity indigenous to biomaterials is based on the intricacies of their molecular architecture. However, it is important to overcome the difficulty of the utilization of biomass caused by the above structural intricacies in order to maintain the sustainable developments that can keep the rich and convenient life developed by science. Recently, it has strongly been recommended to convert biomass into industrial products and to develop biobased industrial products that are cost-competitive[1].

It can be considered that the compounds produced through biosynthesis can be used as raw materials for the synthesis of useful plastics and materials in human life. Major plant components, such as saccharides and lignin, contain highly reactive hydroxyl groups, which can be used as reactive chemical reaction sites. As shown in Figure 1, it is possible to

Biodegradable Polymers and Plastics, Edited by Chiellini and Solaro
Kluwer Academic/Plenum Publishers, New York, 2003

convert saccharides and lignin to useful industrial products such as packaging, insulation and construction materials, films for agricultural purposes, composites, and engineering plastics[2-30].

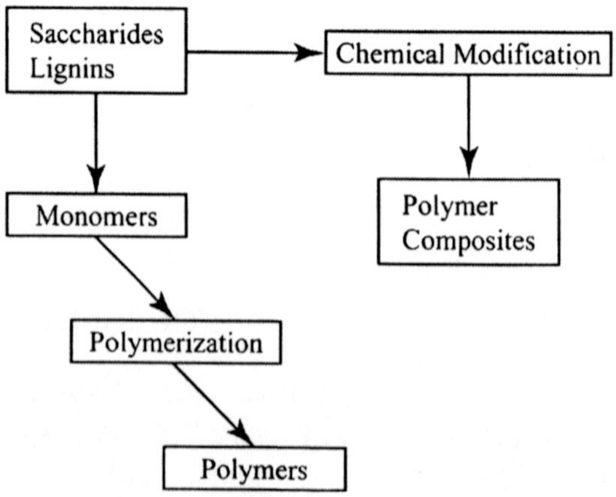

Figure 1. Conversion of polysaccharides and lignin to industrial products.

This paper concerns the preparation and the thermomechanical properties of environmentally compatible polymers derived from saccharides and lignins at our laboratory. The above research results have been obtained over the last several years. The environmentally compatible polymers include polyurethane (PU) and poly(ε-caprolactone) (PCL) derivatives. PU derivatives were prepared from saccharides and lignins. PCL derivatives were synthesized from lignins, saccharides, cellulose and cellulose acetate. The thermal properties of the above polymers were studied by differential scanning calorimetry (DSC), thermogravimetry (TG) and TG-Fourier transform-infrared spectrometry (FTIR). Mechanical properties were measured by mechanical testing.

2. METHODS OF CHARACTERISATION

2.1 Differential Scanning Calorimetry (DSC)

DSC was carried out on all prepared samples using a Seiko DSC 220C in N_2 atmosphere. The scanning rate was 10 °C min^{-1}. Before carrying out the measurements, the samples were dried in an air-oven at 120 °C for 2 h. T_g was defined as the temperature at the point of intersection between the tangents drawn at the point of inflection of the transition and at the flat part of the curve before the transition[31]. Heat capacity difference at T_g was measured. Phase transition temperatures such as T_c and T_m were also defined.[31] The enthalpy of the first order phase transitions was estimated.

2.2 Thermogravimetry (TG)

TG curves of all samples were recorded using a Seiko TG/DTA (differential thermal analyzer) 220. The curves were obtained at a heating rate of 10 °C min^{-1} in nitrogen atmosphere (30 ml min^{-1}). In order to examine the effect of atmosphere on decomposition behaviour, the TG curves were also obtained in air. The thermal degradation temperature (T_d) is defined as the temperature at the point of intersection of the tangents drawn from a point before the main decomposition step (i.e. where the curve is horizontal) and from the point of inflection of the main step.[31]

2.3 TG-Fourier Transform Infrared Spectrometry (TG-FTIR)

TG-FTIR was performed using a Seiko TG 220 instrument equipped with a JASCO FTIR 7000 spectrometer. TG measurements were carried out at a heating rate of 20 °C min^{-1}. Nitrogen and air flow-rates were controlled at 100 ml min^{-1}. The evolved gases during thermal degradation were simultaneously analyzed by FTIR. In order to obtain one spectrum, data of ten scans were accumulated at 1 s intervals. Each spectrum was recorded every 30 sec. The spectral resolution was 1 cm^{-1}.

3. SACCHARIDE- AND LIGNIN-BASED PU DERIVATIVES

PU derivatives (PU's) were prepared according to the process shown in Figure 2. Due to the limited solubility of saccharides such as glucose, fructose, sucrose and molasses in polyols such as polyethylene glycol (PEG), triethylene glycol (TEG) and diethylene glycol (DEG), the saccharides were dissolved in the above polyols at 50 or 60 °C. Prior to reaction with diphenylmethane diisocyanate (MDI), small amounts of water, surfactant and catalyst were added to the polyol solution. The NCO/OH (moles of isocyanate group/moles of OH groups) ratio was changed from 1.0 to 1.2, depending on the required physical properties of the prepared PU's. Molasses were obtained from Syonan Sugar Manufacture Co. Ltd. A schematic chemical structure of the PU sample with saccharide structure is shown in Figure 3.

Various kinds of industrial lignins such as Kraft lignin (KL), sodium lignosulfonate (LS), alcoholysis lignin (AL) and solvolysis lignin (SL) were used as raw materials for the preparation of lignin-based PU's at our laboratory. KL was provided by Westvaco Co. Ltd. LS was provided by Nippon Paper Industries Co., Ltd. AL was provided by Repap Co. Ltd. SL was obtained as a by-product in organosolve pulping of Japanese beech (*Fagus crenata*) with aqueous cresol. SL was provided by the Japan Pulp and Paper Research Institute Co. Ltd.

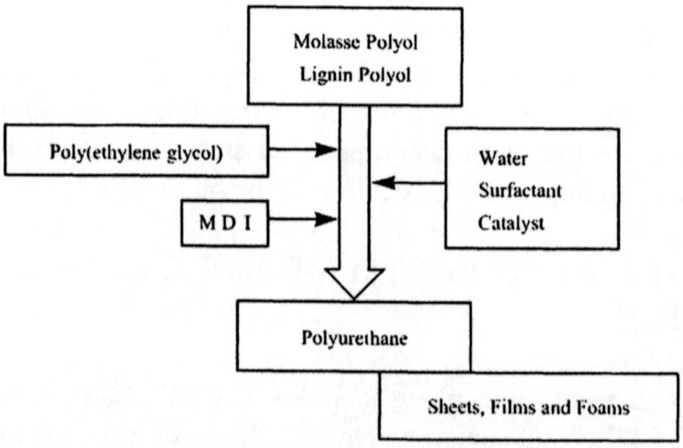

Figure 2. Preparation of biobased polyurethanes.

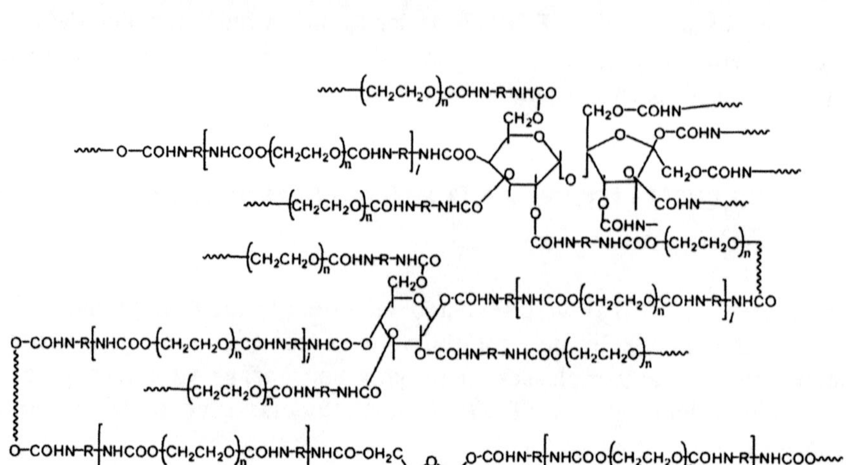

Figure 3. Schematic chemical structure of saccharide-based PU.

The following methods are examples of preparation of lignin-based PU's. Prior to obtaining PU's, KL, LS, AL and SL were dissolved in polyols. In order to prepare PU foams, the polyol solution was mixed first with surfactant (silicone oil) and a catalyst, then MDI was added. This mixture was vigorously stirred with a trace amount of water, which was added as a foaming agent. In the above processes, the NCO/OH ratio was decided according to the required physical properties of prepared PU foams.

A schematic chemical structure of the lignin-containing PU sample is shown in Figure 4.

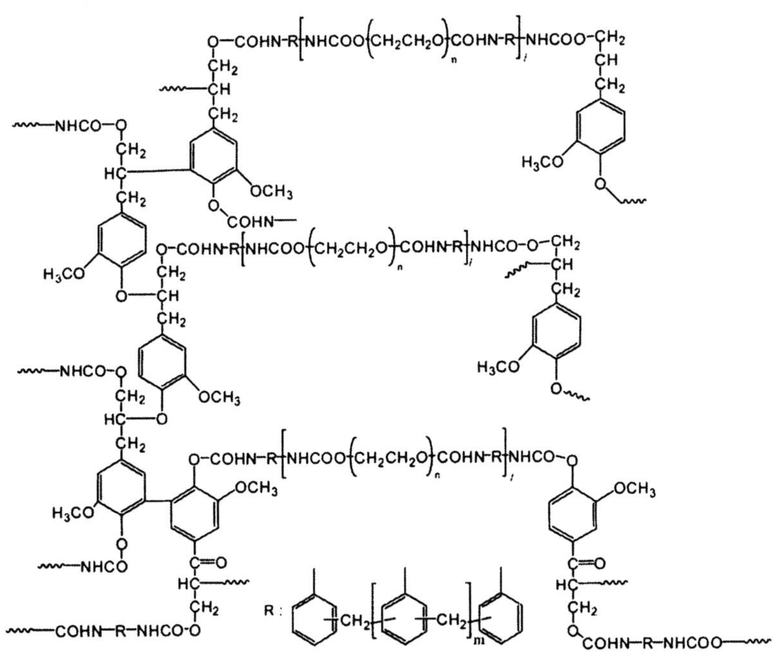

Figure 4. Schematic chemical structure of lignin-based PU.

3.1 Thermal Properties of PU's

It is generally recognized that PU is one of the most useful three-dimensional polymers, since PU has unique features: for example, various materials such as sheets, foams, adhesives, and paints can be obtained from PU, and their physical properties can easily be controlled.

Concerning saccharide-based PU's, their preparation and physical properties have already been reported elsewhere[23,24,26-29]. Accordingly, the recent development of lignin-based rigid PU foams is discussed here.

Figure 5 shows the relationship between T_g and the LS contents in PEG, TEG and DEG from which LS-based rigid PU foams were prepared, where

LSPPU, LSTPU, and LSDPU are PU from LS in PEG, TEG, and DEG solution, respectively. T_g increases almost steadily with the LS content for all PU foams. The incorporation of LS into the PU structure leads to an increase in crosslinking density due to the large number of hydroxyl groups per molecule of lignin. The higher the crosslinking density, the more restricted is the main chain motion and the higher is T_g. In addition of the large effect on the crosslinking density, lignin also acts as hard segment that causes an increase in T_g. It is also noteworthy to consider the effect of large molecular weight of lignin on the segmental motion of PU. The glass transition occurring at higher temperature (T_{gh}) seems to correspond with the glass transition temperature of lignin, since it does not depend on the glycol type (Figure 5).

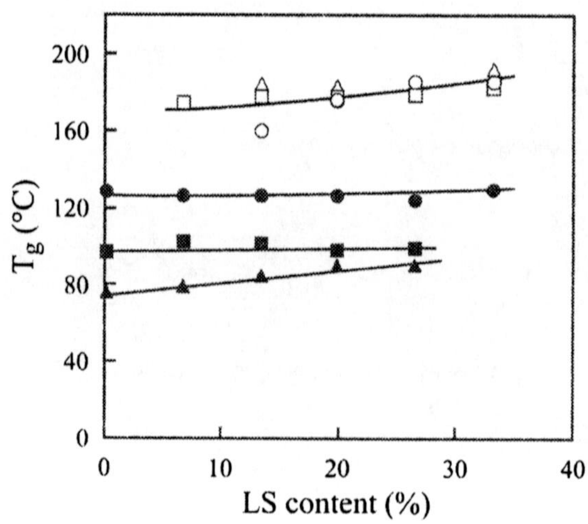

Figure 5. Dependence of T_g (closed symbols) and T_{gh} (open symbols) on the LS content of PU's prepared from LS-DEG-MDI (LSDPU: circles), LS-TEG-MDI (LSTPU: squares), and LS-PEG200-MDI (LSPPU: triangles) systems.

Figure 6 shows TG and DTG curves of LSPPU with various LS contents in PEG. The sample without LS decomposes in one stage according to the observation of derivative TG (DTG) curve and T_d is about 305 °C. DTG curves of LSPPU with lignin in the molecular chain show two-stage decomposition.

Figure 7 shows that T_d of LSPPU foams decreases slightly with the lignin content. Usually, the dissociation of urethane bonds formed between isocyanate and phenolic hydroxyl groups occurs in a temperature range lower than that of urethane bonds formed between isocyanate and alcoholic hydroxyl groups[19]. Accordingly, the increase of lignin content in polyol seems to result in the slight decrease of T_d of LSPPU foams. However, this decrease of T_d for LSTPU and LSDPU is not recognizable. This may be caused by the short length of TEG and DEG crosslinks that makes the PU matrix rigid. This rigid structure makes difficult the diffusion of degradation gas within the PU matrix.

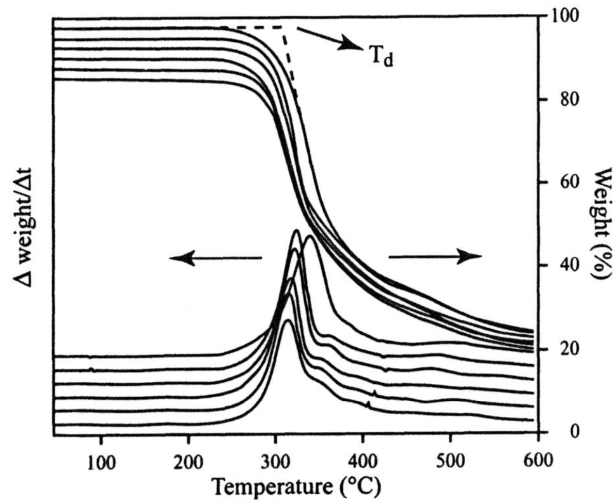

Figure 6. TG and DTG curves of LSPPU with various LS contents in PEG feed; LS content (from top to bottom): 0, 6.6, 13.2, 19.8, 26.4, and 33.0%.

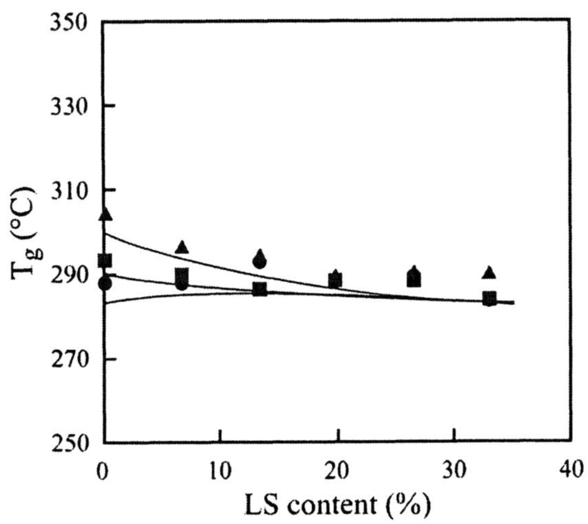

Figure 7. Dependence of T_d on the LS content of PU's prepared from LS-DEG-MDI (LSDPU: ●), LS-TEG-MDI (LSTPU: ■), and LS-PEG200-MDI (LSPPU: ▲) systems.

Figures 8 and 9 show representative stacked three-dimensional diagram showing the relationship between IR intensity, wavenumber and temperature in TG-FTIR. Wavenumbers range from 600 to 4000 cm^{-1} and temperatures range from 40 to 600 °C. As shown in the diagram, IR absorption bands can mainly be observed in the temperature range from 250 to 450 °C. The main peaks are at 1128 (C-O-C), 1620 (C=C), 2277 (NCO), 2358 (CO_2), 2920 (C-H), and 3700 cm^{-1} (H_2O).

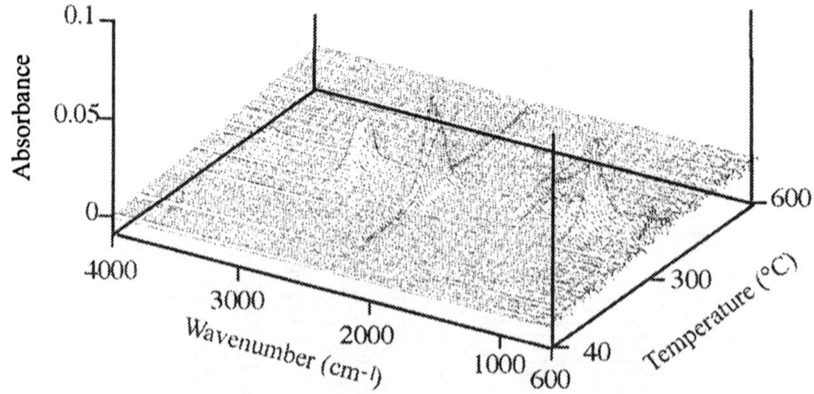

Figure 8. Stacked three-dimensional TG-FTIR diagram for PU prepared from the LS-PEG200-MDI system (LS content = 0).

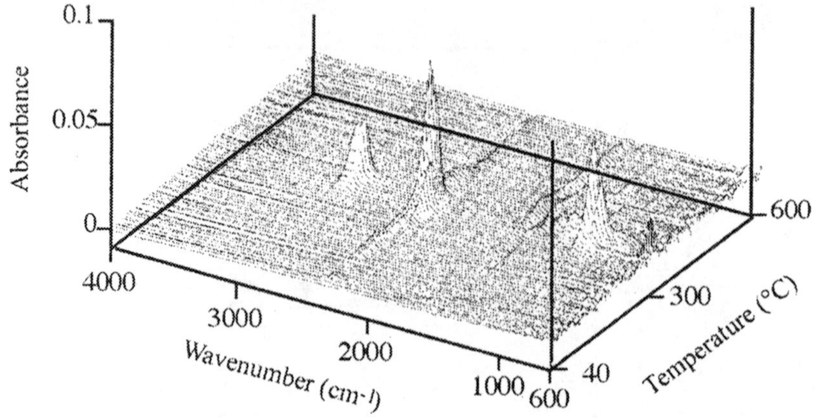

Figure 9. Stacked three-dimensional TG-FTIR diagram for PU prepared from the LS-PEG200-MDI system (LS content = 13.2 %).

Figures 10 and 11 show the changes of C-O-C, C=C, NCO, CO_2, CH, and OH peak intensities in the temperature range from at 300 to 450 °C. As shown in the above figures the intensities of C-O-C, NCO, CO_2, CH, and OH signals decrease with increasing temperature, suggesting that the evolved gases are formed by the degradation of PU's. However, the peak intensities of the evolved gases from PU not containing LS (Figure 10) indicate that gas evolution occurs between 300 and 400 °C, whereas the evolution of gases from PU containing LS ceases at about 350 °C (Figure 11) and the amount of evolved gases is about 2/3 of that from PU without LS. This indicates that the LS containing PU forms a more rigid matrix that may interfere with PU thermal degradation and blocks the gas evolution.

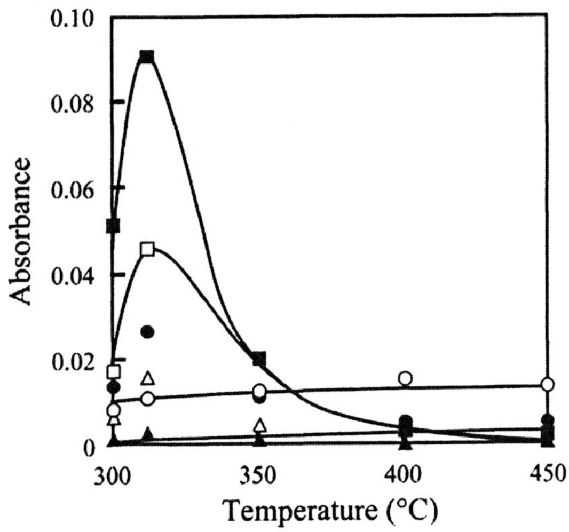

Figure 10. Changes of C-O-C (●), C=C (○), NCO (■), CO_2 (□), O-H (▲), and C-H (△) peak intensities in the temperature range from 300 to 450 °C (LS content 0%).

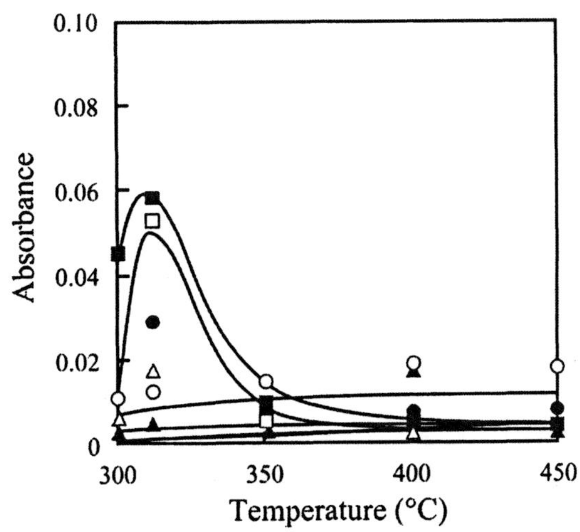

Figure 11. Changes of C-O-C (●), C=C (○), NCO (■), CO_2 (□), O-H (▲), and C-H (△) peak intensities in the temperature range from 300 to 450 °C (LS content 13.2%).

4. SACCHARIDE AND LIGNIN-BASED PCL DERIVATIVES

4.1 Preparation of Saccharide-based and Lignin-based PCL Derivatives

Figure 12 shows the preparation scheme of polycaprolactone derivatives (CAPCL's) based on cellulose acetate (CA). As shown in the scheme, CAPCL's were synthesized from cellulose acetate by reaction with ε-caprolactone by using dibutyltin dilaurate (DBTDL) as catalyst.

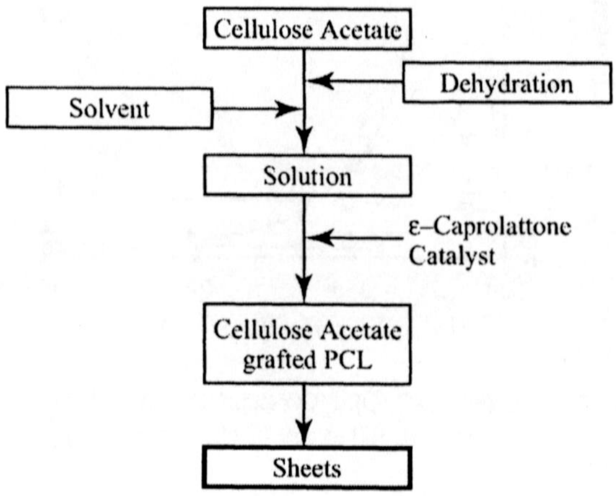

Figure 12. Preparation scheme of CA-based PCL's (CAPCL's).

Lignin-based PCL's (LigPCL's) were synthesized by polymerization of CL initiated by lignin hydroxyl groups. The amount of CL was varied from 1 to 25 molecules per each lignin OH group (CL/OH molar ratio = 1, 2, 3, 4, 5, 10, 15, 20, and 25). Polymerization experiments were carried out in the presence of a small amount of DBTDL. LigPCL sheets were prepared by heat-pressing the synthesized polymers at 160-180 °C at about 10 MPa. Figure 13 shows a schematic representation of LigPCL chemical structure.

4.2 Thermal Properties of CAPCL and LigPCL Polymers

Figure 14 shows the dependence of the T_g of CAPCL samples on the CL/OH molar ratio. It is known that dry cellulose does not present glass transition in the temperature range from 20 °C to the thermal decomposition temperature.[32] When cellulose is acetylated, the glass transition is observed at about 150 °C (Figure 14). These data suggest that the T_g of glucopyranose chains should be observed after introduction of large side-chain substituents such as PCL.

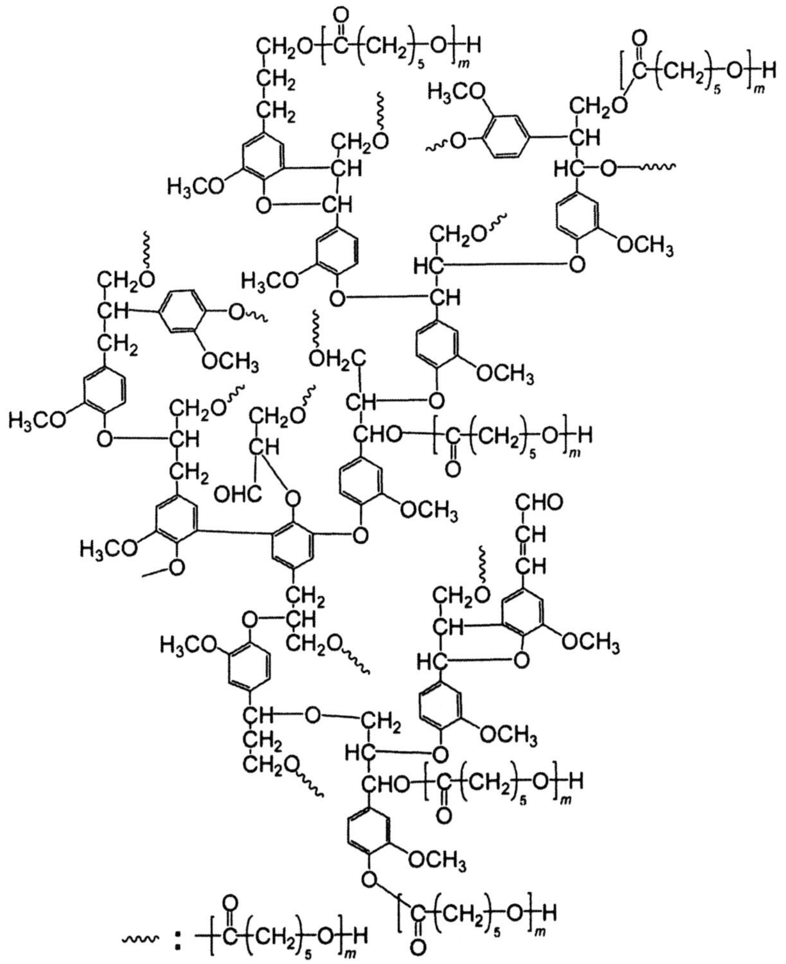

Figure 13. Schematic representation of LigPCL chemical structure.

When the DSC curves were magnified, the two baseline deviations due to the glass transitions of CA and PCL were clearly observed. The glass transition of CA can be observed at low CL/OH molar ratio and becomes very difficult to detect when this ratio exceeds 8. The T_g of CA decreases on increasing the CL content, very likely because the introduction of large side chain molecules gives rise to intermolecular distances expansion; the increased free space enhances the main chain motions. The glass transition of PCL is observed at about –50 °C; the T_g decreases at low CL/OH molar ratios and then increases slightly after reaching a minimum at about [CL]/[OH] = 10. The T_g increase observed for the sample with CL/OH molar ratio = 20 suggests that the molecular motions of PCL random chains are restricted by the presence of crystalline regions.

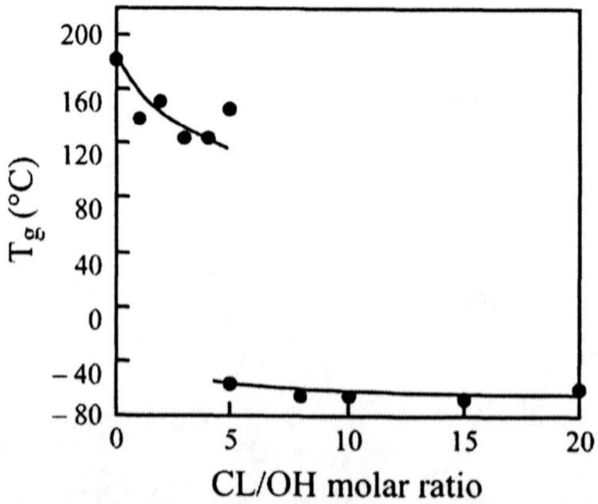

Figure 14. Dependence of the T_g of CAPCL samples on the CL/OH molar ratio.

The melting peak of PCL chains was observed for samples with CL/OH molar ratio = 10, 15, and 20. A broad exothermic peak due to cold crystallization (T_c) at -30 °C was also observed for the sample with CL/OH molar ratio = 15. Figure 15 shows a schematic model for the higher order molecular structure of CAPCL. As shown in Figure 14, T_g of CA main chain is observable at low CL/OH molar ratios, since the cellulose main chain becomes mobile because of the increased intermolecular distances. When the CL/OH molar ratio becomes over 10, enough long chains are assembled to form a crystalline region and the melting of PCL chains become increasingly observable with the growth of crystalline regions.

Figure 16 shows the phase diagrams of kraft lignin (KLPCL) and alcoholysis lignin (ALPCL) grafted PCL. In general, no significant difference is observed between KLPCL and ALPCL samples. T_g decreases with increasing CL/OH molar ratio from 2 to 10, since PCL chains act as soft segments in the lignin molecular network. However, a melting endotherm is clearly observed when the CL/OH molar ratio is 10 to 25, in agreement with the presence of crystalline regions. Noteworthy, the T_g increases also with increasing the CL/OH molar ratio above 10. This behaviour suggests that the long PCL side-chains form crystalline domains that restrict the motion of the amorphous PCL regions. The reported data indicate that the structural differences between AL and KL do not affect markedly the thermal behaviour of KLPCL and ALPCL, indicating that both lignins can be used as raw materials for the preparation of PCL derivatives.

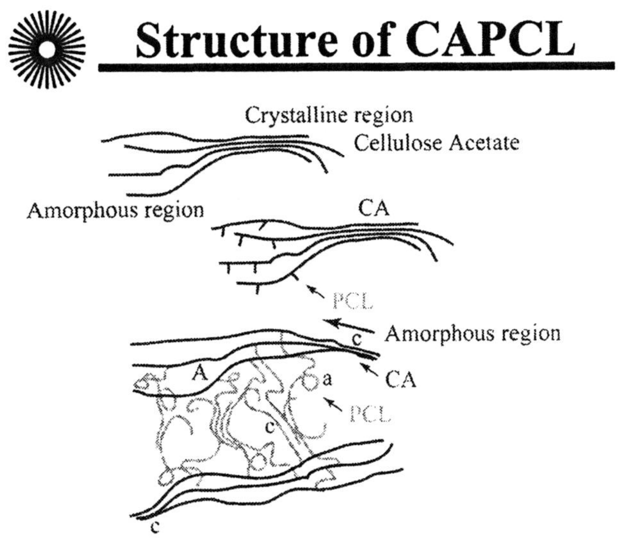

Figure 15. Schematic model for the higher order structure of CAPCL.

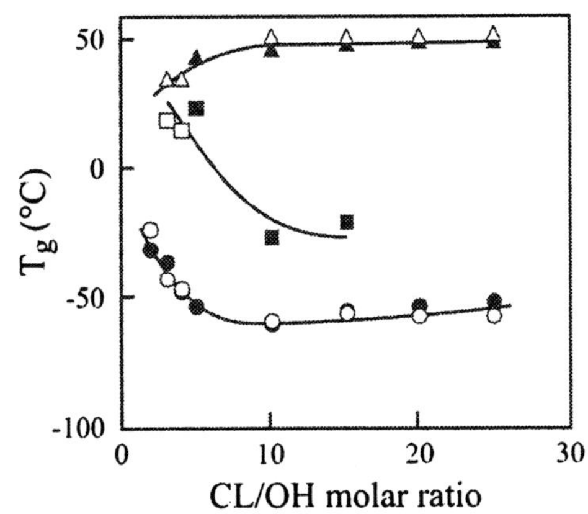

Figure 16. Phase diagrams of KLPCL (close symbols) and ALPCL (open symbols): T_g (circles), T_c (squares), and T_m (triangles).

5. POLYURETHANES FROM SACCHARIDE AND LIGNIN BASED PCL'S

5.1 Preparation of Polyurethanes

Saccharide-based polyurethanes (SacPCLPU) were obtained from saccharide-grafted PCL's by the following procedure. Saccharide-grafted

PCL's were dissolved in THF and reacted with MDI. The obtained prepolymer was cast on a glass plate, the solvent was evacuated under dry conditions and the sheets were cured at 120 °C. The core structure of the prepared polyurethanes consists of saccharide-grafted PCL molecules linked by a three dimensional network of urethane bonds. Figure 17 presents a schematic representation of the chemical structure of PU derived from SucPCL.

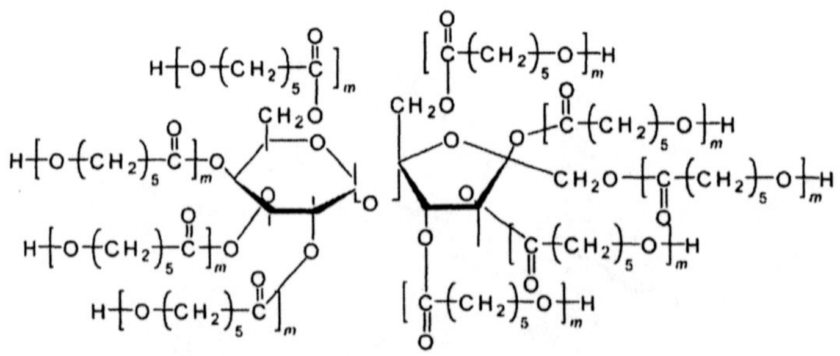

Figure 17. Schematic chemical structure of polyurethanes derived from SucPCL.

Lignin-based polyurethanes (LigPCLPU) were obtained by the following procedure. Lignin-grafted PCL' was dissolved in THF and reacted with MDI. The obtained polymer was cast on a glass plate, and the solvent was evacuated under dry conditions. The resulting ALPCL- and KLPCL-based polyurethane sheets were cured at 120 °C. Figure 18 shows a schematic representation of the chemical structure of LigPCLPU samples.

5.2 Thermal Properties of LigPCLPU's

Figure 19 shows the phase diagram of PU's derived from KLPCL and ALPCL with CL/OH molar ratios varying from 2 to 25. The observed thermal behaviour is almost independent of the lignin type. As shown in the figure, T_g decreases with increasing CL/OH ratio, indicating that PCL chains act as soft segments in PU networks. The DSC curves of PU's with CL/OH molar ratio of 15 exhibit an exothermic peak due to cold-crystallization; also a large peak due to the melting of crystalline region is observed at CL/OH ratios over 15. T_c and T_m slightly increase with increasing the CL/OH ratio over 15. These results indicate the presence of crystalline regions in polyurethanes derived from KLPCL and ALPCL with CL/OH ratios larger than 15.

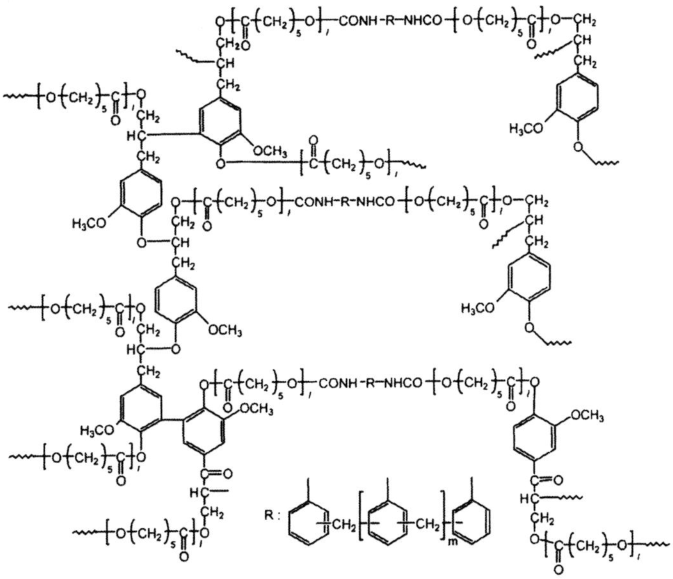

Figure 18. Schematic chemical structure of polyurethanes derived from lignin-grafted PCL.

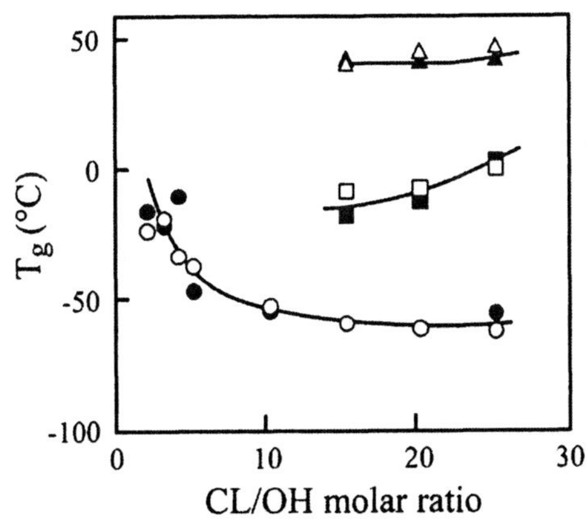

Figure 19. Phase diagram of polyurethanes derived from KLPCL (open symbols) and ALPCL (close symbols): T_g (circles), T_c (squares), and T_m (triangles).

6. CONCLUSIONS

From the reported results, we may conclude that saccharides, polysaccharides, and lignins can be used as useful components of biobased polymeric materials such as poly(ε-caprolactone) and polyurethanes. Saccharides and lignins efficiently act as hard segments in the above polymers. It is possible to control the flexibility and stiffness of the prepared polymers by changing the chain length that connects saccharide and lignin components in the polymer networks. It can be concluded that the conversion of biomass into industrial products and the development of biobased industrial products that are cost-competitive can be realized by the use of the above technologies.

REFERENCES

1. For example, *US Public Law 106-224* "Biomass Research and Development Act of 2000".
2. Hatakeyama, H., Hayashi, E., and Haraguchi, T., 1977, Biodegradation of poly(3-methoxy-4-hydroxy styrene). *Polymer* **18**: 759-763.
3. Hatakeyama, T., Nakamura, K., and Hatakeyama, H., 1978, Differential thermal analysis of styrene derivatives related to lignin. *Polymer* **19**: 593-594.
4. Nakamura, K., Hatakeyama, T., and Hatakeyama, H., 1983, Effect of substituent groups on hydrogen bonding of polyhydroxystyrene derivatives. *Polym. J.* **15**: 361-366.
5. Hirose, S., Hatakeyama, H., and Hatakeyama, T., 1983, Synthesis and thermal analysis of polyacylhydrazones having guaiacyl units with alkylene groups. *Sen-i Gakkaishi* **39**: 496-500.
6. Saraf, V. P., and Glasser, W. G., 1984, Engineering plastics from lignin. III. Structure property relationships in solution cast polyurethane films. *J. Appl. Polym. Sci.* **29**: 1831-1846.
7. Saraf, V. P., and Glasser, W. G., 1985, Engineering plastics from lignin. VI. Structure property relationships of PEG-containing polyurethane networks. *J. Appl. Polym. Sci.* **30**: 2207-2224.
8. Nakamura, K., Hatakeyama, T., and Hatakeyama, H., 1986, DSC studies on hydrogen bonding of poly(4-hydroxy-3, 5-dimethoxystyrene) and related derivatives. *Polym. J.* **18**: 219-225.
9. Yoshida, H., Morck, R., Kringstad, K. P., and Hatakeyama, H., 1987, Kraft lignin in polyurethanes I. Mechanical properties of polyurethanes from a kraft lignin-polyethers triol-polymeric MDI system. *J. Appl. Polym. Sci.* **34**: 1187-1198.
10. Hirose, S., Yano, S., Hatakeyama, T., and Hatakeyama, H., 1989, Heat-resistant Polyurethanes from Solvolysis Lignin. *ACS Symp. Ser.* **397**: 382-389.
11. Hatakeyama, H, Hirose S., and Hatakeyama, T., 1989, High-Performance Polymers from Lignin Degradation Products, *ACS Symp. Ser.* **397**: 205-218.
12. Hatakeyama, T., Nakamura, K., Yoshida, H., and Hatakeyama, H., 1989, Mesomorphic properties of highly concentrated solutions of polyelectrolytes from saccharides, Food Hydrocolloids. *Food Hydrocolloid* **3**: 301-311.
13. Hirose, S., Nakamura, K., and Hatakeyama, H., 1989, Design of Linear Aromatic Polymers Derived from Phenols Related to Lignin. In *Cellulose and Wood* (C. Schuerch, ed.), John Wiley and Sons, New York, p. 1133-1144.
14. Nakamura, K., Morck, R., Reimann, A., Kringstad, K., and Hatakeyama, H., 1989, Compression Properties of Polyurethane Foam Derived from Kraft Lignin. In *Wood Processing and Utilization* (J. F. Kennedy, P. A. Williams and G. O. Phillips, eds.), Ellis Horwood, Chichester, pp. 175-180.
15. Yoshida, H., Morck, R., Kringstad, K. P., and Hatakeyama, H., 1990, Kraft lignin in polyurethanes. II. Effects of the molecular weight of kraft lignin on the properties of

polyurethanes from a kraft lignin -polyether triol-polymeric MDI system. *J. Appl. Polym. Sci.* **40**: 1819-1832.
16. Nakamura, K., Hatakeyama, T., and Hatakeyama, H., 1991, Formation of the glassy state and mesophase in the water-sodium alginate system. *Polym. J.* **23**: 253-258.
17. Morck, R., Reimann, A., Kringstad, K., and Hatakeyama, H., 1991, Mechanical properties of solvolysis lignin derived polyurethanes. *Polym. Adv. Technol.* **2**: 41-47.
18. Hatakeyama, T., and Hatakeyama, H., 1992, Molecular Relaxation of Cellulosic Polyelectrolytes with Water. *ACS Symp. Ser.* **489**: 329-340.
19. Hirose, S., Kobashigawa, K., Izuta, Y. and Hatakeyama, H., 1998, Thermal degradation of polyurethanes containing lignin studied by TG-FTIR. *Polym. Int.* **41**: 247-256.
20. Hatakeyama, H., and Hatakeyama, T., 1998, Interaction between water and hydrophilic polymers. *Thermochim. Acta* **308**: 3-22.
21. Tanaka, R., Hatakeyama, T., and Hatakeyama, H., 1998, Formation of locust bean g um hydrogel by freezing thawing. *Polym. Int.* **45**: 118-126.
22. Hatakeyama, H., Hirose, S., Nakamura, K., and Hatakeyama, T., 1993, New Types of Polyurethanes Derived from Lignocellulose and Saccharides. In *Cellulosics: Chemical, Biochemical and Material Aspects* (J. F. Kennedy, G. O. Phillips and P. A. Williams, eds.), Ellis Horwood, Chichester, pp. 524-536.
23. Morohoshi, N., Hirose, S., and Hatakeyama, H., Tokashiki, T. and Teruya, K., 1995, Biodegradability of polyurethane foams derived from molasses. *Sen-i Gakkaishi* **51**: 143-149.
24. Hatakeyama, H., Hirose, S., Hatakeyama, T., Nakamura, K., Kobashigawa, K., and Morohoshi, N., 1995, Biodegradable polyurethanes from plant components. *J. Macromol. Sci., Pure Appl. Chem.* **A32**: 743-750.
25. Donnely, M. J., 1995, In Vitro enzymic synthesis of polymers containing saccharides, lignins, proteins or related Components: a Review. *Polym. Int.* **47**: 257-266.
26. Nakamura, K., Nishimura, Y., Zetterlund, P., Hatakeyama, T., and Hatakeyama, H., 1996, TG-FTIR Studies on biodegradable polyurethanes containing mono- and disaccharides Components. *Thermochim. Acta* **282/283**: 433-441.
27. Zetterlund, P., Hirose, S., Hatakeyama, T., Hatakeyama, H. and Albertsson, A-C., 1997, Thermal and mechanical properties of polyurethanes derived from mono- and disaccharides. *Polym. Int.* **42:** 1-8.
28. Hatakeyama, H., Kobahigawa, K., Hirose, S., and Hatakeyama, T., 1998, Synthesis and physical properties of polyurethanes from saccharide-based polycaprolactones. *Macromol. Symp.* **130**: 127-138.
29. Hatakeyama, T., Tokashiki, T., and Hatakeyama, H., 1998, Thermal propeties of polyurethanes derived from molasses before and after biodegradation. *Macromol. Symp.* **130**: 139-150.
30. Gandini, A., and Belgacem, N. M., 1998, Recent advances in the elaboration of polymeric materials derived from biomass components. *Polym. Int.* **47**: 267-276.
31. Hatakeyama, T. and Quinn, F. X., 1994, *Thermal Analysis, Fundamentals and Applications to Polymer Science*, John Wiley and Sons, Chichester. Pp. 81-87.
32. T. Hatakeyama, T. and Liu, Z., 1998, *Handbook of Thermal Analysis*, John Wiley, Chichester, pp. 206-210.

Biodegradable Kraft Lignin-based Thermoplastics

YAN LI and SIMO SARKANEN
Kaufert Laboratory, University of Minnesota, 2004 Folwell Avenue, St. Paul, MN, USA

1. INTRODUCTION

Lignins are seldom accorded a prominent place in compendia about biodegradable polymeric materials. Yet lignin derivatives are available in huge quantities from plant sources, and they are (albeit slowly) biodegradable. Indeed high (85%) industrial by-product lignin contents in thermoplastics were first reported in 1997[1], and since then a U.S. patent[2] has been issued in response to the discovery of plasticizers for simple derivatives of the same kind of raw materials. Previously it had usually been impossible to incorporate any lignin derivative at levels beyond 25–40% into a polymeric material without sacrificing its mechanical integrity. The present article traces the development of the conceptual basis for the paradigm shift that has occurred in formulating useful lignin-based thermoplastics.

1.1 Lignins

The earliest vascular plants appearing in the fossil record (from the time between the Late Silurian and Middle Devonian periods) appear to possess lignified cell walls. Lignins impart rigidity to the network of cellulose microfibrils that, in being crosslinked through hydrogen bonding to other intervening glycans, form the basis for the structure of modern plant cell walls. The appearance of lignins during evolutionary time is credited with having allowed vascular plants to adopt an upright habit and develop water-conducting cells that prolong metabolic activity in desiccating

Biodegradable Polymers and Plastics, Edited by Chiellini and Solaro
Kluwer Academic/Plenum Publishers, New York, 2003

environments[3]. Thus lignins are of enormous importance to life on Earth and, as a group of biopolymers that represent 15–30% of all vascular plant cell wall materials, they are second in abundance only to cellulose itself.

Lignins have been traditionally portrayed as being constituted from *p*-hydroxyphenylpropane units through random distributions of about 10 different linkages[3] (Figure 1)[4]. Lignin macromolecules are assembled through the dehydrogenative polymerization of one or more among three possible monolignol precursors, namely *p*-hydroxycinnamyl (*p*-coumaryl) alcohol, 4-hydroxy-3-methoxycinnamyl (coniferyl) alcohol, and 4-hydroxy-3,5-dimethoxycinnamyl (sinapyl) alcohol. Roughly half of the inter-unit linkages are of the same 8-*O*-4' alkyl aryl ether type, but there are pronounced variations between the lignins of conifers (gymnosperms), hardwoods (angiosperm eudicotyledons), and cereals and grasses (monocotyledons) in regard to their monomer composition[3].

The primary lignin precursor in conifers is coniferyl alcohol with a smaller contribution originating from *p*-coumaryl alcohol, but in hardwoods sinapyl alcohol is also incorporated at substantial levels into the macromolecular matrix. The lignins of grasses and cereals, however, embody units derived from the three primary monolignols in comparable proportions, while *p*-coumaric, ferulic and sinapic acids may be variously bound to the resulting framework as well[3]. It is worth emphasizing that, in any given plant species, there are also significant variations of lignin configuration with cell type and morphological region within the cell wall. This variability may be one of the reasons why structural formulae proposed for lignin macromolecules are usually very complicated (*cf.* Figure 1). To a great extent they have been based on analyses of degradative reaction products from lignin preparations[3] that were not homogeneous with respect to macromolecular composition.

Moreover, during the past decade 2-D NMR spectroscopic findings have tended to call into question earlier beliefs about the frequencies and configurations of some important inter-unit linkages in lignins. As a result, the most recent structural formula proposed for a representative softwood lignin fragment[5] is far simpler than what prevailing wisdom would have preferred some two decades earlier[4]. The later macromolecular lignin chain is less branched and embodies fewer distinguishable structural features, but a new 8-membered dibenzodioxocin ring has been prominently incorporated into it[5] as a structure accounting for about 8% of all inter-unit linkages[6]. These comparisons are not intended merely for idle curiosity: there are better prospects for developing useful polymeric materials from derivatives of the more simple lignin structure[5] proposed in 1998 than from those of the unattractive predecessor[4] shown in Figure 1.

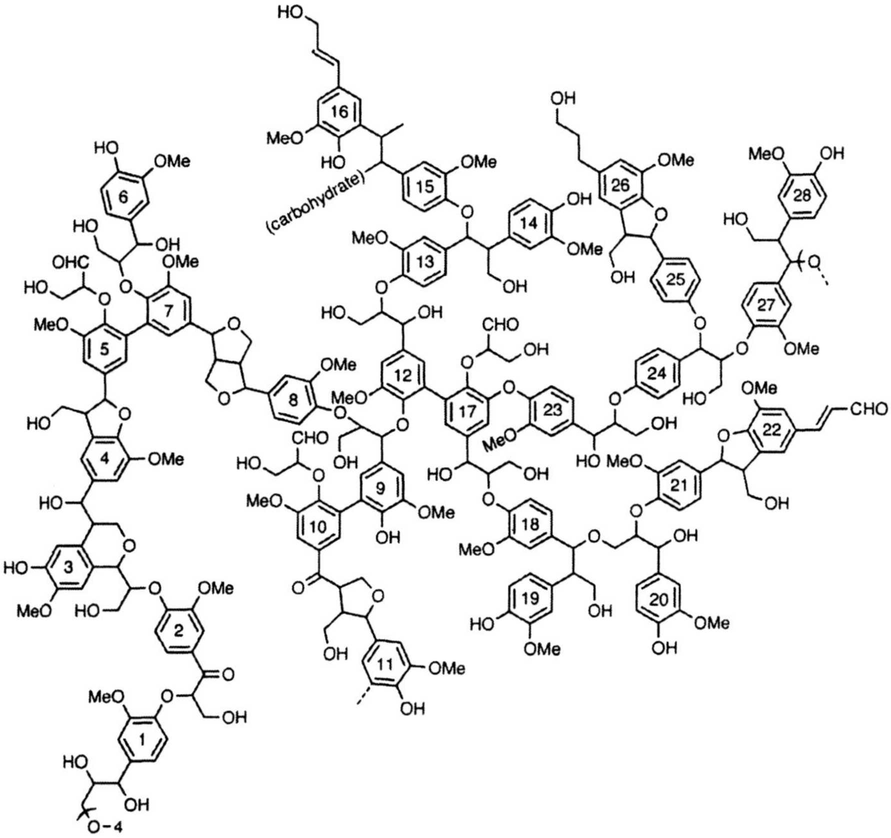

Figure 1. A traditional schematic depiction of the structural features in softwood lignins. (Adapted from reference 4).

1.2 Sources of Industrial Lignin Derivatives

Kraft lignins, the by-products of the kraft process through which wood chips are converted into pulp for manufacturing paper, constitute the most plentiful single type of lignin derivative produced in the world today. The individual molecular kraft lignin components have been extensively modified with respect to the configurations of the native lignin chains from which they are derived[7]. They have been exposed to quite severe conditions (*e.g.* for 2 h or more at 170 °C to aqueous solution containing 45 g L^{-1} NaOH and 12 g L^{-1} Na_2S, in the case of softwood chips) during which many of the inter-unit linkages along the macromolecular lignin chains are cleaved and new covalent bonds can be formed between previously independent molecular species (Figure 2). Further modifications involving, *inter alia*, the introduction of conjugated double bonds into the framework of the transformed components can complicate the range of products formed considerably[7].

Most kraft lignin is burned as fuel in the recovery furnace of the pulp mill, but a surplus can be created if the capacity of the unit is exceeded through attempts to maximize pulp production[8]. In addition to kraft lignins from such recovery loaded mills, it can be anticipated that other kinds of by-product lignin derivatives may be forthcoming in the future as a result of environmentally friendly pulping processes[9] and also bioethanol generation[10] and chemical production[11] from lignocellulose-derived monosaccharides. Moreover, if initiatives to make pharmaceuticals and vaccines in transgenic plants[12] were expanded to include higher volume (bio)chemicals, considerable quantities of new lignin derivatives would begin to accumulate that would have to be processed in an economically profitable way. Certainly transgenic plants (including trees) have been reported with reduced lignin contents, but it is unlikely that lignins can be eliminated from their cell walls altogether: it was the appearance of these biopolymers during the course of evolution that allowed vascular plants to begin to thrive on dry land in the first place (*vide supra*).

It is, however, difficult to predict precisely what the lignin derivatives will be with which industry will have to contend in the future, and therefore it is at present more reasonable to work with the kraft lignins that are currently available. There is every reason to expect that the fundamental insights gained in developing thermoplastics with the highest attainable kraft lignin contents will be applicable to thermoplastics embodying other kinds of lignin derivatives as well. The formulations for thermoplastics based on different lignin derivatives will vary, but the approach to optimizing them will remain essentially the same. Indeed it is evident that a substantial step has already been taken in this direction (*vide supra*).

1.3 Traditional Lignin-Containing Polymeric Materials

In research and development efforts carried out between the mid 1970's and 1990's, it was common for suitable lignin preparations to be incorporated covalently into other kinds of polymeric materials. In this way by-product lignins and their derivatives were successfully introduced into phenol-formaldehyde resins, polyurethanes, epoxies and acrylics[14]. In addition they were used as backbones to which other synthetic polymer chains could be grafted through free radical polymerization reactions[15].

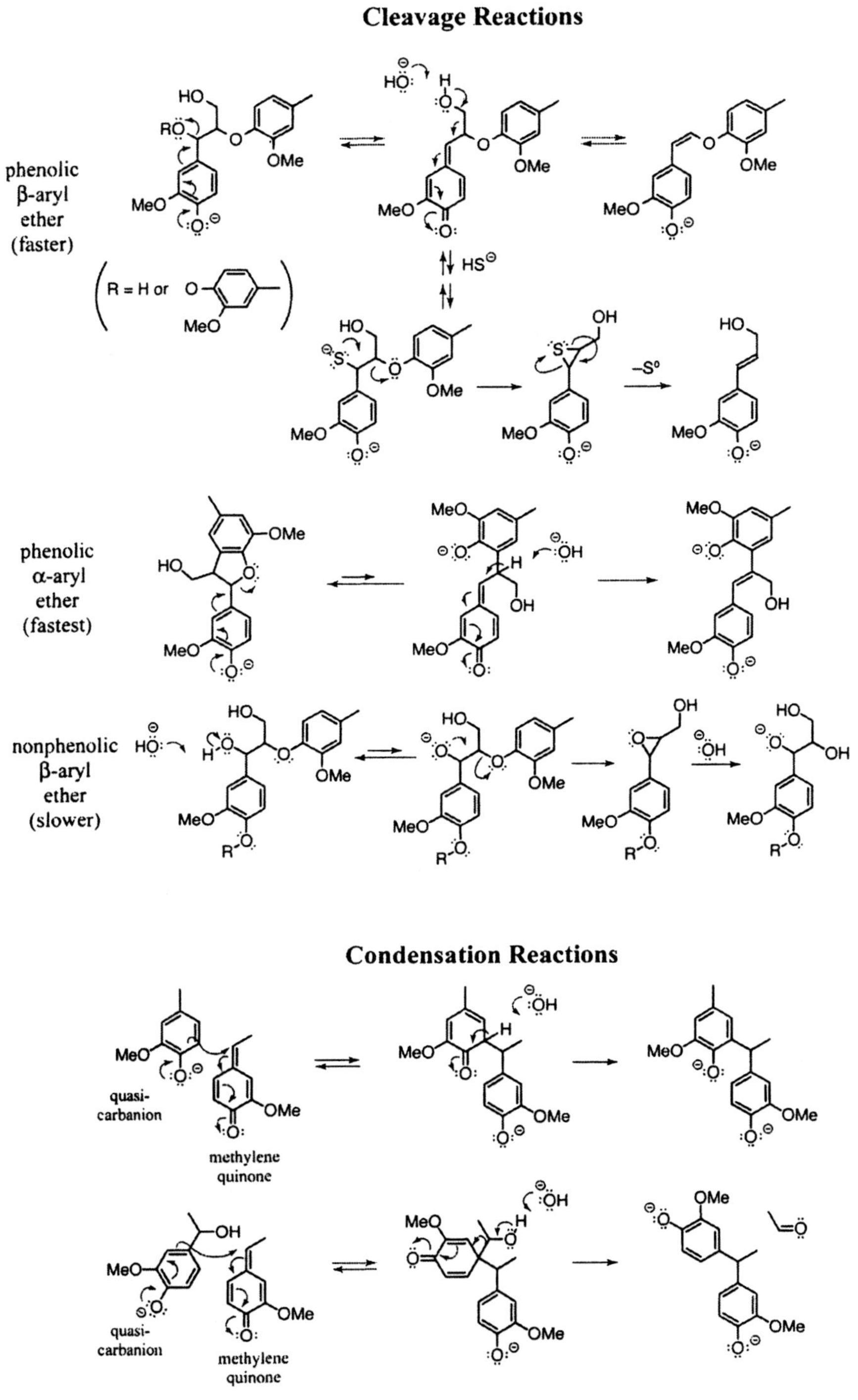

Figure 2. Predominant reactions of lignins during kraft pulping[7].

Alternatively it was suggested that greater flexibility in the properties of the polymeric materials being created could be obtained if the lignin preparations were first degraded into monomers and oligomers, which could then be copolymerized in a variety of ways to yield a much broader range of products[16]. However, the severity of the degradative reaction conditions that would be needed, coupled with the fractionation of the resulting component mixtures, are likely to prove prohibitively costly.

Some quintessential examples of traditional lignin-containing polymeric materials were embodied in a series of kraft lignin–polyether triol–polymeric MDI polyurethanes developed by Hatakeyama, Kringstad *et al.*[17] In a comparative study, these were produced using four kraft lignin fractions derived by solvent extraction from an industrial parent preparation; their apparent number-average molecular weights varied by a factor of 8 on the basis of the profiles generated by their acetylated derivatives upon elution in THF through poly(styrene-divinylbenzene) columns. The formulation employed a relatively low NCO/OH ratio of 0.9 in the expectation of a greater sensitivity on the part of their mechanical behaviour to variations in kraft lignin content.

The tensile strengths of the polyurethanes made with the first three fractions (in order of increasing molecular weight) initially increased with kraft lignin content up to 25–30%, where they levelled off, but the tensile strength of the corresponding material containing the highest molecular weight fraction reached a maximum already at a kraft lignin content of ~10%, where after it decreased markedly (Figure 3). The difference in behaviour was ascribed to the presence of increasing quantities of unreacted kraft lignin particles in the cured material incorporating the highest molecular weight fraction. On the other hand, the polyurethanes produced from all four fractions exhibited Young's moduli that increased with kraft lignin content up to 30–33% where they attained their maximum values. However, the most telling aspect of the reported findings was that the polyurethanes invariably became glassy and rigid whenever the overall proportion of kraft lignin exceeded 30%, regardless of the molecular weight of the fractions involved[17].

These observations aptly exemplified a general concern at the time[18] that "even the most judicious selection of lignin isolation or modification schemes has not allowed researchers to overcome the incorporation limit of 25 to 40 weight percent of lignin as an active component in polyurethanes." Indeed similar limitations have confounded the introduction of lignin derivatives into other polymeric materials as well. This arose from the fact that the physicochemical properties of the lignin components themselves had not been explicitly taken into account: at the molecular level, the kraft lignin species that were incorporated into polyurethanes, for example, had been regarded primarily in terms of their hydroxyl group content[17]. Yet there are particularly strong non-covalent interactions between the individual molecular components in kraft lignin preparations[19]; they will have a pronounced impact upon the mechanical properties of any polymeric materials into which these lignin derivatives have been incorporated as active constituents (rather than fillers).

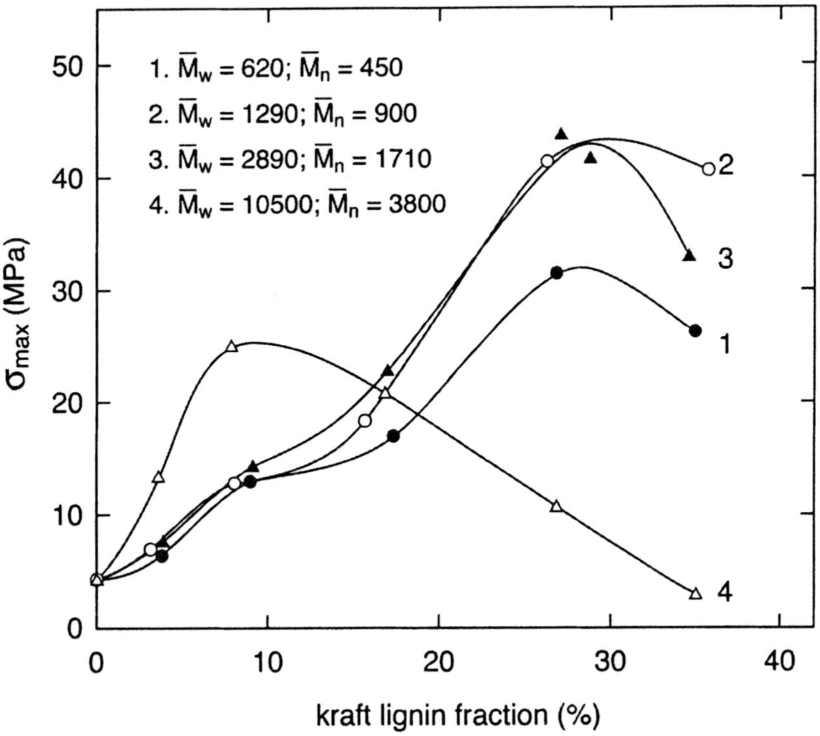

Figure 3. Tensile strengths (σ_{max}) of cured kraft lignin–polyether triol–polymeric MDI polyurethanes, NCO/OH ratio 0.9. Variation with content of softwood kraft lignin fractions characterized by different molecular weights that were isolated from the parent preparation through solvent extraction. (Data from reference 17.).

2. TOWARDS THE FIRST THERMOPLASTICS WITH HIGH LIGNIN CONTENTS

Although traditional views of native lignin configuration leave no room for specific primary structures (*i.e.* sequences of inter-unit linkages) along the macromolecular chains, the non-covalent interactions between the components in (highly degraded) kraft lignin derivatives are quite selective in their effects[19]. The intermolecular associative processes that engender the formation of kraft lignin complexes in solution are clearly nonrandom[20]. This suggests that specific intermolecular interactions are likely to have a direct influence upon the mechanical behaviour of materials that are fundamentally based on kraft lignins[21] (and indeed on other lignin derivatives also).

2.1 Physicochemical Properties of Kraft Lignins

In aqueous alkaline solutions, the upper limit to the molecular weight distributions of softwood kraft lignin preparations is decidedly below 1.0×10^5, as far as the vast majority of the individual molecular species are concerned[22]. Under these conditions, incubation at high kraft lignin concentrations (~200 g L^{-1}) engenders a spontaneous increase in the proportion of higher molecular weight species through a reversible time-dependent process; upon dilution, the associated complexes that have been formed undergo dissociation[23] (albeit quite slowly).

Most, but not all, of the individual molecular components in softwood kraft lignins participate in associated complex formation; those that do not have been covalently modified to such an extent that vital structural information has been compromised. The highest attainable degrees of association in aqueous alkaline solutions are limited by the negative charge densities on the polyphenolic kraft lignin components, but these restrictions are no longer operative around neutral pH (in the presence of a cosolvent like dioxane to maintain sufficient solubility). Under such circumstances all of the kraft lignin components that can contribute to associated complex formation take part in assembling these large non-covalently bonded species. Those components that are unable to participate (for reasons of effective stoichiometry or structure) can be readily separated from the associated complexes chromatographically[20].

The associative processes are not governed by hydrogen-bonding: after acetylation (Ac_2O/pyridine) and methylation (CH_2N_2), multimodal distributions of supramacromolecular kraft lignin complexes extending to very high molecular weights (around 1.0×10^8) are observed in polar solvents such as DMF[21,24] (Figure 4). Transmission electron micrographs reveal that the largest entities appear to have dimensions of about 0.25 μm across[21]. Moreover, for each sample the relative proportions of the species above 2×10^6 in apparent molecular weight are systematically related to the degree of association exhibited in aqueous alkaline solution[21,24] (Figure 4).

This, together with the fact that these huge supramacromolecular kraft lignin complexes are remarkably few in number, indicates that the associative processes occur in a specific manner. Intermolecular registration in the resulting domains is therefore likely to be quite well defined and hence the complexes would be expected to play an important part in determining the mechanical behaviour of materials that embody these enormous entities as active constituents[21].

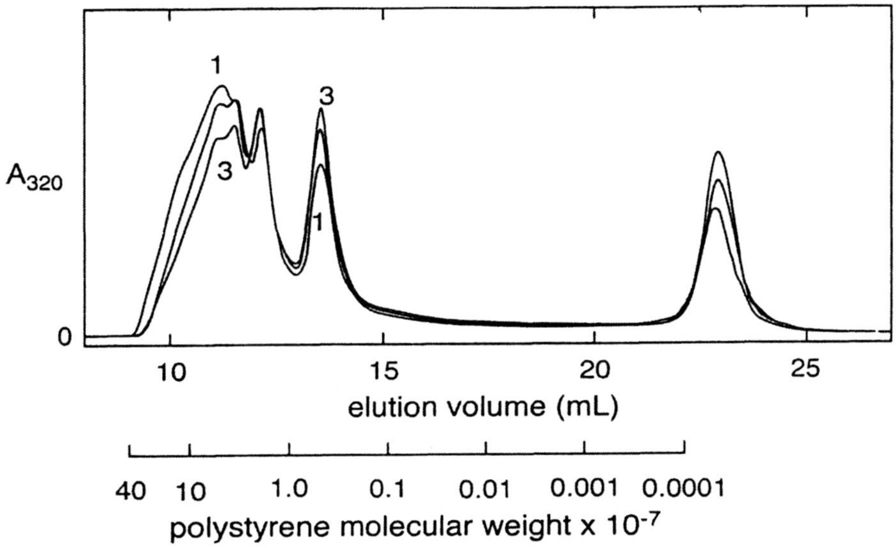

Figure 4. Apparent molecular weight distributions in DMF of kraft lignin samples acetylated and methylated following fractionation through Sephadex LH20 in aqueous 35% dioxane after association for (1) 6740 h, (2) 3910 h and (3) 1630 h at 195 g L^{-1} aqueous 1.0 M ionic strength 0.40 M NaOH. Profiles from 10^7 Å pore-size poly(styrene-divinylbenzene) column monitored at 320 nm[24].

2.2 Advent of 85% Kraft Lignin Based Thermoplastics

Inevitably the creation of the first series of thermoplastics with very high kraft lignin contents was the product of far-ranging exploratory studies. There was no precedent for these materials; they were arguably the first thermoplastics ever made that were genuinely lignin-based. They were formulated by blending 85% underivatized kraft lignin with poly(vinyl acetate) in the presence of (1.6%) diethyleneglycol dibenzoate and (0.8%) indene[1]. The degree of association of the kraft lignin preparations had been modulated by prior incubation in aqueous alkaline solutions; the effect upon the tensile behaviour of the resulting thermoplastics was considerable[1].

The material based upon the most associated kraft lignin preparation was quite rigid, but the one containing the most dissociated kraft lignin evaded fracture until its elongation had exceeded 65% (Figure 5). Since the two thermoplastics differed *only* in the inherent degree of association between the *same* set of individual kraft lignin components, the dramatic contrast between them clearly demonstrated, for the first time, what was necessary for plastic deformation to occur in true lignin-based materials. Evidently the macromolecular chains can be drafted past one another in response to mechanical stress providing they are free of the well-defined domains that maintain associated complex integrity (and in the process restrict molecular motion).

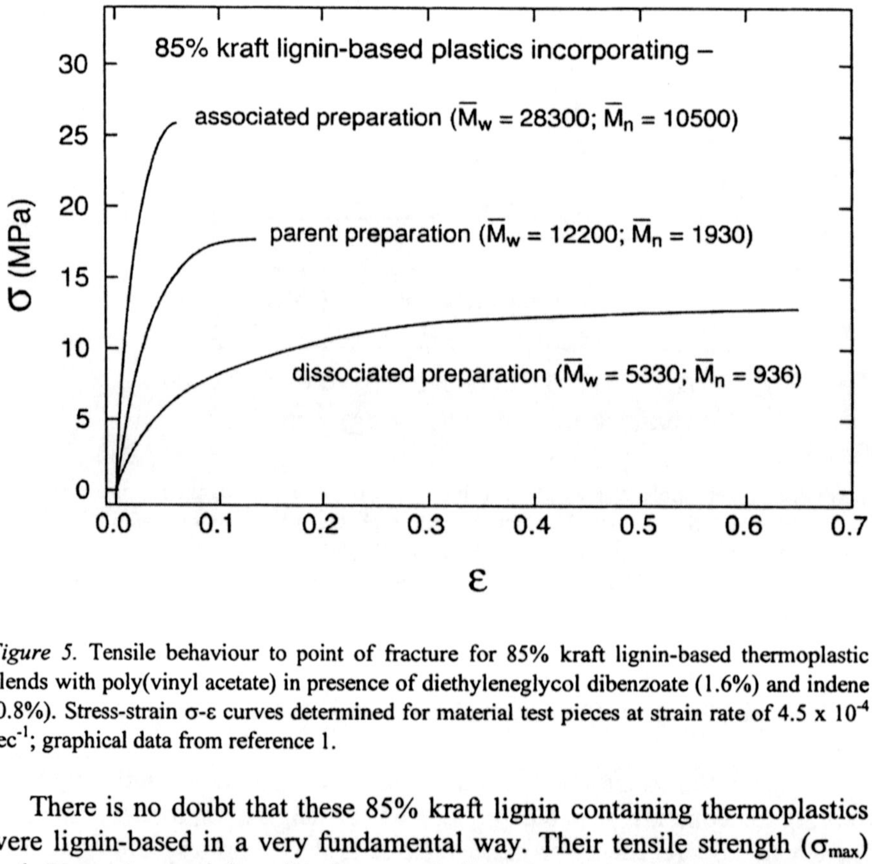

Figure 5. Tensile behaviour to point of fracture for 85% kraft lignin-based thermoplastic blends with poly(vinyl acetate) in presence of diethyleneglycol dibenzoate (1.6%) and indene (0.8%). Stress-strain σ-ε curves determined for material test pieces at strain rate of 4.5×10^{-4} sec^{-1}; graphical data from reference 1.

There is no doubt that these 85% kraft lignin containing thermoplastics were lignin-based in a very fundamental way. Their tensile strength (σ_{max}) and Young's modulus (E) increased linearly with the weight-average molecular weight ($\overline{M}_w$) of the kraft lignin preparation that had been incorporated into the blend (Figure 6); the two parameters attained values of 25 MPa and 1.5 GPa, respectively, in the material made from the most associated preparation. The T_gs of these homogeneous thermoplastics were close to room temperature, and their melt-flow indices were sufficient for extrusion-moulding purposes[1].

Be that as it may, some important insight at the molecular level into the constitution of the 85% kraft lignin-based thermoplastics was furnished by a small modification to the formulation that had been developed for these blends. As already mentioned (see section 2.1), kraft lignin preparations are generally composed of individual molecular components and associated macromolecular complexes (the latter having been assembled from the former species). The majority of the individual kraft lignin components can participate in the formation of associated complexes but some, on account of a stoichiometric constraint or structural modification, are unable to do so[20,21]. After these non-associating components had been removed through a chromatographic fractionation step, the resulting 85% kraft lignin-based blends (embodying exactly the same formulation as before) produced polymeric materials that were too weak for their mechanical properties to be measured. In the blends involving the unfractionated kraft lignin, associated complexes and individual components were both dispersed within the poly(vinyl acetate), but in those based on the fractionated kraft lignin, only the associated complexes were present. Hence it was the individual kraft lignin components that were responsible for promoting compatibility between the associated kraft lignin complexes and poly(vinyl acetate).

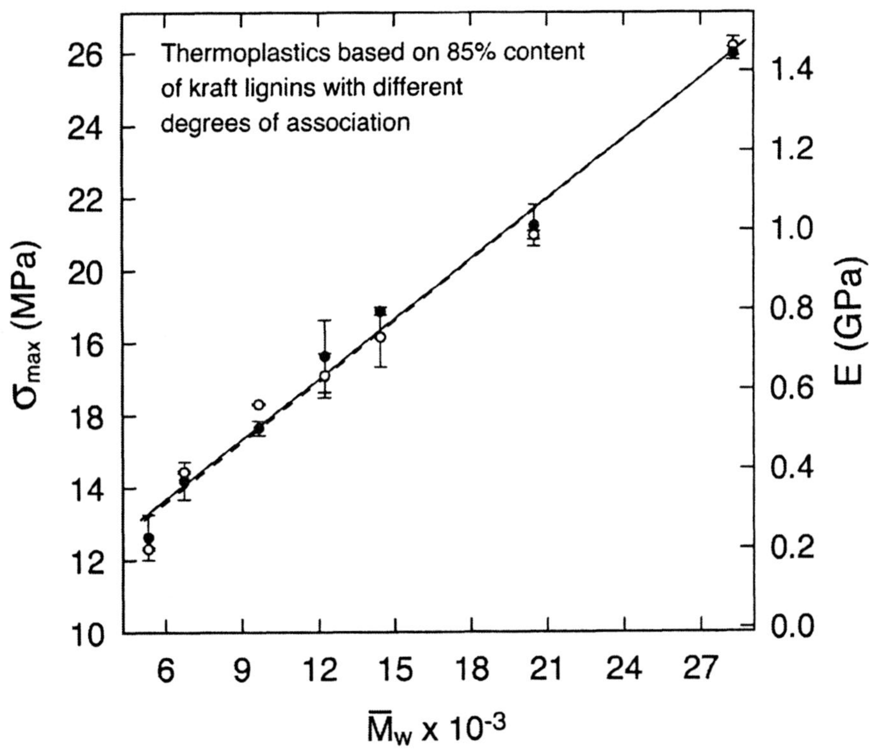

Figure 6. Variation of (●) tensile strength (σ_{max}) and (O) Young's modulus (E) with weight-average molecular weight ($\overline{M}_w$) of kraft lignins incorporated at 85% (w/w) levels in thermoplastic blends with same composition (data from reference 1).

3. ALKYLATED KRAFT LIGNIN-BASED THERMOPLASTICS

The advent of 85% kraft lignin-based thermoplastics was an important development since it had been assumed until then that polymeric materials with such high lignin contents would be extremely weak — if, indeed, they were capable of retaining any measurable cohesiveness at all. However, the kraft lignin components in these new thermoplastics were still partially soluble in aqueous alkaline solution, and thus alternative formulations were sought to obviate dissolution in aqueous solutions of any kind.

3.1 Alkylated 100% Kraft Lignin-Based Thermoplastics

Soon polymeric materials made solely of ethylated methylated kraft lignin were successfully fabricated for the first time[25]. A softwood parent preparation (identical to what had already been used for producing 85% kraft lignin-based thermoplastics—section 2.2) and a related higher molecular weight fraction (retained by a 10,000 nominal molecular weight cutoff membrane during ultrafiltration in aqueous 0.10 M NaOH) were chosen as starting materials for these ground-breaking studies. Both were ethylated ($(EtO)_2SO_2$ in aqueous 60% dioxane at pH 11–12) and then methylated (CH_2N_2 in $CHCl_3$), the resulting derivatives being respectively solvent cast from DMSO into the forms of test pieces.

The tensile parameters (Young's modulus, tensile strength and elongation to break) for the polymeric material made from the ethylated methylated higher molecular weight fraction were between 1.2- and 1.5-fold greater than those for the material based on the corresponding derivative of the parent kraft lignin. This is perhaps not surprising since there was a substantially larger proportion of lower molecular weight components in the latter than in the former alkylated kraft lignin preparation. The primary significance of these new materials lay in the fact that they were the first ever made exclusively from any simple by-product lignin derivative. Furthermore they were capable of exhibiting liquid flow behaviour at elevated temperatures: powdered samples would spontaneously coalesce into globules on heating under the right conditions.

3.2 Alkylated Kraft Lignin–Aliphatic Polyester Blends

With an elongation to break of about 2%, the ethylated methylated 100% kraft lignin-based materials are quite brittle, and so adequate plasticization is essential if alkylated kraft lignins are going to be employed in producing components that are tough enough for practical use. It was actually first disclosed in 1999 that miscible aliphatic polyesters can plasticize alkylated 100% kraft lignin-based polymeric materials quite well[26]. The effect is illustrated in Figure 7 for the higher molecular weight kraft lignin fraction after it had been methylated successively with $(MeO)_2SO_2$ and CH_2N_2 (*cf.* section 3.1). Before blending with poly(butylene adipate)[26], this methylated higher molecular weight kraft lignin-based polymeric material was quite brittle and tended to undergo fracture prematurely during tensile tests.

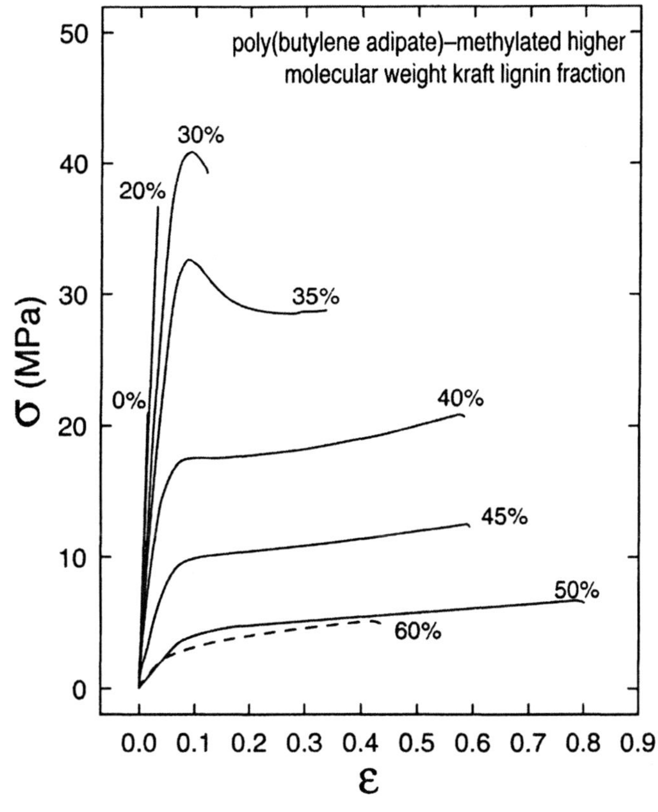

Figure 7. Progressive plasticization of methylated higher molecular weight kraft lignin-based polymeric material by poly(butylene adipate). Stress-strain σ-ε curves determined for blends at strain rate of 9.3×10^{-5} sec^{-1}.

Indeed its tensile strength in blends containing 20% poly(butylene adipate) increased 2-fold, but there was no sign of plastic deformation in this methylated 80% kraft lignin-based material. The first indication of stress yielding actually appeared (between 39 and 41 MPa) in blends with 30% poly(butylene adipate), while blends containing 35% levels of the aliphatic polyester exhibited some strain softening before the onset of substantial plastic deformation. However blend compositions involving 40–50%

poly(butylene adipate) displayed detectable strain hardening after the onset of plastic deformation (Figure 7).

The polymeric material made entirely of the methylated parent kraft lignin preparation was so brittle that its tensile behaviour could not be documented. Yet the threshold in poly(butylene adipate) content for blends showing the first signs of plastic deformation seemed to be somewhat smaller than for those based on the methylated higher molecular weight kraft lignin fraction. After plasticization had been achieved, the elongation to break tended to be larger for blends (with a particular composition) of the methylated parent kraft lignin than those of the higher molecular weight fraction, while the ultimate stresses were appreciably smaller. The broad similarity in tensile behaviour observed for the blends based on the two rather different lignin samples arose from the fact that in each the predominant species at the molecular level were the customarily huge supramacromolecular associated kraft lignin complexes (section 2.1).

It has been found that a range of aliphatic polyesters possessing methylene/carboxylate ester group ratios of 2.0–4.0 form homogeneous blends with alkylated kraft lignins; in doing so they all act as effective plasticizers for these simple kraft lignin derivatives. The strength of intermolecular attraction between the aliphatic polyester and alkylated kraft lignin components (relative to the polyester–polyester and lignin–lignin interactions) plays a central role in determining plasticizer efficacy. This was reflected in the way in which the glass transition temperature (T_g) of the blend varied with its composition. The gradual blending of more alkylated kraft lignin with the polyester brought about an increase in the T_g of the resulting material, as exemplified in Figure 8 by the effect of the methylated higher molecular weight kraft lignin fraction upon poly(trimethylene adipate) and poly(trimethylene glutarate), respectively[27].

It was difficult to measure the T_gs of blends with alkylated kraft lignin contents above 70% (w/w) reliably. The T_gs of the unplasticized higher molecular weight alkylated kraft lignin fraction and the corresponding parent preparation could not be observed at all because the majority of the individual components were incorporated within supramacromolecular associated complexes where molecular motion is restricted. The value of 160 °C employed in Figure 8 was estimated by extrapolation from the T_g values successfully determined for a series of lower molecular weight oligodisperse alkylated kraft lignin fractions.

Some 10 years ago, Lu and Weiss derived a general expression with no adjustable parameters for the T_gs of homogeneous binary polymer blends in terms of the T_gs of the constituent polymers[28]. Any attempt to fit the Lu-Weiss expression to the data in Figure 8 was beyond the scope of the present work, but under certain conditions reasonable approximations may be furnished by the Gordon-Taylor equation[29] (1) or Kwei's equation[30] (2), respectively:

$$T_g = \frac{w_1 T_{g1} + k w_2 T_{g2}}{w_1 + k w_2} \quad (1)$$

$$T_g = \frac{w_1 T_{g1} + k w_2 T_{g2}}{w_1 + k w_2} + q w_1 w_2 \quad (2)$$

Here w_i is the weight fraction and T_{gi} the glass transition temperature of polymer$_i$.

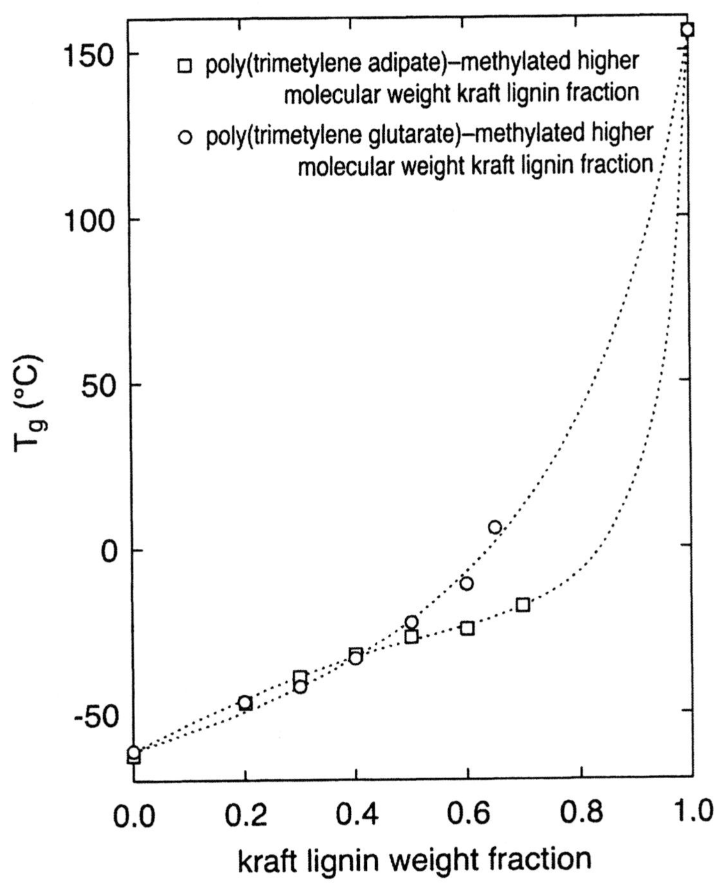

Figure 8. Dependence of T_g on composition of blends involving the methylated higher molecular weight kraft lignin fraction and either (□) poly(trimethylene adipate) or (O) poly(trimethylene glutarate) (Data from reference 27).

Some auspicious attempts have been made to correlate *k* in the Gordon-Taylor equation[31] and *q* in Kwei's equation[32] empirically with the strengths of the intermolecular interactions in binary polymer blends. When these equations were curve-fitted to the T_g data for the various miscible blends of aliphatic polyesters with ethylated and/or methylated kraft lignin preparations, the largest respective values of *k* and *q* were obtained when the methylene/carboxylate ester group ratio for the former fell in the 2.5–3.0 interval[33]. Thus the intermolecular interactions between the aliphatic polyester and alkylated kraft lignin components were the strongest near the middle of the miscibility range.

Under these circumstances there was a greater tendency for the supramacromolecular kraft lignin complexes to undergo dissociation and thereby release larger numbers of individual alkylated kraft lignin components. As a result, the T_gs of the blends tended to increase more rapidly with alkylated kraft lignin content, as exemplified in Figure 8 for the materials incorporating poly(trimethylene glutarate). Consequently somewhat more polyester was needed to plasticize the alkylated kraft lignin, an effect that preferentially involves the molecules in the peripheral regions, rather than the interiors, of the supramacromolecular complexes. The greater plasticizer requirement arose from the fact that a larger proportion was consumed counterproductively in dismantling the associated kraft lignin complexes when the intermolecular interactions between the polyester and lignin components were the strongest.

Support for the foregoing analysis has been forthcoming from X-ray powder diffraction studies of the alkylated kraft lignin–aliphatic polyester blends[33]. Thus the characteristic reflections from poly(1,4-butylene adipate) (with a methylene/carboxylate ester group ratio of 4.0) were no longer visible in blends containing ~45% (w/w) of the methylated higher molecular weight kraft lignin fraction. On the other hand, as far as poly(trimethylene succinate) was concerned (for which the methylene/carboxylate ester group ratio is 2.5), the distinctive X-ray powder diffraction pattern had already disappeared in blends involving 30% (w/w) of the same methylated higher molecular weight kraft lignin fraction. Such comparisons are certainly consistent with the trends encountered in the T_g–composition data for these alkylated kraft lignin-based thermoplastics.

4. CONCLUSIONS

The blends formulated from simple alkylated kraft lignin derivatives and aliphatic polyesters seem to have opened the door to a range of potentially versatile thermoplastics with high lignin contents. Blend miscibility is preserved when the methylene/carboxylate ester group ratio for the polyester falls between 2.0 and 4.0. The predominant alkylated kraft lignin species in the blends are huge supramacromolecular associated complexes. As an efficacious plasticizer, the aliphatic polyester should interact productively with the individual kraft lignin components in the peripheral domains of these complexes without disturbing their inner structures appreciably.

ACKNOWLEDGEMENTS

The work from our laboratory was supported by the United States Department of Agriculture (Grant 98-35103-6730), the United States Environmental Protection Agency through the National Center for Clean Industrial and Treatment Technologies (although it does not necessarily reflect the views of the Agency or Center, so no official endorsement should be inferred), the Vincent Johnson Lignin Research Fund, and the Minnesota Agricultural Experiment Station (through Project No. 43-68, maintained by Hatch Funds).

REFERENCES

1. Li, Y., Mlynár, J., and Sarkanen, S., 1997, The first 85% kraft-lignin-based thermoplastics. *J. Polym. Sci., Part B: Polym. Phys.* 35: 1899-1910.
2. Sarkanen, S., and Li, Y., January 9, 2001, Compositions Based on Lignin Derivatives. *US Pat.* 6172204.
3. Lewis, N. G., Davin, L. B., and Sarkanen, S., 1999, The Nature and Function of Lignins. In *Comprehensive Natural Products Chemistry* (D. H. R. Barton, K. Nakanishi and O. Meth-Cohn, eds.), Elsevier Science, Oxford, Vol. 3 (B. M. Pinto, Vol. ed.), pp. 617-745.
4. Sakakibara, A., 1980, A structural model of softwood lignin. *Wood Sci. Technol.* **14**: 89-100.
5. Brunow, G., Kilpeläinen, I., Sipilä, J., Syrjänen, K., Karhunen, P., Setälä, H., and Rummakko, P., 1998, Oxidative Coupling of Phenols and the Biosynthesis of Lignin. In *Lignin and Lignan Biosynthesis* (N. G. Lewis and S. Sarkanen, eds.), ACS Symposium Series No. 697, American Chemical Society, Washington, D.C., pp. 131-147.
6. Argyropoulos, D. S., Jurasek, L., Kristofová, L., Xia, Z., Sun, Y., and Palus, S., 2002, Abundance and reactivity of dibenzodioxocins in softwood lignin. *J. Agric. Food Chem.* **50**: 658-666.
7. Gierer, J., 1980, Chemical aspects of kraft pulping. *Wood Sci. Technol.* **14**: 241-266.
8. Kirkman, A. G., Gratzl, J. S., and Edwards, L. L., 1986, Kraft lignin recovery by ultrafiltration—economic feasibility and impact on the kraft recovery system. *Tappi J.* **69** (5): 110-114.
9. Hergert, H. L., 1998, Developments in Organosolv Pulping—An Overview. In *Environmentally Friendly Technologies for the Pulp and Paper Industry* (R. A. Young and M. Akhtar, eds.), Wiley, New York, pp. 5-67.

10. Vinzant, T. B., Ehrman, C. I., Adney, W. S., Thomas, S. R., and Himmel, M. E., 1997, Simultaneous saccharification and fermentation of pretreated hardwoods—effect of lignin content. *Appl. Biochem. Biotechnol.* **62**: 99-104.
11. Dorsch, R., and Nghiem, N., 1998, Specialty chemicals with emphasis on environmentally benign products and processes. *Appl. Biochem. Biotechnol.* **70-72**: 843-844.
12. Holzman, D., 1999, Monsanto moves into the contract production arena. *Gen. Eng. News* **19** (4): 1, 8, 35.
13. Anterola, A. M., and Lewis, N. G., 2002, Trends in lignin modification—a comprehensive analysis of the effects of genetic manipulations/mutations on lignification and vascular integrity. *Phytochemistry* **61**: 221-294.
14. *Lignin—Properties and Materials* (W. G. Glasser and S. Sarkanen, eds.), 1989, ACS Symposium Series No. 397, American Chemical Society, Washington, D.C.
15. Meister, J. J., 1986, Review of the Synthesis, Characterization, and Testing of Graft Copolymers of Lignin. In *Renewable-Resource Materials—New Polymer Sources* (C. E. Carraher, Jr. and L. H. Sperling, eds.), Plenum Publishing Corp., New York, pp. 305-322.
16. Lindberg, J. J., Kuusela, T. A., and Levon, K., 1989, Specialty Polymers from Lignin. In *Lignin—Properties and Materials* (W. G. Glasser and S. Sarkanen, eds.), ACS Symposium Series No. 397, American Chemical Society, Washington, D.C., pp. 190-204.
17. Yoshida, H., Mörck, R., Kringstad, K. P., and Hatakeyama, H., 1990, Kraft lignin in polyurethanes; II: Effects of the molecular weight of kraft lignin on the properties of polyurethanes from a kraft lignin-polyether triol-polymeric MDI system. *J. Appl. Polym. Sci.* **40**: 1819-1832.
18. Kelley, S. S., Glasser, W. G., and Ward, T. C., 1989, Effect of Soft-Segment Content on the Properties of Lignin-Based Polyurethanes. In *Lignin—Properties and Materials* (W. G. Glasser and S. Sarkanen, eds.), ACS Symposium Series No. 397, American Chemical Society, Washington, D.C., pp. 402-413.
19. Sarkanen, S., Teller, D. C., Stevens, C. R., and McCarthy, J. L., 1984, Associative interactions between kraft lignin components. *Macromolecules* **17**: 2588-2597.
20. Dutta, S., Garver, T. M., Jr., and Sarkanen, S., 1989, Modes of Association between Kraft Lignin Components. In *Lignin—Properties and Materials* (W. G. Glasser and S. Sarkanen, eds.), ACS Symposium Series No. 397, American Chemical Society, Washington, D.C., pp. 155-176.
21. Dutta, S., and Sarkanen, S., 1990, A New Emphasis in Strategies for Developing Lignin-Based Plastics. In *Materials Interactions Relevant to the Pulp, Paper, and Wood Industries* (D. F. Caulfield, J. D. Passaretti and S. F. Sobczynski, eds.), MRS Symposium Proceedings Vol. 197, Materials Research Society, Pittsburgh, pp. 31-39.
22. Mlynár, J., and Sarkanen, S., 1996, Renaissance in Ultracentrifugal Sedimentation Equilibrium Calibrations of Size Exclusion Chromatographic Elution Profiles. In *Strategies in Size Exclusion Chromatography* (M. Potschka and P. L. Dubin, eds.), ACS Symposium Series No. 635, American Chemical Society, Washington, D.C., pp. 379-400.
23. Garver, T. M., Jr., Iwen, M. L., and Sarkanen, S., 1989, The Kinetics of Macromolecular Kraft Lignin Complex Dissociation. In *Fifth International Symposium on Wood and Pulping Chemistry*, TAPPI Proceedings, TAPPI, Atlanta, Vol. I, pp. 113-119.
24. Li, Y., 1995, Noncovalent Interactions between Kraft Lignin Components. *MS Thesis*, University of Minnesota.
25. Li, Y., and Sarkanen, S., 2000, Thermoplastics with Very High Lignin Contents. In *Lignin—Historical, Biological, and Materials Perspectives* (W. G. Glasser, R. A. Northey and T. P. Schultz, eds.), ACS Symposium Series No. 742, American Chemical Society, Washington, D.C., pp. 351-366.

26. Sarkanen, S., and Li, Y., 1999, Plasticizers that Transform Alkylated Kraft Lignins into Thermoplastics. In *Biomass—A Growth Opportunity in Green Energy and Value-Added Products* (R. P. Overend and E. Chornet, eds.), Proceedings of the 4th Biomass Conference of the Americas, Elsevier Science Ltd., Oxford, Vol. 1, pp. 533-539.
27. Li, Y., and Sarkanen, S., 2001, Kraft lignin-based thermoplastic polymer blends. In *11th International Symposium on Wood and Pulping Chemistry*, ATIP, Paris, Vol. I, pp. 75-78.
28. Lu, X., and Weiss, R. A., 1992, Relationship between the glass transition temperature and the interaction parameter of miscible binary polymer blends. *Macromolecules* **25**: 3242-3246.
29. Gordon, M., and Taylor, J. S., 1952, Ideal copolymers and the second-order transitions of synthetic rubbers. I: Non-crystalline copolymers. *J. Appl. Chem.* **2**: 493-500.
30. Kwei, T. K., 1984, The effect of hydrogen-bonding on the glass transition temperatures of polymer mixtures. *J. Polym. Sci., Polym. Lett.* **22**: 307-313.
31. Prud'homme, R. E., 1982, Miscibility phenomena in polyester/chlorinated polymer blends. *Polym. Eng. Sci.* **22**: 90-95.
32. Lin, A. A., Kwei, T. K., and Reiser, A., 1989, On the physical meaning of the Kwei equation for the glass transition temperature of polymer blends. *Macromolecules* **22**: 4112-4119.
33. Li, Y., and Sarkanen, S., 2002, Alkylated kraft lignin based thermoplastic blends with aliphatic polyesters. *Macromolecules* **35**: 9707-9715.

Biodegradable Hybrid Polymeric Materials Based on Lignin and Synthetic Polymers

ANDREA CORTI, FEDELE CRISTIANO, ROBERTO SOLARO, and EMO CHIELLINI*
Department of Chemistry and Industrial Chemistry, University of Pisa, via Risorgimento 35, 56126 Pisa, Italy

1. INTRODUCTION

Environmentally compatible polymeric materials based on the combination of natural components and synthetic polymers are attracting increasing attention. Several contributions have been reported on the preparation of blends and composites based on biodegradable synthetic polymers such as poly(vinyl alcohol) (PVA) and poly(ε-caprolactone) (PCL) with starch[1-3], cellulose[4,5], and protein materials[6,7]. Starch gives substantially immiscible blends with PVA[8,9], nevertheless a significant improvement in the mechanical properties of the natural polysaccharide has been recorded, thus indicating at least a partial compatibility between starch and the synthetic polymer[10,11].

Lignin, which constitutes the major source of natural aromatic carbon, despite its structural and compositional complexity, represents an attractive material by considering its role in the biogeochemical cycles of the organic matter, as well as its character of cheap by-product of paper, pulps, and tannery industries. Lignin is subjected to relatively slow degradation processes mediated by actinomycetes and fungi leading to the production of proteic derivatives and humic substances in natural soils[12], with benign effects on their structure and bio-fertility[13].

Substantially immiscible blends of lignin with PCL, poly(vinyl chloride), and PVA[14-16] have been produced and the presence of lignin induced only negligible effects on the thermal properties of the synthetic components. Nevertheless, the mechanical properties of poly(vinyl acetate) based thermoplastics containing a very large amount (85%) of kraft lignin can be effectively modulated by varying the degree of association of the kraft lignin used in the preparations[17].

Lignin modification by alkylation reactions was also utilized to enhance the compatibility with synthetic polymers[18] as well as to produce thermoplastics based only on alkylated kraft lignin[19]. Moreover, the chemical structure and functional groups of lignin represent an attractive tool for the synthesis of lignin-based copolymers. Indeed, graft polymerization of vinyl monomer, such as acrylamide, and styrene[20], onto lignin by free radical reactions has been repeatedly reported. Redox initiators based either on 2-hydroperoxy-1,4-dioxycyclohexane and Ce(IV) salts, or on hydrogen peroxide and $CaCl_2$ mediate this reaction[21].

Recently, the chemo-enzymatic grafting of acrylamide onto lignin by using *laccase*, which is a lignin-degrading enzyme of fungal origin has been also investigated for the production of bio-based polymeric materials[22].

In the present contribution, we report on the results of an investigation aimed at the formulation of biodegradable hybrid matrices based on synthetic polymers and lignin. Poly(vinyl acetate) (PVAc), poly(vinyl alcohol), and poly(ε–caprolactone) were chosen as synthetic matrices for the production of biodegradable hybrid formulations to be used in agricultural and packaging applications. Organosolv lignin (OL), a lignin derivative resulting from more eco-compatible pulping processes was utilized as natural component. In fact, wood extraction with a mixture of organic solvent (usually ethanol or acetone) and water at relatively high temperatures (135-190 °C)[23] gives practically untarnished lignin macromolecules, thus allowing for better utilization in the production of lignin-based thermoplastics.

Hybrid materials containing lignin covalently bound to the synthetic matrices were prepared by graft polymerization of selected monomers onto organosolv lignin. For comparison, hybrid materials were also prepared by solution casting of blends of PVAc, PVA, and PCL with 10-30% by weight of lignin. All blends and copolymers were characterized by NMR spectroscopy and then submitted to respirometric biodegradation tests performed by using mature compost as incubation medium.

2. MATERIALS AND METHODS

2.1 Synthesis of Lignin Copolymers

2.1.1 Grafting of Vinyl Acetate on Lignin

Polymerization experiments were carried out in round bottom flasks under nitrogen atmosphere, by adding organosolv lignin (OL) and $CaCl_2$ in that order to a known volume of DMSO. After the reagents dissolved completely, H_2O_2 was added and the reaction mixture was bubbled with nitrogen for 10 min. Freshly distilled vinyl acetate (Vac) was added drop-wise and nitrogen was bubbled again in the solution for 10 min. The reaction mixture was maintained under stirring at 30-60 °C for 48 hours, the resulting slurry was then poured in water and the coagulated polymer was recovered.

Experiments were performed by varying both redox system concentration and reaction temperature (Table 1). Homopolymerization of vinyl acetate (Run VA0) was also carried out under the same experimental conditions in the absence of lignin. Moreover, in order to get an insight on the free-radical grafting mechanism, the polymerisation of vinyl acetate was performed under the same conditions in the presence of 2-methoxy-4-methylphenol (MAF), a model compound of lignin functional groups.

Table 1. Graft polymerization of vinyl acetate (VAc) with organosolv lignin (OL) and 2-methoxy-4-methylphenol (MAF)

Run	VAc (g)	OL[a] (g)	H_2O_2 (ml)	$CaCl_2$ (g)	T (°C)	Yield (%)
LVA0	9.2	1.0	1.0	1.0	30	44
LVA1	9.2	1.0	0.1	0.1	30	26
LVA45	9.2	1.0	1.0	1.0	45	29
LVA60	9.2	1.0	1.0	1.0	60	27
MAFVA0	9.2	0.8[b]	1.0	1.0	30	29
MAFVA1	4.6	2.4[b]	1.0	1.0	30	20
VA0	9.2	-	1.0	1.0	30	0

[a] Content of organosolv lignin, if not otherwise stated. [b] MAF content.

2.1.2 Methanolysis of Lignin-grafted Vinyl Acetate Polymers

Hydrolysis reactions were carried out in methanol solution in the presence of NaOH (1 mmol/g polymer) at 40 °C. The reaction mixture was neutralized with acetic acid and the resulting polymeric material was purified by two precipitations in acetone. The amount of lignin in the hydrolyzed copolymers was evaluated from the UV absorbance at 280 nm, by using different basic solution of lignin for calibration.

2.1.3 Grafting of ε-Caprolactone on Lignin

PCL/lignin copolymer (Sample LCL0) was prepared by bulk polymerization of ε-caprolactone (CL) in the presence of the potassium salt of organosolv lignin at 160 °C for 12 hours, by using 9:1 CL/OL weight ratio. The OL potassium salt was obtained by dissolving OL in 1M KOH solution followed by freeze-drying. The reaction product was dissolved in THF and coagulated in *n*-hexane to give 97% polymer yield. Anionic polymerisation of ε-caprolactone in the presence of 2-methoxy-4-methylphenol (MAF) and KOH, was carried out under the same conditions, to give 83% polymer yield. After purification, all reaction products were characterized by spectroscopic analysis (FT-IR, ^{1}H-NMR) as well as by Size Exclusion Chromatography (SEC).

2.2 Preparation of Solution Cast Lignin/Synthetic Polymer Blends

Poly(vinyl acetate)/lignin and poly(ε-caprolactone)/lignin blends were prepared by casting chloroform solutions of PVAc (Mw 140 kDa) and PCL (Mw 80 kDa) mixtures containing 9-30% by weight of lignin (Table 2).

Poly(vinyl alcohol)/lignin (PVAL) blends were prepared by casting aqueous solution of 88% hydrolyzed PVA (PVA88) and basic OL, in order to attain a lignin content in the resulting blends ranging between 8 and 30% by weight (Table 2).

Table 2. Composition of cast films of lignin/synthetic polymer blends

Sample	PVA (%)	PVAc (%)	PCL (%)	OL (%)
PVAs	100	-	-	-
PVAL10s	90	-	-	10
PVAL20s	79	-	-	21
PVAL30s	71	-	-	29
PVAcs	-	100	-	-
PVAcL10s	-	91	-	9
PVAcL20s	-	80	-	20
PVAcL30s	-	72	-	28
PCLs	-	-	100	-
PCLL10s	-	-	91	9
PCLL20s	-	-	78	22
PCLL30s	-	-	70	30

2.3 Respirometric Biodegradation Tests

The "Biometer flask" test procedure[24], partially modified[25] for testing polymeric materials with low-to-moderate propensity to biodegradation was used. Biodegradation tests were carried out in glass vessels containing a multilayer substrate in which a mixture of mature compost with perlite, a chemically inert heat expanded aluminium-silicate rock, was sandwiched between two perlite layers (Figure 1). Test polymer samples were placed in the middle layer and the vessels were kept in the dark at 55 °C. The CO_2 evolved from samples and blanks was trapped in beakers containing 40-50 ml of 0.05 N KOH solution placed inside the test vessels. Every few days, this solution was back titrated with 0.1 N HCl after addition of 4-5 ml of 0.25 N $BaCl_2$, and then replaced with fresh solution.

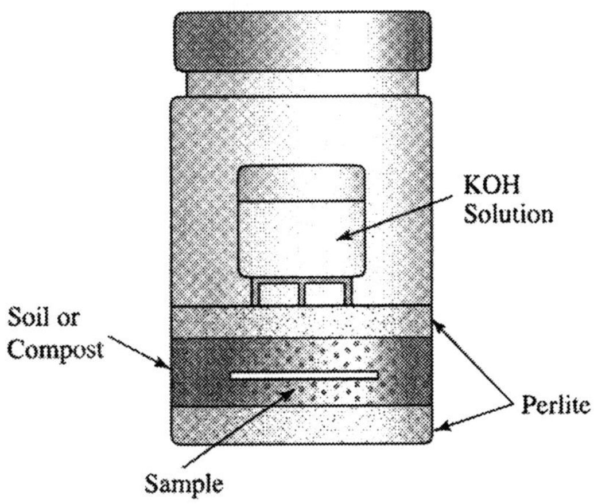

Figure 1. Culture set-up for biodegradation tests with soil or mature compost inocula.

3. RESULTS AND DISCUSSION

Free radical grafting of vinyl acetate to lignin was successfully accomplished by using the $H_2O_2/CaCl_2$ redox system to promote the formation of free radicals in the reaction mixture. ^{1}H-NMR analysis of the purified reaction products evidenced the presence of the typical signals of poly(vinyl acetate) as well as those of lignin (Figure 2). Interestingly, the polymerization of vinyl acetate carried out in the absence of lignin under the same conditions did not afford any polymeric product, clearly indicating the involvement of lignin in the initiation step of the reaction.

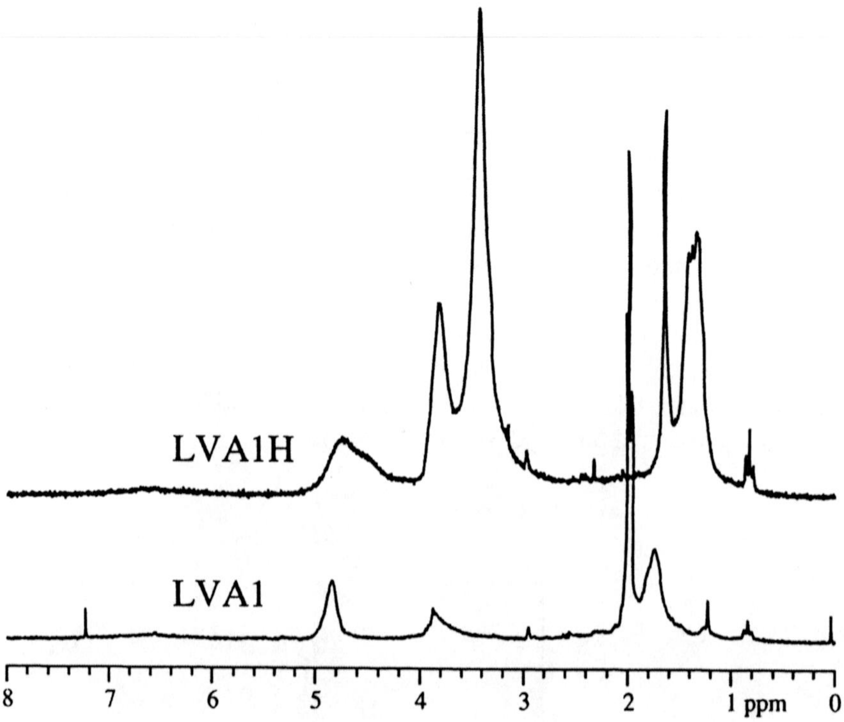

Figure 2. ^{1}H-NMR spectra of lignin-vinyl acetate (LVA1) and lignin-vinyl alcohol (LVA1H) graft copolymers.

This indication was supported by SEC analysis carried out by using both UV and RI detectors. In fact, polymeric fractions having molecular weight higher than lignin were clearly detected at 280 nm (Figure 3), whereas poly(vinyl acetate) does not show any significant absorption at this wavelength.

The collected data evidenced that the reaction temperature as well as the monomer-catalyst molar ratio affect both yield (Table 1) and molecular weight of the resulting polymeric materials (Table 3). No appreciable influence on the copolymer composition was detected.

Similar results were obtained in the polymerization of vinyl acetate in the presence of 2-methoxy-4-methylphenol (MAF), which was selected as low molecular weight model compound of lignin structural units. The obtained polymeric materials, which showed a sharp absorption in the UV region, were analysed by SEC (Figure 4) after purification by semi-preparative TLC. The average molecular weight was significantly affected by the vinyl acetate/MAF molar ratio utilized in the polymerisation reactions, the higher was the ratio and the higher was the molecular weight (Table 3), in agreement with the mechanism of free radical polymerisation. On the basis of these results, it seems reasonable that also the polymerization of vinyl acetate in the presence of lignin, by using the $CaCl_2/H_2O_2$ redox initiator, should match the mechanism previously proposed for other vinyl monomers[21].

Alkaline methanolysis of graft vinyl acetate-lignin copolymers (Samples LVA0, LVA1, and LVA60) led to the corresponding water-soluble vinyl alcohol copolymers, whose lignin content, as determined by the absorption at 280 nm, resulted to be substantially higher than in the parent VAc copolymer samples hydrolyzed samples, respectively, thus demonstrating however the formation of copolymer macromolecules.

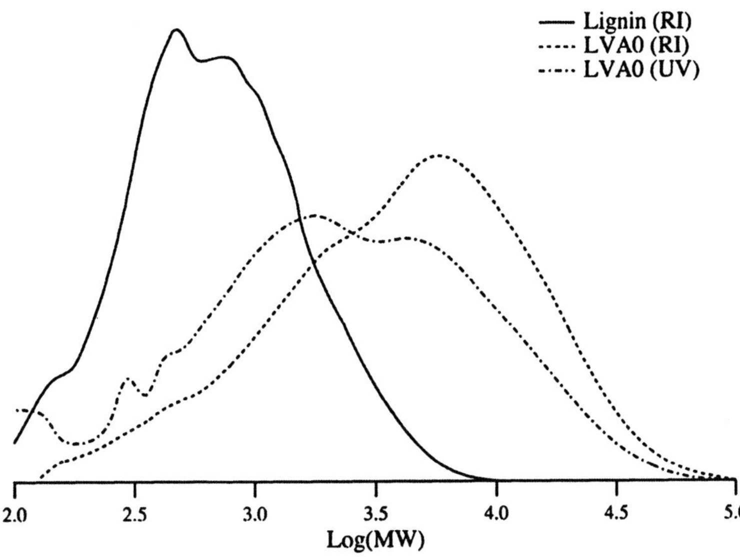

Figure 3. Molecular weight distributions of lignin and lignin-vinyl acetate graft copolymers.

Table 3. Molecular weight and molecular weight distribution of lignin copolymers

Run	Lignin [a] (% wt)		T (°C)	Mw (kDa)	Mw/Mn
	Feed	Polymer			
LVA0	11	11	30	7.4	3.8
LVA1	11	9	30	5.9	4.2
LVA45	11	10	45	9.4	6.6
LVA60	11	11	60	2.9	2.7
MAFVA0	9 [b]	0.1 [b]	30	10.1	1.2
MAFVA1	53 [b]	4 [b]	30	6.2	1.6
LCL0	11	6	160	4.5	2.6
MAFCL	5 [b]	1 [b]	160	6.3	2.0
OL	-	100	-	0.9	1.9

[a] Content of lignin, if not otherwise stated. [b] MAF content.

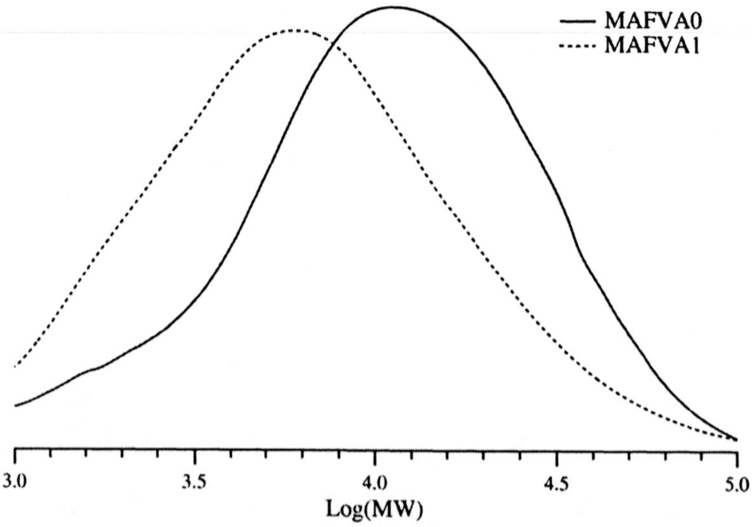

Figure 4. Molecular weight distributions of MAF-vinyl acetate graft copolymers.

Comparison of the thermal properties of PVAc/OL cast blends and OL-VAc graft copolymers evidenced a clear difference in the glass transition temperature (T_g) of the two types of hybrid materials. In fact, the T_g of the blends was only slightly larger than that of pure PVAc (Table 4). On the contrary, LVA1 copolymer displayed a marked drop of the T_g, very likely because of the short length of VAc chain branches grafted onto lignin and their ability to act as a sort of lignin plasticizer.

Table 4. DSC thermal characteristics of PVAc/OL blends and OL-VAc graft copolymers

Sample	Lignin content (% wt)	T_g (°C)	ΔC_p ($J\ g^{-1}K^{-1}$)
PVAc	-	42	0.52
LVA1	9	23	1.04
PVAcL10s	9	45	0.50
PVAcL20s	21	46	0.50
PVAcL30s	28	48	0.48
OL	100	91	0.65

Almost quantitative yields were obtained in the anionic bulk polymerization of ε-caprolactone carried out in the presence of either the potassium salt of lignin or MAF. The formation of polyester chains was clearly evidenced by ^{1}H-NMR spectroscopy, even though the absence of typical absorption signals of aromatic protons did not allow demonstrating clearly that the polyester chains were covalently bound to the aromatic substrates. Nevertheless, SEC analysis of the resulting polymeric materials showed an evident UV absorption at 280 nm, thus indicating that the polymerization of ε-caprolactone should have been initiated by the phenol groups of lignin (Figure 5).

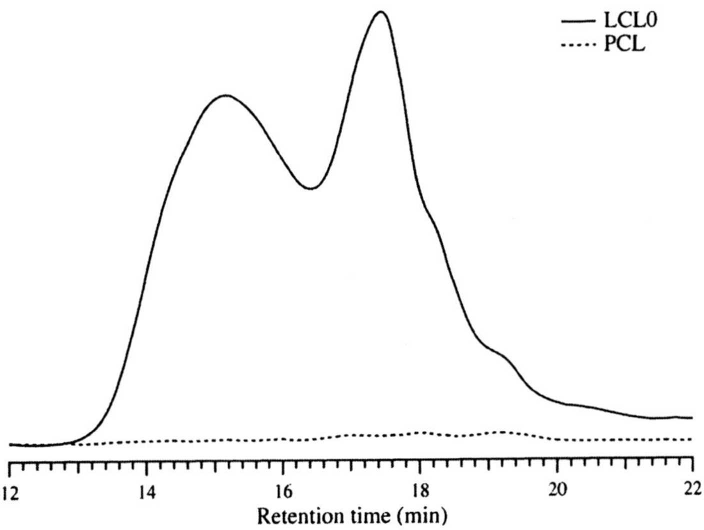

Figure 5. SEC chromatograms (UV detector) of lignin/ε-caprolactone graft copolymer and of poly(ε-caprolactone).

The potential biodegradation behaviour of lignin-based graft copolymers and blends was tested in respirometric experiments in the presence of mature compost as incubation medium.

PVAc containing hybrid materials experienced very limited biodegradation that did not exceed 20% after 110 days of incubation. However, a slight positive effect exploited by lignin was evidenced, particularly in the case of the graft copolymer (LVA0 run) (Figure 6).

A pronounced positive influence of lignin on the rate and extent of PVA biodegradation, whose recalcitrance to the biological attack in solid matrices such as soil and compost has been repeatedly reported[26-30], was detected in the case of the graft copolymer (LVAH), that approached 45% biodegradation, whereas in the relevant PVA/lignin blends the natural component did not affect significantly the biodegradation behaviour of the vinyl polymer (Figure 7).

On the other hand, lignin seems to have a negative effect on the biodegradation process of PCL based blends, while inducing only a delay in the case of the corresponding copolymer (LCL0) that however, approached the same extent (60%) of biodegradation of pure PCL only after longer incubation time (Figure 8).

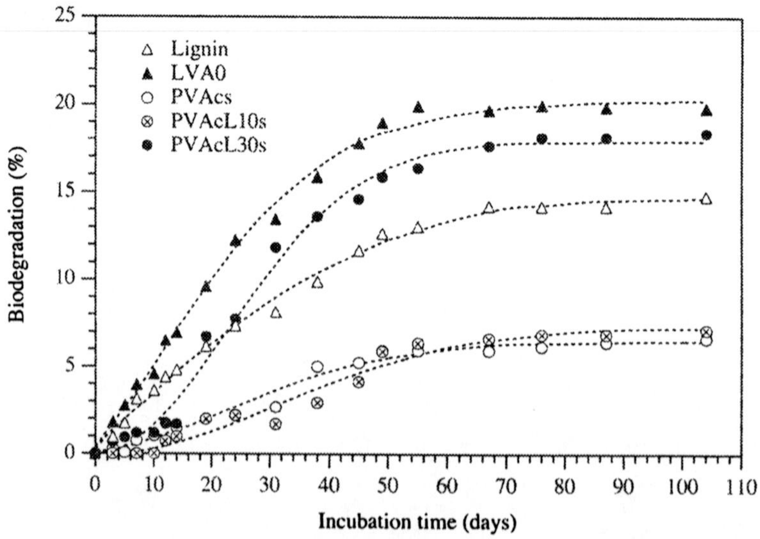

Figure 6. Biodegradation profiles of PVAc-lignin blends (PVAcL10s, PVAcL30s) and graft copolymer (LVA0) in mature compost respirometric tests.

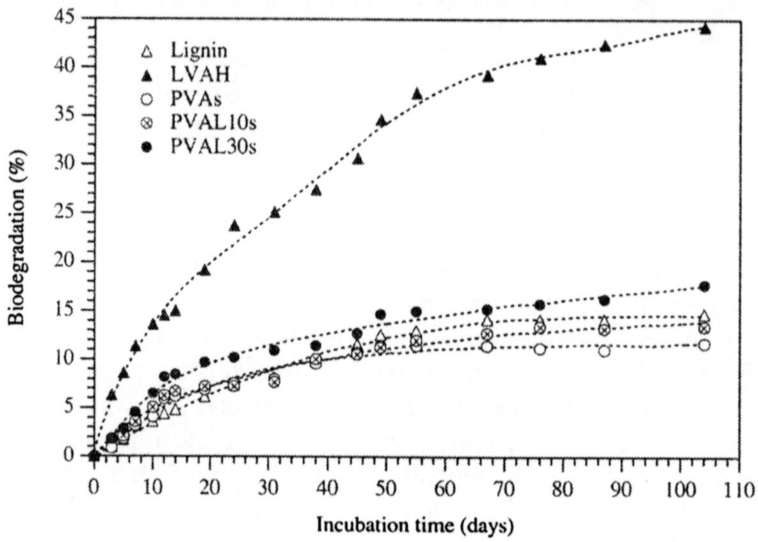

Figure 7. Biodegradation profiles of PVA-lignin blends (PVAL10s, PVAL30s) and graft copolymer (LVAH) in mature compost respirometric tests.

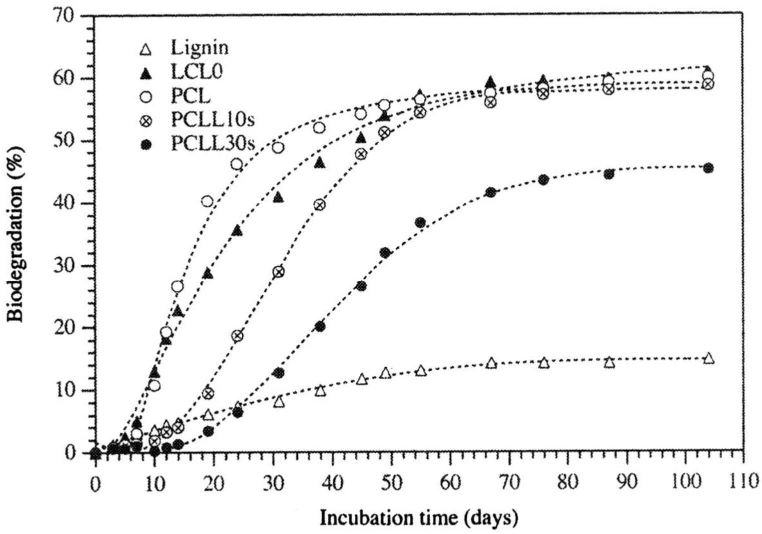

Figure 8. Biodegradation profiles of PCL-lignin blends (PCLL10s, PCLL30s) and graft copolymer (LCL0) in mature compost respirometric tests.

4. CONCLUSIONS

Hybrid polymeric materials containing PVAc, PVA, and PCL as synthetic component and lignin, a cheap by-product of paper, pulp, and tannery industry, as natural component were prepared. In particular, lignin-PVAc graft copolymers were prepared by free radical polymerization in the presence of $H_2O_2/CaCl_2$ redox initiator, whereas PCL was grafted onto lignin by anionic bulk polymerization. Lignin-PVA copolymers were obtained by methanolysis of the corresponding lignin-PVAc graft copolymers

In any case, the synthetic polymer chains are covalently bound to lignin, as indicated by both spectroscopic and SEC characterizations.

The biodegradation behaviour of the analyzed polymeric hybrid materials clearly indicates the positive effect of lignin when chemically bound to vinyl alcohol polymer chains, whereas in the case of the lignin-PCL copolymer only a delay in the biodegradation process was observed.

In spite of the preliminary character of the reported investigation, the results so far obtained evidence the possibility to synthesize in a relatively simple way, low cost, potentially biodegradable polymeric hybrid materials containing a significant amount of lignin, with obvious beneficial economic and environmental effects.

REFERENCES

1. Young, A. H., 1967, Polyvinyl alcohol plasticized amylose compositions. *US Pat.* 3,312,641.
2. Otey, F. H., Mark, A. M., Mehltretter, C. L., and Russell, C. R., 1974, Starch-based film for degradable agricultural mulch. *Ind. Eng. Chem., Prod. Res. Develop.* **13:** 90-92.
3. Averous, L., Moro, L., Dole, P., and Fringant, C., 2000, Properties of thermoplastic blends: starch-polycaprolactone. *Polymer* **41**: 4157-4167.
4. Nishio, Y., and Manley, R. StJ., 1988, Cellulose/poly(vinyl alcohol) blends prepared from solutions in N,N-dimethylacetamide-lithium chloride. *Macromolecules* **21:** 1270-1277.
5. Nishio Y., Haratani, T., and Takahashi, T., 1989, Cellulose/poly(vinyl alcohol) blends: an estimation of thermodynamic polymer-polymer interaction by melting point depression analysis. *Macromolecules* **22**: 2547-2549.
6. Chiellini, E., Cinelli, P., Fernandes, E.G., Kenawy, E-R., and Lazzeri, A., 2001, Gelatin based blends and composites. Morphological and thermal mechanical characterization. *Biomacromolecules* **2**: 806-811.
7. Chiellini, E., Cinelli, P., Corti, A., and Kenawy E-R., 2001, Composite films based on waste gelatin: thermal-mechanical properties and biodegradation testing. *Polym. Degr. Stab.* **73**: 549-555.
8. Mayer, J. M., and Kaplan, D. L., 1994, Biodegradable materials: balancing degradability and performance. *Trends Polym. Sci.* **2:** 227-235.
9. Okaya, T., Kohno, H., Terada, K., Sato, T., Maruyama, H., and Yamauchi, J., 1992, Specific interaction of starch and polyvinyl alcohols having long alkyl-groups. *J. Appl. Polym. Sci.***45**: 1127-1134.
10. Otey, F. H., and Mark, A. M., 1976, Degradable starch-based agricultural mulch film. *US Pat.* 3,949,145.
11. Shogren, R. L., Lawton, J. W., Tiefenbacher, K. F., and Chen, L., 1998, Starch-poly(vinyl alcohol) foamed articles prepared by a baking process. *J. Appl. Polym. Sci.* **68**: 2129-2140.
12. Stevenson, F. J., 1982, Humus chemistry. Wiley, New York.
13. Kononova, M. M., 1961, *Soil organic matter, its nature, its role in soil formation and in soil fertility*. Pergamon Press, Oxford.
14. Li, J. C., He, Y., and Inoue, Y., 2001, Study on thermal and mechanical properties of biodegradable blends of poly(epsilon-caprolactone) and lignin. *Polymer J.* **33**: 336-343.
15. Feldman, D., and Banu, D., 1997, Contribution to the study of rigid PVC polyblends with different lignins. *J. Appl. Polym. Sci.* **66**: 1731-1744.
16. Corradini, E., Pineda, E. A. G., and Hechenleitner, A. A. W., 1999, Lignin-poly(vinyl alcohol) blends studied by thermal analysis. *Polym. Degr. Stab.* **66**: 199-208.
17. Li, Y., Mlynár, J., and Sarkanen, S., 1997, The first 85% kraft-lignin based thermoplastics. *J. Polym. Sci., Part B: Polym. Phys.* **35**: 1899-1910.
18. Ciemnecki, S. L., and Glasser, W. G., 1989, Lignin properties and materials. *ACS Symp. Ser.* **397**: 452-459.
19. Li, Y., and Sarkanen, S., 2000, Thermoplastics with Very High Lignin Contents. *ACS Symp. Ser.* **742**, 351-366.
20. Meister, J. J., and Patil, D. R., 1985, Solvent effects and initiation mechanisms for graft polymerization of pine lignin. *Macromolecules* **18**: 1559-1564.
21. Meister, J. J., and Chen, M. J., 1991, Graft 1-phenylethylene copolymers of lignin. 1. Synthesis and proof of copolymerization. *Macromolecules* **24**: 6843-6848.
22. Mai, C., Milstein, O., and Huttermann, A., 2000, Chemoenzymatical grafting of acrylamide onto lignin. *J. Biotechnol.* **79**: 173-183.
23. Hergert, H. L., 1998, Developments in Organosolv Pulping-An Overview. In *Environmentally Friendly Technologies for the Pulp and Paper Industry* (R. A. Young and M. Akhtar, eds.), Wiley, New York, pp. 5-67.

24. Yabannavar, A. V., and Bartha, R., 1994, Methods for assessment of biodegradability of plastic films in soil. *Appl. Environ. Microbiol,.* **60**: 3608-3614.
25. Solaro, R., Corti, A.,and Chiellini, E., 1998, A new respirometric test simulating soil burial conditions for the evaluation of polymer biodegradation. *J. Environ. Polym. Degr.* **6**: 203-208.
26. Chiellini, E., Corti, A., and Solaro, R., 1999, Biodegradation of poly(vinyl alcohol) based blown films under different environmental conditions. *Polym. Degr. Stab.* **64**: 305-312.
27. Bloembergen, S., David. J., Geyer. D., Gustafson, A., Snook, J., and Narayan, R., 1994, Biodegradation and composting studies of polymeric materials. In *Biodegradable Plastics and Polymers* (Y. Doi and K. Fukuda, eds.): Elsevier, Amsterdam, pp. 601-609.
28. Krupp, L. R., and Jewell, W. J., 1992, Biodegradability of modified plastic films in controlled biological environments. *Environ. Sci. Technol.* **26**: 193-198.
29. Sawada, H., 1994, Field testing of biodegradable plastics In *Biodegradable Plastics and Polymers* (Y. Doi and K. Fukuda, eds.): Elsevier, Amsterdam, pp. 298-310.
30. Kimura, M., Toyota, K., Iwatsuki, M., and Sawada, H., 1994, Effects of soil conditions on biodegradation of plastics and responsible microorganisms. In *Biodegradable Plastics and Polymers* (Y. Doi and K. Fukuda, eds.): Elsevier, Amsterdam, pp. 92-106.

Production and Applications of Microbial Polyhydroxyalkanoates

GUO-QIANG CHEN
Department of Biological Science and Biotechnology, Tsinghua University, Beijing 100084, China

1. INTRODUCTION

Polyhydroxyalkanoates (PHA) have shown extensive structure variety[1]. Depending on growth substrates and types of organisms used, the side chain R can change from a simple methyl group to functional structures containing unsaturated double or triple bonds, halogens, benzyl, cyanide or epoxy groups (Fig. 1)[2]. Both the monomer structure and content affect PHA physical properties. PHA can be very brittle, such as polyhydroxybutyrate (PHB) produced by many bacteria; they can also be flexible, such as PHBV copolyesters consisting of 3-hydroxybutyrate (HB) and 3-hydroxyhexanoate (HHx), and PHBHx copolyesters consisting of HB and 3-hydroxyvalerate (HV). PHA can be elastic too when their monomers are 3-hydroxyoctanoate or 3-hydroxydecanoate (Table 1)[3].

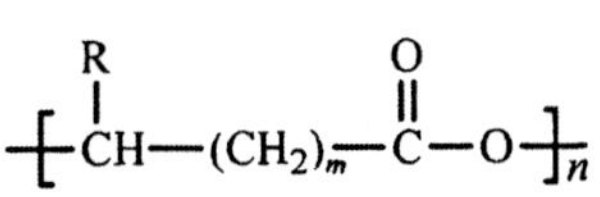

Figure 1. General molecular structure of polyhydroxyalkanoates. m = 1, 2, 3, yet m = 1 is most common, n can range from 100 to several thousands. R is variable. When m = 1, R = CH_3, the monomer structure is 3-hydroxybutyrate, while m = 1 and R = C_3H_7, it is a 3-hydroxyhexanoate monomer. If $R \geq C_3H_7$, the PHA are called medium-chain-length PHA; If $R < C_3H_7$, the PHA are referred to as short-chain-length PHA.

Biodegradable Polymers and Plastics, Edited by Chiellini and Solaro
Kluwer Academic/Plenum Publishers, New York, 2003

There are many bacteria capable of producing PHA. Among PHA, PHB is most commonly found in many bacteria. In order to find bacteria able to synthesize non-PHB polyesters, screening process will have to be carried out.

Although many PHA have been found, only three of them were produced in large scale for commercial exploitation, these are PHB[4], PHBV[5] and PHBHx[6]. There are still a lot of unknown for the production of these unique polyesters.

The cost for the production of PHA is still too high for application for biodegradable packaging. High value added applications, especially biomedical application and fine chemical application, may be realistic for the current PHA applications. Many efforts have been made in this area.

Table 1. Physical properties of various PHA in comparison with conventional plastics[3]

Samples	T_m (°)	T_g (°)	Tensile strength (MPa)	Elongation at break (%)
PHB	177	4	43	5
P(HB-*co*-10% HV)	150	-	25	20
P(HB-*co*-20% HV)	135	-	20	100
P(HB-*co*-10% HHx)	127	-1	21	400
P(HB-*co*-17% HHx)	120	-2	20	850
Polypropylene	170	-	34	400
Polystyrene	110	-	50	-

HB: 3-hydroxybutyrate; HV: 3-hydroxyvalerate; HHx: 3-hydroxyhexanoate

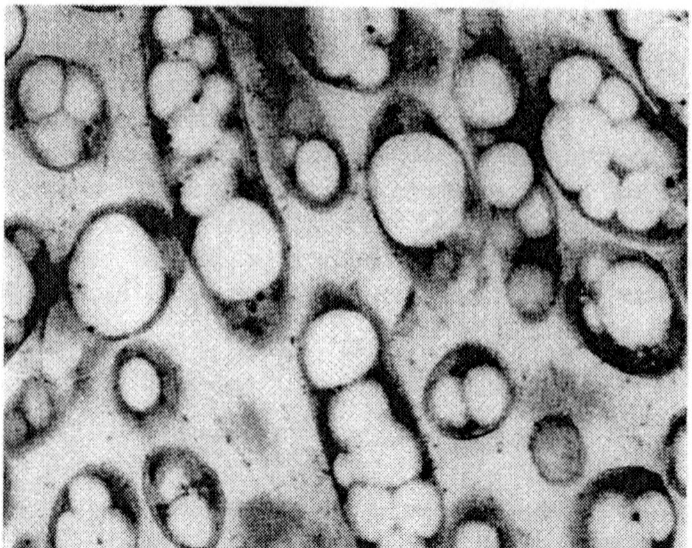

Figure. 2. Granules of copolyesters consisting of 3-hydroxybutyrate and 3-hydroxyhexanoate (PHBHx) produced by *Aeromonas hydrophila.* 20,000 X magnification.

2. PRODUCTION OF POLYHYDROXYALKANOATES (PHA)

To exploit the application of PHA, large quantity of PHA has to be supplied. Although 90 PHA with various monomer units were reported in 1991 and this number is still increasing, application research on PHA can only be conducted with a handful of PHA that can be produced in sufficient quantity. The high cost associated with finding the right organism and developing an industrial PHA production process has contributed to the slow development of PHA production technology, this further leads to the high production cost for PHA. Thus, large-scale application of PHA as environmentally friendly materials has been discouraged partially by the high cost. On the other hand, mechanical properties of PHA play an important role in their applications. Many approaches have been adopted to improve the flexibility of this unique material.

2.1 Screening for Industrial PHA Producing Microorganisms

Many bacteria are able to produce PHA, especially PHB[1,7,8,9]. However, very few of them can be used for industrial production purposes. As an industrial PHA production strain, the microorganism should satisfy the following requirements: rapid growth in cheap carbon sources, high PHA accumulation in the cells, high transformation efficiency of substrate to product, harmless to animals, human and the environment, large in size for separation purposes, and easily lysable for PHA extraction. Ideally, the bacteria should also have a low oxygen demand so that cells can be grown to high density without encountering oxygen limitation; at the same time, the cells should also produce PHA with high molecular weight for application purposes. Due to the difficulty to isolate bacteria that satisfy so many requirements, only *Alcaligenes latus*[10,11] and *Rastonia eutropha*[5] (Formerly called *Alcaligenes eutrophus*) were used for PHB and PHBV industrial production. Other bacteria, such as *Pseudomonas oleovorans* and *Aeromonas hydrophila*, which produce medium chain length PHA and copolyesters PHBHx, respectively, do not meet all above requirements[6,12]. However, they are still used for industrial production due to the lack of alternative strains. Therefore, it is extremely important to develop a rapid screening method that will allow for the discovery of a suitable industrial PHA production strain from the many available bacterial strains.

Normally, the screening process of PHA producing strains can be divided into the following steps: isolation and purification of a single clone, followed by growing the strain in shake flasks; centrifugation to collect biomass after cell growth, followed by freeze dry the cells; PHA extraction out of the cells; gas chromatographic study of PHA monomer structures; GC-MS (Gas chromatography and mass spectroscopy) study to confirm the PHA monomer structures. The entire process can last months before the ability of PHA production and the types of PHA produced by certain bacteria grown on certain substrate are known. This approach is labour intensive and time consuming. Although it has a lot of disadvantages, by many researchers worldwide practiced this method. Obviously, this lengthy screening process will not be effective for screening a large number of PHA producing microorganisms.

At least two methods were developed for rapid screening of PHA producing strains. One is the Nile red method, another is the FT-IR approach. Using the Nile red method, all PHA polyesters show similar fluorescence behaviour, revealing a clear fluorescence maximum at an excitation wavelength between 540 nm and 560 nm and an emission wavelength between 570 nm and 605 nm. The examination of native PHB granules isolated from cells of *Ralstonia eutropha* H16 showed that the addition of 6.0 μg Nile red is necessary for total staining of 1.0 mg granules. The fluorescence intensity at an excitation wavelength of 550 nm and an emission wavelength of 600 nm showed high correlation to the PHB concentration of granule suspensions at different granule concentrations. These results and the staining of cell suspensions during cultivation experiments revealed that Nile red has a high potential for the quantitative determination of hydrophobic bacterial PHA[13]. The Nile red staining method was successfully applied to distinguish PHA producing strains from non-PHA producers among many clones grown on Petri dishes[14]. However, the Nile red staining approach cannot tell the PHA structure. If one observes the bacterial clones stained with Nile red on the Petri dish, it will not be possible to estimate the PHA contents and the types of synthesized PHA. Therefore, an improved method needs to be developed to overcome the drawbacks of Nile red staining method.

A FT-IR method for fast screening of PHA producers was developed[15]. The method is rapid, convenient, non-invasive, combined with the possibility to distinguish short-chain-length and medium-chain-length PHA, as well as quantitatively assay the intracellular PHA content[16].

The FT-IR spectra of pure PHA containing short-chain-length monomers, such as hydroxybutyrate (HB), medium-chain-length hydroxyalkanoate (mclHA) monomers including hydroxyoctanoate (HO) and hydroxy-decanoate (HD), or both HB and mclHA monomers, show their strong characteristic band at 1728, 1740, and 1732 cm^{-1} respectively. Other accompanying bands near 1280 and 1165 cm^{-1} help identify the types of PHA. The intensity of the methylene band near 2925 cm^{-1} provides additional information for PHA characterization. In comparison, bacterial cells accumulating the above PHA also showed strong marker bands at 1732, 1744, and 1739 cm^{-1}, corresponding to intracellular PHB, mclPHA, and P(HB + mclHA), respectively. The accompanying bands of pure PHA were also observable in intact cells. Therefore, by scanning the bacterial cells, it will be possible to know the approximate PHA content in the cells and the types of PHA synthesized by the cells within 10 seconds. Thus, the FT-IR technique will allow the rapid screening of PHA producing strains from among a large number of bacterial colonies.

A broad screening process using the FT-IR technique was carried out[17]. Samples were collected from various geological locations around China. The FT-IR method proved very effective. It was found that the PHA composition depends very much on the geological location. In some locations, bacteria mainly synthesized short-chain-length PHA, in other locations, medium-chain-length PHA were accumulated by inhabiting bacteria. Additionally, the synthesis of blend polymers consisting of PHB, short-chain-length PHA and mclPHA is a common phenomenon among the studied bacteria. 40% of the 371 strains cultivated on six substrates were able to synthesize PHA, with many of them making blends of PHB and mclPHA. This result will help polymer researchers to identify sources of PHA synthesizing bacteria.

2.2 Production of Polyhydroxybutyrate (PHB)

2.2.1 PHB Production by *Bacillus* spp.

Bacillus spp. were among the very first to be reported as PHB producers[18]. However, we were surprised to learn that no PHB production research was conducted with this organism in terms of process development, although *Bacillus* spp. have long been known to grow rapidly; they are also capable of using various cheap carbon sources for growth and they are very resistant to contamination by other bacteria. Chen *et al.* reported that the production of PHB from 11 *Bacillus* spp. randomly selected from German Culture Collection (DSM) never exceeded 50% when growth was conducted in shake flasks[8].

To investigate the possibility for PHB production using *Bacillus* spp., a *Bacillus* strain isolated from molasses contaminated soil was used as a model[19]. It appeared that PHB formation was growth associated; factors that normally promote PHB production including high ratios of carbon to nitrogen, carbon to phosphorus and low oxygen supply, did not lead to high PHB production. Instead, these factors resulted in sporulation, which further leads to reduced PHB contents and cell dry weight. The molecular weights of PHB produced by this *Bacillus* sp. were all low. The competition of PHB synthesis and sporulation seemed to be the reason for low PHB production. Therefore, *Bacillus* spp. may not be a suitable PHB industrial production strain. Furthermore, the thick Gram-positive cell wall will make the breakage of cells and PHB extraction difficult.

2.2.2 PHB Production by *Alcaligenes latus*

Alcaligenes latus is one of the strains that satisfy the requirement for industrial PHB production[20]. The strain grows rapidly in sucrose, glucose and molasses. PHB accumulation can be as high as over 90% of the cell dry weight[21].

Chemie Linz AG/Austria (later btf Austria) produced PHB in a quantity of 1000 kg/week in a 15 m^3 fermentor using *Alcaligenes latus* DSM 1124[4]. The cells were grown in mineral medium containing sucrose as carbon source. The PHB produced by *Alcaligenes latus* has been used to make sample cups, bottles, and syringes for application trials. Biomer in Germany now owns the PHB production and processing technology. Different products including combs, pens, and bullets have been made from PHB produced by *Alcaligenes latus*.

2.2.3 PHB Production by Ralstonia eutropha

Ralstonia eutropha was used to investigate the PHB production in 1 m^3 fermentor under the joint action of Institute of Microbiology affiliated to the Chinese Academy of Sciences and Tianjin Northern Food Inc./China. Growth was carried out for 48 h in glucose mineral medium. At the end of the cell growth, cell density reached 160 g/l. The cells produced 80% PHB in their dry weight. Most surprisingly, the strain grown to such a high density did not require oxygen-enriched air. This was perhaps the highest cell density for PHB production achieved in pilot scale production.

2.2.4 PHB Production by *Azotobacter vinelandii*

Azotobacter vinelandii strain UWD was demonstrated to grow rapidly in molasses medium[22]. The strain has a large size, ranging from 1 to 8 μm. It can produce PHB up to 90% of cell dry weight. At the same time, the strain produces PHB with a molecular weight ranging from 1 to 4 million Dalton[23]; this is rarely seen with any microorganism. PHB production could be promoted by lower aeration, therefore, PHB production can be separated into two-stage: one for cell growth under high aeration and another for PHB accumulation under lower aeration[24]. In a small-scale lab top fermentor, 36 g/l PHB were produced from molasses after 48 h growth.

Collaboration between the Microbiology Lab at Tsinghua University and Guangdong Jiangmen Center for Biotech Development/China for pilot PHB production by *A. vinelandii* UWD was carried out on molasses medium. The pilot study was done in 4 m^3 fermentor without automatic oxygen supply control. After 48 h growth, the cells reached a density of 75-80 g/l. The PHB content in the cells was as high as 72% of the cell dry weight. The cell size was at least 6 μm in diameter. Due to the high PHB accumulation efficiency and the large cell size, separation of biomass from the fermentation broth using continuous disk centrifuges was convenient. At the same time, the cells were easily broken in 0.2% SDS solution at 60 °C for 2 h, making the downstream processing relatively easy. Major problem with this strain has been the difficulty to grow the cells to high density, as this strain requires high dissolved oxygen concentration for high-density growth. The supply of oxygen enriched air for industrial fermentation is impossible due to its explosive danger and the high cost of pure oxygen.

PHB produced by this strain is now under study by the Institute of Polymer Sciences and Engineering at Tsinghua University. Major efforts have been focused on improving the mechanical strength and on the exploitation of tissue engineering application for this polyester.

2.3 Production of Copolyesters Consisting of 3-Hydroxybutyrate and 3-Hydroxyvalerate (PHBV)

ICI Bioproducts & Fine Chemicals (now Zeneca) was the first to really produce PHA in large scale, namely, copolyesters (PHBV) of 3-hydroxybutyrate (HB) and 3-hydroxyvalerate (HV)[5]. The production strain is *Ralstonia eutropha*; the strain is able to grow on glucose and produce the copolymer PHBV to a density as high as 70-80 g/l after over 70 h growth. Shampoo bottles were produced from PHBV (Trademarked as BIOPOL) and were available in supermarkets in Europe. However, due to the economic reason, the Biopol products did not succeed and the PHBV patents were sold to Monsanto.

Hangzhou Glutamate Ltd./China, in collaboration with the Institute of Microbiology affiliated to the Chinese Academy of Sciences, has developed a model process that can produced PHBV in high efficiency. Without supply of pure oxygen, *R. eutropha* grew to a density of 160 g/l cell dry weight within 48 h in a 1000 L fermentor. The cells accumulated 80% of PHBV with a production efficiency of 2.5 g/h/l. The HV content in the copolymer ranged from 8 to 10%. This process can significantly reduce the production cost for PHBV.

PHBV can become economically competitive only by achieving high growth rate, high PHBV production efficiency and high cell and polymer densities. We assume that PHBV or other PHA can become cost effective after extensive improvement in fermentation process and downstream process.

2.4. Production of Copolyesters Consisting of 3-Hydroxybutyrate and 3-Hydroxyhexanoate (PHBHx)

Recently, Tsinghua University in Beijing/China, in collaboration with Guangdong Jiangmen Center for Biotech development/China, KAIST/Korea, and Procter & Gamble in USA has succeeded in producing PHBHx by *Aeromonas hydrophila* grown in 20 cubic meter fermentor[6]. The PHBHx production was carried out on glucose and lauric acid for about 60 h. Cell dry weight reached 50 g/l; only 50% of PHBHx was produced in the cell dry weight. The extraction of PHBHx was a very complicated process involving the use of ethyl acetate and hexane, which dramatically increased the production cost. PHBHx produced by Jiangmen/China is now been exploited for application in areas of flushable, non-wovens, binders, films, flexible packaging, thermoformed articles, coated paper, synthesis paper, coating systems, and medical devices (www.nodax.com). Copolymers consisting of HB and medium-chain-length HA have been trademarked by P&G as NODAX.

The current production cost for PHBHx is still too high for real commercial application. However, many efforts have been made to improve the PHBHx production process including the downstream process technology. Most efforts have been focused on increasing the cell density and simplifying the downstream process. A better production strain able to utilize glucose will be one of the most important issues for reducing the PHBHx production costs.

2.5 Production of Copolyesters Consisting of Medium-Chain-Length Hydroxyalkanoates

Medium-chain-length PHA can be produced by *Pseudomonas oleovorans* and *Pseudomonas stutzeri* as well as by other *Pseudomonas* spp. It was reported that mclPHA could be produced at costs below US$10/kg if production scale is 1000 tones/year by using the *P. oleovorans* grown on octane. However, mclPHA made up only less than 40% of the cell dry weight. It would be very important if a strain could produce at least over 50% mclPHA.

Strain *P. stutzeri* 1317 isolated from oil-contaminated soil was found to grow on a variety of carbon source including glucose and soybean oil[26]. The strain produced over 63% mclPHA when grown on soybean oil. On glucose, 51% mclPHA was synthesized by this organism. The strain is currently under intensive investigation due to the possibility to increasing the mclPHA production level.

3. APPLICATION OF POLYHYDROXYALKANOATES AS BIOMATERIALS FOR TISSUE ENGINEERING

Applications for PHA can be found in areas of flushable, non-wovens, binders, films, flexible packaging, thermoformed articles, coated paper, synthesis paper, and coating systems (www.nodax.com). However, current PHA production cost is still too high to satisfy such low added value demand. Therefore, we believe that high added value application should be more realistic. As PHA are biodegradable, and possibly biocompatible, their application as biomaterials or tissue engineering materials should be very attractive.

To test the biocompatibility of PHA, three polymers were selected, namely, PHB, PHBHx and poly(L-lactic acid) (PLA). Mouse fibroblast cell line L929 was incubated on films made of PHB, PHBHx and their blends, as well as PLA. Results showed that PHBHx had the best biocompatibility, followed by PHBHx/PHB blend; PLA was the least biocompatible polymer[27,28]. Since the mechanical strength of PHBHx is much better than that of PHB and PLA, it is expected that PHBHx possessing better biocompatibility and mechanical strength will have a promising future in tissue engineering application.

To test this promise, polymer scaffolds consisting of poly(hydroxy-butyrate-*co*-hydroxyhexanoate)/polyhydroxybutyrate (PHBHx/PHB) blends were investigated for possible application as a matrix for the three-dimensional growth of chondrocytes[29]. PHBHx/PHB blends were fabricated into three- dimensional porous scaffolds by the salt-leaching method. Chondrocytes isolated from rabbit articular cartilage (RAC) were seeded on the scaffolds and incubated for 28 days, by replacing the culture medium every 4 days. PHB scaffold was taken as control. Methylthiazoltetrazolium (MTT) (3-[4,5-dimethylthiazol-2-yl]-2,5-diphenyltetrazolium bromide) assay was used to quantitatively examine the proliferation of chondrocytes. Results showed that chondrocytes proliferated better on PHBHx/PHB scaffolds than on PHB. The maximal cell densities were observed after 7 days of incubation. As far as the blend composition is concerned, cells grew better on scaffolds consisting of 2:1 and 1:2 PHBHx/PHB blends than they did on 1:1 PHBHx/PHB. Scanning electron microscopy (SEM) also showed that large quantities of chondrocytes grew initially on the surface of the scaffold. After 7 days, they further grew into the open pores of the blend polymer scaffolds. Morphologically, cells found on the surface of the scaffold exhibited fibril appearance and slowly formed confluent cell multilayers starting from 14 to 28 days of growth. In contrast, cells showed rounded morphology, formed aggregates and islets inside the scaffolds. In addition, chondrocytes proliferated on the scaffold and preserved their phenotype for up to 28 days. This further showed that PHBHx is a good candidate for tissue engineering application.

4. CONCLUSIONS

Microbial production of PHA has been developed over the past two decades. PHB and PHBV fermentation technology has been very well exploited. Production costs for these two types of polymers have been significantly reduced. Better performing polymers, especially copolyesters consisting of HB and mclHA, should be the focal point for development. However, current industrial production strains for this type of polymers suffer from slow growth, low substrate to product transformation efficiency, and dependence on expensive substrates. Efforts should be made to isolating or engineering high productive strains and to improving fermentation and downstream processing technology.

Before the PHA production cost can be brought to the point where it can compete with conventional plastics, high value added application in medical fields should be the real destination of PHA. As these polymers possess not only excellent mechanical properties and processability but also good biocompatibility, their application as tissue engineering materials looks very promising.

ACKNOWLEDGEMENTS

The financial support by Natural Science Foundation of China Grant No. 30170017 and 20074020, Tsinghua University 985 project, State 9th Five-Year R&D Project, and by Procter & Gamble is gratefully acknowledged.

REFERENCES

1. Steinbüchel, A., 1991, Polyhydroxyalkanoic acids. In *Biomaterials: Novel Materials from Biological Sources* (D. Byrom, ed.), Stockton, New York, pp124-213.

2. Chen, G.Q., Wu, Q., Zhao, K., and Yu, P.H., 2000, Functional polyhydroxyalkanoates synthesized by microorganisms. *Chinese J. Polym. Sci.* **5**: 389-396.
3. Doi, Y., Kitamura, S., and Abe, H., 1995, Microbial synthesis and characterization of poly(3-hydroxybutyrate-*co*-3-hydroxyhexanoate). *Macromolecules* 28: 4822-4828.
4. Hrabak, O., 1992, Industrial production of poly-β-hydroxybutyrate. *FEMS Microbiol. Rev.* **103**: 251-256.
5. Byrom, D., 1992, Production of poly-β–hydroxybutyrate and poly-β-hydroxyvalerate copolymers. *FEMS Microbiol. Rev.* **103**: 247-250.
6. Chen, G.Q., Zhang, G., Park, S. J., Lee S. Y., 2001, Industrial Production of Poly(hydroxybutyrate-*co*-hydroxyhexanoate). *Appl Microbiol Biotechnol* **57**: 50-55.
7. Anderson, A. J., Dawes, E. A., 1990, Occurrence, metabolism, metabolic role and industrial uses of bacterial polyhydroxyalkanoates. *Microbiol. Rev.* **45**: 450-472.
8. Chen, G. Q., Koenig, K. H., Lafferty, R. M., 1991, Occurrence of poly-D (-)-3-hydroxyalkanoaies in the genus *Bacillus*. *FEMS Microbiol. Lett.* **84**: 173-176.
9. Haywood, G. W., Anderson, A. J., Dawes, E. A., 1989, A survey of the accumulation of novel polyhydroxyalkanoates by bacteria. *Biotechnol. Lett.* **7**: 471-476.
10. Chen, G. Q., Koenig, K. H. Lafferty, R. M., 1991, Production of poly-D(-)-3-hydroxybutyrate and poly-D(-)-3-hydroxyvalerate by strains of *Alcaligenes latus*. *Anton. Leeuw.* **60**: 61-66.
11. Hanggi U. J., 1990, Pilot scale production of PHB with *Alcaligenes latus*. In: *Novel Biodegradable Microbial Polymers.* (E. D. Dawes ed), Kluwer Academic Publishers, Netherlands, pp. 65-70.
12. Brandl, H., Cross, R. A., Lenz, R. W., Fuller, C., 1988, *Pseudomonas oleovorans* as a source of poly(β-hydroxyalkanoates) for potential applications as biodegradable polyesters. *Appl. Environ. Microbiol.* **54**: 1977-1982.
13. Gorenflo, V., Steinbüchel, A., Marose, S., Rieseberg, M., Scheper, T., 1999, Quantification of bacterial polyhydroxyalkanoic acids by Nile red staining, *Appl. Microbiol. Biotechnol.* **51**: 765-772.
14. Tian, W.D., 1998, Screening and study of novel polyhydroxyalkanoate producing microorganisms. *M.Sc. Thesis*, Tsinghua University, Beijing/China.
15. Hong, K., Sun, S.Q., Tian, W.D., Chen, G.Q., 1999, A rapid method for detecting bacterial PHA in intact cells by FT-IR. *Appl. Microbiol. Biotechnol.* **51**: 523-526.
16. Kansiz, M., Billman-Jacobe, H., McNaughton, D., 2000, Quantitative determination of the biodegradable polymer poly(β-hydroxybutyrate) in a recombinant *Escherichia coli* strain by use of mid-infrared spectroscopy and multivariative statistics. *Appl. Environ. Microbiol.* **66**: 3415-3420.
17. Wu, Q., Sun, S.Q., Yu, P.H.F., Chan, A.X.Z., Chen, G.Q., 2000, Environmental dependence of microbial synthesis of polyhydroxyalkanoates. *Acta Polym. Sin.* **6**: 751-756.
18. Macrae, R.M., Wilkinson, J.F., 1958, Poly-beta-hydroxybutyrate metabolism in washed suspensions of *bacillus cereus* and *Bacillus megaterium*. *J. Gen. Microbiol.* **19**: 210-220.
19. Wu, Q., Huang, H.H., Hu, G.H., Chen J.C., Ho, K.P., Chen, G.Q., 2001, Production of poly-3-hydroxybutyrate by Bacillus sp. JMa5 cultivated in molasses media. *Anton. Leeuw.* **80**:111-118.
20. Küng, W., 1982, Wachstum und Poly-D(-)-3-Hydroxybuttersäure Akkumulation bei *Alcaligenes latus*. *M.Sc. Thesis*, Graz University of Technology, Graz/Austria.
21. Chen, G.Q., 1989, Produktion von Poly-D(-)-3-Hydroxybuttersäure und Poly-D(-)-3-Hydroxyvaleriansäure mit Einzel- und Mischpopulation von *Alcaligenes latus* DSM 1122, 1123 bzw. 1124. *Ph.D. Thesis*, Graz University of Technology, Graz/Austria.

22. Page, W.J., 1992, Production of polyhydroxyalkanoates by *Azotobacter vinelandii* UWD in beet molasses culture. *FEMS Microbiol. Rev.*. **103**: 149-158.
23. Chen, G.Q., Page, W.J., 1994, The effect of substrate on the molecular weight of poly-β-hydroxybutyrate produced by *Azotobacter vinelandii* UWD. *Biotechnol. Lett.* **16**: 155-160.
24. Chen, G.Q., Page, W.J., 1997, Production of poly-β-hydroxybutyrate by *Azotobacter vinelandii* UWD in a two-stage fermentation process. *Biotechnol. Tech.* **11**: 347-350.
25. Hazenberg, W., Witholt, B., 1997, Efficient production of medium-chain-length poly(3-hydroxyalkanoates) from octane by *Pseudomonas oleovorans*: economic considerations. *Appl. Microbiol. Biotech.* **48**: 588-596.
26. He, W.N., Tian, W.D., Zhang, G., Chen, G.Q., Zhang, Z.M., 1998, Production of novel polyhydroxyalkanoates by *Pseudomonas stutzeri* 1317 from glucose and soybean oil. *FEMS Microbiol Lett.* **169**: 45-49.
27. Zhao, K., Yang, X.S., Chen, J.C., Chen, G.Q., 2002, Effect of lipase treatment on the biocompatibility of microbial polyhydroxyalkanoates. *J. Mater. Sci. Mater. Med.* **13**: 1-6.
28. Yang, X.S., Zhao, K., Chen, G.Q., 2002, Effect of Surface Treatment on the Biocompatibility of Microbial Polyhydroxyalkanoates.. *Biomaterials* **23**: 1391-1397.
29. Deng, Y., Zhao, K., Zhang X.F., Hu, P., Chen, G.Q., 2002, Study on the three-dimensional proliferation of rabbit articular cartilage-derived chondrocytes on polyhydroxyalkanoate scaffolds. *Biomaterials* **23**: 4049-4056.

The Solid-State Structure, Thermal and Crystalline Properties of Bacterial Copolyesters of (R)-3-Hydroxybutyric Acid with (R)-3-Hydroxyhexanoic Acid

ZHIHUA GAN[*], KAZUHIRO KUWABARA[†], HIDEKI ABE[*†], and YOSHIHARU DOI[*,†]
[*]Department of Innovative and Engineered Materials and the SORST Group of Japan Science and Technology Corporation (JST), Tokyo Institute of Technology, 4259 Nagatsuta, Midori-ku, Yokohama 226-8502, Japan, and [†]Polymer Chemistry Laboratory, RIKEN Institute, 2-1 Hirosawa, Wako-shi, Saitama 351-0198, Japan

1. INTRODUCTION

Bacterial poly(3-hydroxyalkanoic acid)s (PHAs) are interesting biodegradable and biocompatible thermoplastics produced by a wide variety of microorganisms from various carbon sources.[1,2] The PHAs are known to accumulate inside bacterial body as intracellular storage materials for biological carbon and energy sources[1-4]. Until now PHAs with more than 100 different monomeric units as constituents have been found[5]. These biodegradable PHA thermoplastics have attracted much attention in the recent two decades as they are environmentally friendly materials which can be degraded to carbon dioxide, water and biomass by a wide of microorganisms. Therefore the bacterial PHAs are a prospective candidate to replace the normally used plastics, which result in serious environmental pollution by waste polymers.

Poly(3-hydroxbutyric acid) [P(3HB)] is the most famous member in the family of PHAs. This homopolymer is an optically active polyester of (*R*)-specific units and has a high crystallinity up to 60~70%. The particular advantage is that P(3HB) is a thermoplastic and therefore can be processed by using the existing equipment. However the drawbacks are also serious because P(3HB) materials are rather brittle. One solution to overcome the drawback is to produce copolyesters of 3HB with other (*R*)-hydroxyalkanoic acids by feeding bacteria with relatively complex carbon sources. Poly(3-hydroxybutyric acid-*co*-3-hydroxyvaleric acid) [P(3HB-*co*-3HV)] is successful bacterial copolyester, which is produced commercially under the trade name of Biopol[6]. P(3HB-*co*-3HV) copolyester has been extensively studied on its crystal structure, morphology, crystallization and melting behaviours as well as segmental dynamics[7-17]. It has been found that the polyesters can be improved on flexibility by introducing 3HV units into the P(3HB) backbone, but that the copolyesters still have a high degree of crystallinity (>50%) over the entire composition range due to cocrystallization.

Poly(3-hydroxybutyric acid-*co*-3-hydroxyhexanoic acid) [P(3HB-*co*-3HH)] is another bacterial copolyester. Our laboratory has found that random copolyesters of 3HB with 3HH are produced from plant oils by the *Aeromonas caviae* bacterium isolated from soil[18,19]. The basic characterization on the crystallization and physical properties of P(3HB-*co*-3HH) random copolyesters with 3HH contents varying from 0 to 25 mol% has indicated that 3HH units are excluded from the P(3HB) crystal lattice, and that both the crystallinity and crystallization rate are reduced as 3HH content increases. The elongation at break of the copolyester film is near 10 times that of P(3HB) homopolymer[20]. These results have shown that the mechanical properties of P(3HB-*co*-3HH) random copolyesters can be modified although they have the same crystal structure of P(3HB) homopolymer, and they suggested that the copolyesters have better processability and more extensive applicable areas.

In this paper, we investigated the solid-state microstructure, lamellar crystal morphology and growth process, thermal property, as well as the interrelations between these properties and copolymer composition for P(3HB-*co*-3HH) random copolyesters by using differential scanning calorimetry, atomic force microscopy, x-ray diffraction, and solid-state NMR techniques.

2. EXPERIMENTAL METHODS

2.1 Materials

The bacterial copolyesters of (*R*)-3-hydroxybutyric acid (3HB) and (*R*)-3-hydroxyhexanoic acid (3HH) [P(3HB-*co*-3HH)] were supplied from Procter & Gamble Company. Poly[(*R*)-3-hydroxybutyric acid] was supplied from Monsanto Company. All the samples were purified before investigation. The relative molecular weights and distributions were measured by size exclusion chromatography (SEC) at 40 °C with chloroform as eluent at 0.8 mL/min flow rate. SEC measurements were carried out on a Shimadzu 10A system and 10A refractive index detector by using narrow dispersed polystyrene samples as standards. The content and distribution of 3HH units were determined by ^{1}H NMR in solution. The results indicate that 3HH units are distributed randomly along the 3HB backbone chains for all copolyesters. Some characteristic results are summarized in Table 1.

Table 1. Basic properties for P(3HB-*co*-3HH) copolyesters with different 3HH contents[a]

Polyester	M_w [b]	M_w/M_n	T_g (°C) [c]
P(3HB)	436,000	2.7	1.4
P(3HB-*co*-4.6mol%3HH)	1,005,000	3.2	−2.1
P(3HB-*co*-11.9mol%3HH)	1,134,000	3.8	−2.8
P(3HB-*co*-14.6mol%3HH)	289,000	2.6	−4.4

[a] The 3HH molar contents were determined by ^{1}H NMR. [b] The molecular weight and polydispersity were measured by SEC. [c] P(3HB-*co*-3HH) copolyesters were quenched from the melt to the amorphous state and then heated at a rate of 20 °C/min.

2.2 WAXD Measurement

P(3HB-*co*-3HH) copolyesters were first dissolved in chloroform and then poured into Petri Dish to form cast films after evaporation of the solvent. The stripped cast films were sandwiched between two thick Teflon sheets with a thin Teflon sheet (0.05 mm thickness) as spacer, and then melt-compressed on a mini test press (Toyoseiki) at a temperature about 20 °C above the melting temperature under 150 kg/cm^2 pressure for 1 min. They were then quickly transferred into an oven at a given temperature to crystallize for 3 days. The wide-angle X-ray diffraction (WAXD) patterns of the melt-crystallized P(3HB-*co*-3HH) films were recorded at room temperature on Rigaku RINT2500 system by using nickel-filtered Cu Kα radiation (λ = 0.154 nm, 40 kV and 20 mA) in the 2θ range of 6~60° at a scan speed of 2°/min. The degree of crystallinity of melt-crystallized copolyester films was calculated from the x-ray diffraction intensity according to Vonk's method[21].

2.3 DSC Measurements

About 3 mg of small pieces of P(3HB-*co*-3HH) cast films was encapsulated into DSC aluminum pans. The pans were placed on a hot stage at a temperature of 20 °C above the melting point for 2 minutes for melting, and then transferred to the oven at a given temperature for 3-day crystallization. Then, the melt-crystallized copolyester samples were stored in a refrigerator at a temperature below T_g prior to the DSC analyses. The differential scanning calorimetry (DSC) analyses were carried out on a Perkin-Elmer Pyris differential scanning calorimeter equipped with a CryoFill liquid nitrogen cooling system and operated under nitrogen at a flow rate of 20 mL/min. The temperature was accurately calibrated with indium and zinc by standard procedure. Heating scans were performed at 20 °C/min for all the melt-crystallized copolyester samples, except for studying the influence of heating rate on the melting behaviour. Annealing experiments for the melt-crystallized copolyesters were carried out directly on the DSC sample holder. The melt-crystallized samples were heated from room temperature (~25 °C) to a selected temperature T_a at 60 °C/min and kept there for 10 minutes. After that, the samples were quenched to 25 °C before heating again to the melt. The melting traces were recorded. For the determination of glass transition temperature T_g, the copolyesters were quenched from the melt state to −50 °C and again heated to the melt at 20 °C/min. T_g was calculated as the half C_p extrapolated point.

2.4 NMR Characterization

The compositions and sequence distributions of P(3HB-co-3HH) copolyesters were analyzed by NMR technique. The copolyesters were dissolved in $CDCl_3$ at a concentration of 10 mg/mL and their ^1H and ^{13}C NMR spectra were recorded on a JEOL ALPHA-400 spectrometer. Tetramethylsilane (Me_4Si) was used as internal chemical shift standard.

High-resolution solid-state ^{13}C NMR was utilized to investigate the microscopic motions of polymer chains in both crystalline and amorphous regions. The measurements were carried out at room temperature on a Chemagnetics Infinity-400 spectrometer under a static magnetic field of 9.4 T. The ^1H and ^{13}C field strength $\gamma B_1/2\pi$ were 62.5 kHz. The contact time for the cross-polarization (CP) was 2.0 ms throughout this work. The magic angle spin (MAS) rate for each measurement was set to 8.0 kHz to avoid the overlapping of spinning side bands on other resonance lines. ^{13}C chemical shifts were expressed as values relative to Me_4Si by the CH_3 line at 17.36 ppm of hexamethylbenzene crystals as an external reference. ^{13}C spin-relaxation times (T_{1C}) were measured by the Torchia's CPT1 pulse sequence[22]. All copolyester samples for the solid-state ^{13}C NMR measurements were melt-crystallized at a given temperature for 3 days.

2.5 Crystalline Morphology Observation by AFM

The morphologies of melt-crystallized copolyesters were observed by atomic force microscopy (AFM). Very thin polymer films with a thickness of about 100 nm were first prepared on a cover glass by solution cast method[23] and then melt-crystallized at a given temperature for 1 day, and then the surface morphology was observed on SPI3800/SPA400 AFM (Seiko Instrument Inc.) in tapping mode under ambient conditions. A rectangle cantilever with a length of 200 μm, a spring constant of 12 N/m, and a resonance frequency of 126 kHz was used in measurements.

3. RESULTS AND DISCUSSION

3.1 Solid-State Structure of Melt-Crystallized P(3HB-*co*-3HH) Films

Figure 1 shows the X-ray diffraction patterns of melt-crystallized P(3HB-*co*-3HH) films with 3HH contents from 0 to 14.6 mol%. As compared to P(3HB) homopolymer, all the copolyesters have the same diffraction patterns, except that some diffraction peaks decrease in intensity with increasing the 3HH contents.

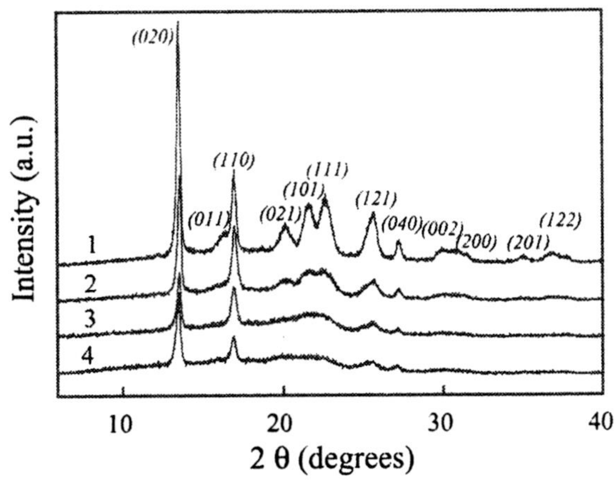

Figure 1. X-ray diffraction patterns of P(3HB-*co*-3HH) copolyesters after melt-crystallization at 90 °C for 3 days. 1: P(3HB); 2–4: P(3HB-*co*-3HH) containing 4.6, 11.9, and 14.6 mol% of 3HH units, respectively.

The calculation on the *d* spacing values for (020), (110) and (002) diffraction peaks given in Table 2 indicates that all the *d* spacing values are almost invariant in 3HH content range from 0 to 14.6 mol%. This result indicates that P(3HB-*co*-3HH) copolyesters containing 4.6, 11.9, and 14.6 mol% 3HH units have the same crystal structure of P(3HB) homopolymer. The 3HH units are excluded completely from the 3HB crystal lattice and act as defects; therefore they greatly decrease the degree of crystallinity. As seen in Table 2 at a crystallization temperature of 90 °C, P(3HB) has a degree of crystallinity of 65%, while introduction of 4.6 mol% 3HH units reduces the crystallinity of copolyester to 41%, and P(3HB-*co*-3HH) with 14.6 mol% 3HH units only has about 24% crystallinity.

Table 2. Crystal structure, *d* spacing, and degree of crystallinity of P(3HB-*co*-3HH) copolyester samples from X-ray diffraction[a]

3HH (mol%)	Crystal Structure	*d* spacing (nm)			Crystallinity (%)
		(020)	(110)	(002)	
0	P(3HB)	0.6568	0.5246	0.2920	65±5
4.6	P(3HB)	0.6539	0.5246	0.2927	41±5
11.9	P(3HB)	0.6568	0.5265	0.2938	27±5
14.6	P(3HB)	0.6578	0.5252	0.2931	24±5

[a] Films after melt-crystallization at 90 °C for three days. The *d* spacings are calculated from the X-ray diffraction peak positions on the basis of Bragg's condition $2d\sin\theta = n\lambda$ (n=1, 2,...) where θ is the half diffraction angle and λ is the X-ray wavelength.

Solid-state ^{13}C NMR results further support the above conclusion that the crystalline cores of P(3HB-*co*-3HH) copolyesters are only composed of 3HB sequences and that the excluded 3HH units are in amorphous regions. Figure 2 shows the CP/MAS ^{13}C NMR spectra of melt-crystallized P(3HB-*co*-3HH) samples containing 4.6, 11.9, and 14.6 mol% of 3HH units between 12 and 30 ppm.

The resonance lines at 20–22 and 14–15 ppm are attributed to the CH_3 carbons of 3HB and 3HH units, respectively. For 3HB units, both resonance lines assigned to crystalline and amorphous components are observed. In contrast, only a single peak ascribed to the amorphous component is observed for 3HH units of all samples. These results clearly indicate that the crystalline regions of P(3HB-*co*-3HH) copolyesters are composed only of 3HB sequences, while that 3HH units are excluded from the crystalline region and located in a single (amorphous) phase.

3.2 Melting Behaviour

As shown in Table 1, the glass transition temperatures of P(3HB-*co*-3HH) copolyesters slightly decrease from 1.4 to −4.4 °C as the contents of 3HH units increase from 0 to 14.6 mol%. When the copolyesters were heated at a rate of 20 °C/min from amorphous to melt state, P(3HB) and P(3HB-*co*-4.6mol%3HH) show the melting peaks, while the other two copolyesters with 3HH contents as 11.9 mol and 14.6 mol% do not present any melting peak, indicating that the copolyesters with relatively higher 3HH contents have a slow overall crystallization rate.

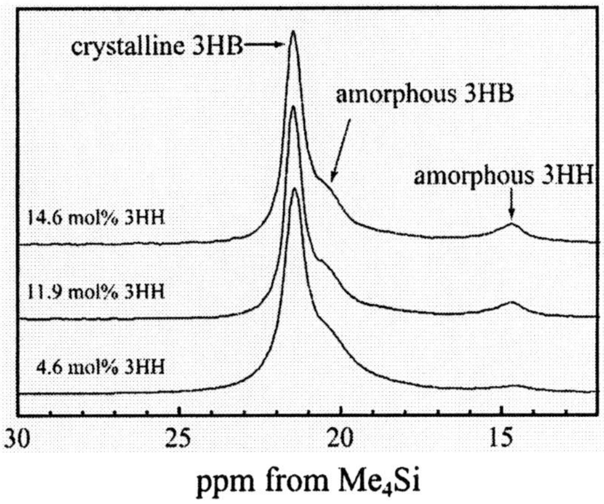

Figure 2. CP/MAS solid-state ^{13}C NMR spectra of P(3HB-*co*-3HH) samples having different contents of 3HH units between 12 and 30 ppm.

Figure 3 shows the melting behaviour for P(3HB) homopolymer and P(3HB-*co*-3HH) copolyesters after isothermal melt-crystallization at a given temperature for 3 days. P(3HB) shows only a single and sharp melting peak at all crystallization temperatures, whereas the melting curves of the three copolyesters show complicated multiple melting peaks, marked as T_{s2}, T_{s1}, T_{m1} and T_{m2} from low to high temperature. It can be observed that: 1) copolyesters have a wide melting range, starting from a temperature T_{s1} just about 10 °C above the crystallization temperature T_c; 2) for the melting peak T_{s1}, its magnitude increases not only with 3HH content, but also with the crystallization temperature; 3) the melting peak T_{m1} is observed for the three copolyesters and shifts to higher temperature as the crystallization temperature increases; 4) the magnitude of the melting peak T_{m2} decreases with increasing crystallization temperature, and finally disappears; 5) the melting peak T_{s2} appears at temperatures below the crystallization temperature T_c for the samples with higher 3HH contents or crystallized at higher temperatures.

Multiple-melting behaviour has been reported for many crystalline polymers; it may originate from different mechanisms, depending on both polymer chemical structure and crystallization conditions. Although several models have been proposed by many researchers[24], divergences among these models still exist. Melting-recrystallization is one of the explanations for the multiple-melting behaviour, and generally it must be first considered and examined by DSC experiment. Our results have confirmed that melting-recrystallization occurs for three P(3HB-*co*-3HH) copolyesters during DSC heating process, and that the melting peak T_{m2} corresponds to crystals formed by partial melting and recrystallization. Our X-ray diffraction results also exclude the possibility that the multiple melting peaks arise from different crystal structures.

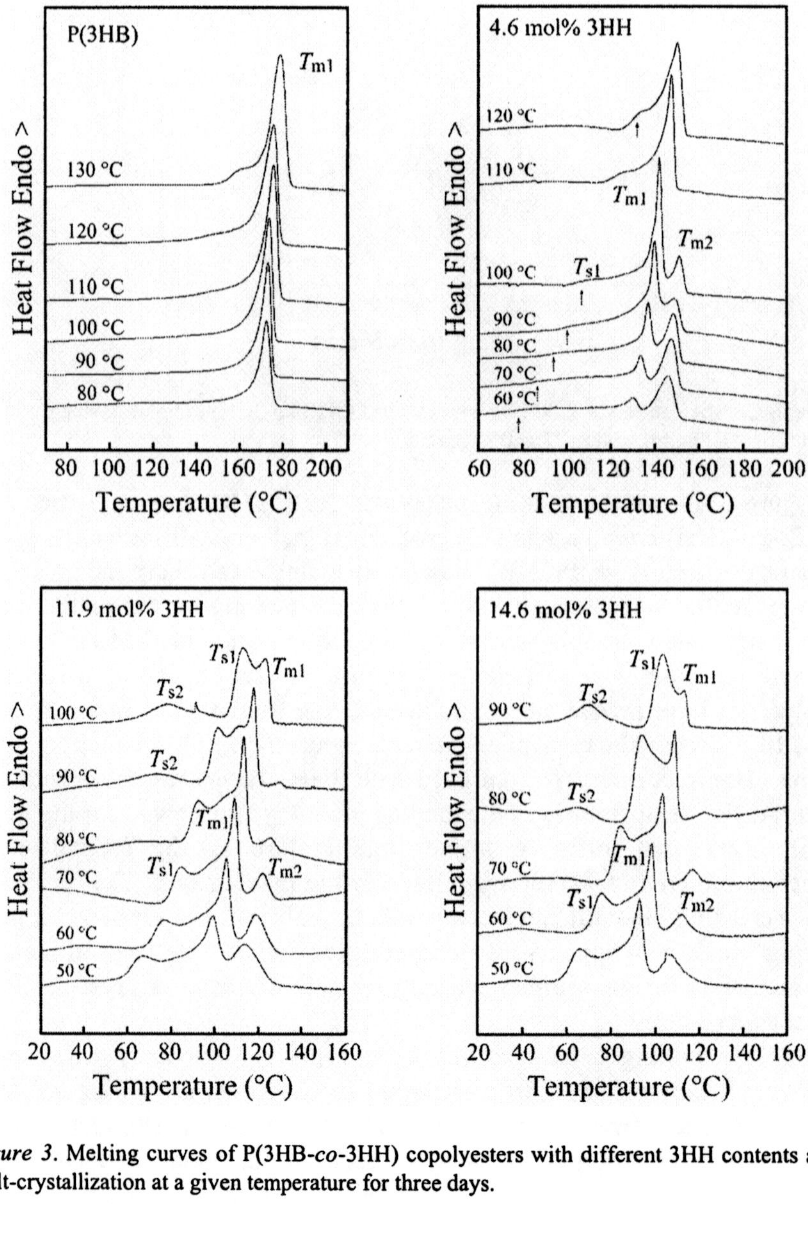

Figure 3. Melting curves of P(3HB-*co*-3HH) copolyesters with different 3HH contents after melt-crystallization at a given temperature for three days.

The melting peak T_{ml} arises from the original crystals, which are formed at the crystallization temperature T_c. Figure 4 shows the dependence of T_{ml} on T_c for P(3HB-*co*-3HH) copolyesters with different 3HH contents. T_{ml} increases linearly with T_c for all the samples, and this temperature decreases with increasing 3HH contents at the same crystallization temperature. The decrease of the melting temperature can be attributed to the random distribution of 3HH units, which are excluded from the P(3HB) sequence crystals and hence reduce the lamellar thickness.

The melting peak T_{sl} in Figure 3 is hardly observed in P(3HB), but it is more evident for P(3HB-*co*-3HH) copolyesters especially at higher 3HH contents and crystallization temperatures. Figure 4 shows that the melting temperature T_{sl} of P(3HB-*co*-3HH) copolyesters at different crystallization temperatures is always about 10 °C above the crystallization temperature T_c, indicating that crystals with melting temperature T_{sl} grow at the crystallization temperature but do not form at room temperature after cooling. Another interesting feature is that T_{sl} is independent of the 3HH content, it depends only on the crystallization temperature. All the T_{sl} data can be fitted by the same straight line which is parallel to the equilibrium line $T_m = T_c$. This kind of melting peak, which is just above the crystallization temperature has been reported as "annealing peak" in other polymers and widely studied[24-27]. Although several suggestions were proposed, debate still exists and direct evidences were seldom found to support their conclusions. In the case of random P(3HB-*co*-3HH) copolyesters, the melting peak T_{sl} is considered to be related to the length of crystallizable P(3HB) sequence in copolyesters.

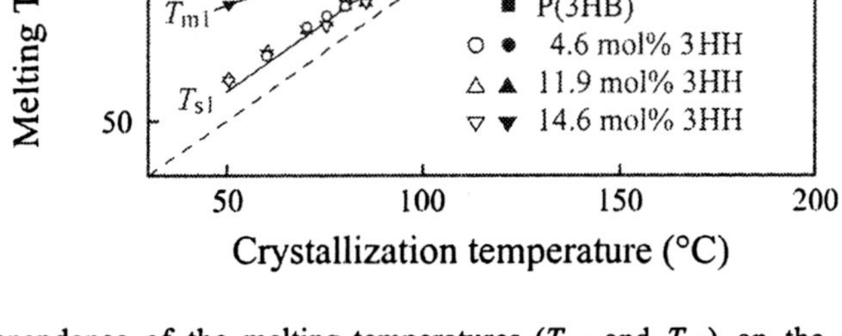

Figure 4. Dependence of the melting temperatures (T_{ml} and T_{sl}) on the crystallization temperature of P(3HB-*co*-3HH) copolyesters with different 3HH contents. The dash line is the equilibrium line of $T_m = T_c$.

As stated by Flory's equilibrium model[28] for a random A-B type copolymer where the non-crystallizable B is totally excluded from the lattice of crystallizable polymer A, at any given crystallization temperature T_c there is a critical length $\zeta(T_c)$. Only the fraction of A sequences with length longer than $\zeta(T_c)$ can participate to the crystallization process, while A sequences with length shorter than $\zeta(T_c)$ will be excluded from the lamellar crystal regions.

Application of the Flory's model to P(3HB-*c o*-3HH) random copolyesters, which are composed of crystallizable 3HB units and non-crystallizable 3HH units, suggests the existence of a critical length $\zeta(T_c)$. Only 3HB sequences with length longer than $\zeta(T_c)$ are able to crystallize to form the regular crystals with melting temperature T_{ml}, while 3HB sequences with length shorter than $\zeta(T_c)$ and some 3HB sequences with length longer than $\zeta(T_c)$ are excluded from the T_{ml} crystals. The excluded sequences may be long enough to form locally ordered imperfect crystals with melting temperature T_{sl}. Both crystals with melting temperatures T_{sl} and T_{ml} grow at the crystallization temperature T_c but not at room temperature or during the cooling process. The crystals formed at room temperature have a melting temperature T_{s2} always lower than the crystallization temperature. Such crystals with melting peak T_{s2} have been previously revealed by DSC and atomic force microscopy for poly[(R)-3-hydroxybutyric acid-*co*-6-hydroxy hexanoic acid] copolyesters[29].

Based on the above discussion, it can be concluded that $\zeta(T_c)$ may be regarded as a scale to judge whether 3HB sequences can participate to the crystallization and what kind of crystals (i.e. crystals with melting temperatures T_{ml} or T_{sl}) will be formed. Therefore, the change of the relative intensity of melting peaks T_{ml} and T_{sl} with temperature and composition as shown in Figure 3 is well explained by the relative amount of 3HB sequences with longer or shorter length than $\zeta(T_c)$. As the crystallization temperature T_c increases for a given copolyester, the lamellar thickness increases, and then a longer critical length $\zeta(T_c)$ is necessary for crystallization. It means that the amount of P(3HB) sequences with length larger than $\zeta(T_c)$ decreases and that a number of P(3HB) sequences shorter than $\zeta(T_c)$ are excluded from lamellar crystals. On the other hand, as the contents of 3HH units along the P(3HB) chains increase, the fraction of P(3HB) sequences longer than the critical length $\zeta(T_c)$ at a given crystallization temperature also decreases. As a result, the number of P(3HB) sequences taking part in the crystallization process to form lamellar crystals with melting temperature T_{ml} decreases. This is the reason why the relative magnitude of melting peaks T_{sl} increases at high crystallization temperatures and at higher 3HH contents, in comparison with that of melting peaks T_{ml}, as shown in Figure 3.

It is clear from Figure 4 that T_{m1} depends on both the crystallization temperature and copolyester composition; while T_{s1} only depends on the crystallization temperature. However, the crystallization mechanism of T_{s1} crystals is still unclear. Investigations on location and size of T_{s1} crystals and their relations to T_{m1} crystals are thought to be difficult. However, some helpful information can be obtained by annealing treatment on the melt-crystallized P(3HB-*co*-3HH) samples. Figure 5 shows the melting behaviour of melt-crystallized P(3HB-*co*-3HH) copolyesters containing 14.6 mol% before and after annealing treatment. After 10 min annealing, only the clear and sharp melting peaks T_{m1} emerge, T_{s1} peaks disappear and the melting peaks T_{m2} caused by reorganization are hardly detectable. Unexpectedly, the positions and the intensities of melting peaks T_{m1} remain almost unchanged before and after annealing. Similar results were also observed for the P(3HB-*co*-3HH) copolyester containing 11.9 mol% (data not shown). This result implies that the annealing treatment only removes crystals which are located outside the T_{m1} lamellar crystal region, and that the removed crystals correspond to the melting peaks T_{s1}. Therefore, the melting peaks T_{s1} may arise from small size crystalline domains located between or outside the regular crystals with melting temperature T_{m1}.

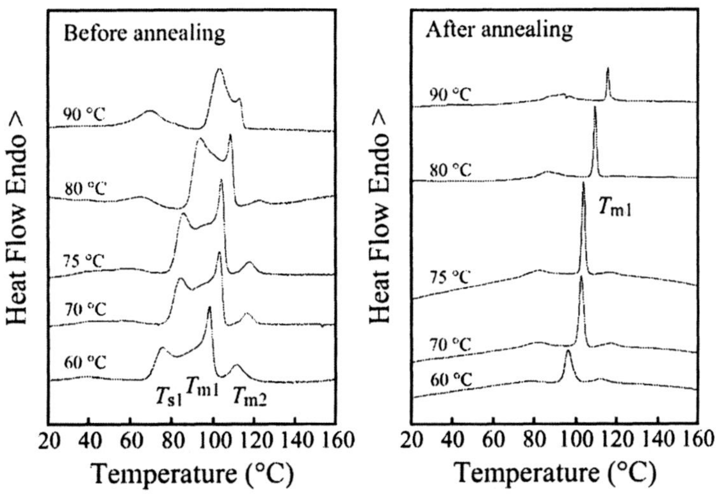

Figure 5. Influence of the annealing treatment on melting curves of the P(3HB-*co*-3HH) copolyester containing 14.6 mol% HH units after isothermal melt-crystallization for 3 day at different temperatures. Annealing temperature: $T_a = (T_{s1}+T_{m1})/2$; time: 10 min; heating rate: 20 °C/min.

Another feature of Figure 5 is the disappearance of melting peaks T_{m2} of recrystallized crystals after the annealing treatment. It has been known that annealing treatment only removes T_{s1} crystals, but it has no influence on T_{m1}

crystals. Therefore, Figure 5 seems to indicate that recrystallization of P(3HB-*co*-3HH) copolyesters has a close relation with T_{s1} crystals, which have poor crystal structures. As it can be seen in Figure 3, P(3HB) homopolymer only shows a single and sharp melting peak, indicating that it is hard to recrystallize during the heating process due to the higher crystallinity (>60%). On the other hand, the introduction of 3HH units along the backbone of P(3HB-*co*-3HH) copolyesters reduces the crystallinity and causes the complicated crystallization behaviour of 3HB sequences, such as the formation of crystals with T_{m1} and T_{s1} melting temperatures. During the heating cycle of DSC measurements, the crystals recrystallized due to their poor crystal structure and the corresponding melting peaks can be observed in Figure 3 for P(3HB-*co*-3HH) copolyesters. However the recrystallization of P(3HB-*co*-3HH) copolyesters is more complicated; not only it is affected by the composition, but also by the crystallization temperature. As shown in Figures 3, the melting peaks of crystals recrystallized at higher temperature are hard to be observed although the relative magnitude of melting peaks T_{s1} is larger than that of melting peaks T_{m1}. Therefore a complete understanding of the crystallization behaviour of P(3HB-*co*-3HH) copolyesters needs further investigation.

3.3 Crystal Growth Kinetics and Morphology

Figure 6 shows the dependence of crystal growth rates G of P(3HB-*co*-3HH) samples on the crystallization temperature T_c. It can be observed that the crystal growth rate decreases more than two orders of magnitude as the 3HH content increases from 0 to 11.9 mol%. No data was shown for P(3HB-*co*-3HH) sample containing 14.6 mol% 3HH units due to the very low growth rate and to the difficulty of measuring it accurately. Data reported in Figure 6 suggest that the presence of 3HH units along the P(3HB) backbone reduces the concentration of crystallizable 3HB sequences and subsequently hinders them from depositing and arranging in front of the growing crystals. Another feature can be observed in Figure 6 is the shift of the maximum crystal growth rate to the lower temperature region as the 3HH content increases from 0 to 11.9 mol%. This is because of the decrease of the melting temperature of P(3HB-*co*-3HH) copolyesters.

The crystal growth kinetics of P(3HB-*co*-3HH) copolyesters was analysed by using the secondary nucleation theory[30], which is written as:

$$G = G_0 \exp[-\frac{U^*}{R(T_c - T_\infty)}] \exp[-\frac{K_g}{T_c(\Delta T) f}] \qquad (1)$$

where G_0 is a constant, U^* is the activation energy for transporting polymer chains to the crystallization site, R is the gas constant, T_∞ is the temperature below which the polymer chain movement ceases, ΔT is the supercooling given as $T_m^0 - T_c$ where T_m^0 is the equilibrium melting temperature, f is a factor accounting for the variation in fusion enthalpy ΔH_f, given as $2T_c/(T_m^0+T_c)$, and K_g is a nucleation constant whose value is:

$$K_g = \frac{nb_0 \sigma \sigma_e T_m^0}{\Delta h_f k} \quad (2)$$

where σ and σ_e are the lateral and surface free energies, respectively, b_0 is the molecular thickness, and the n values are 4, 2 and 4 depending on the regime of crystallization.

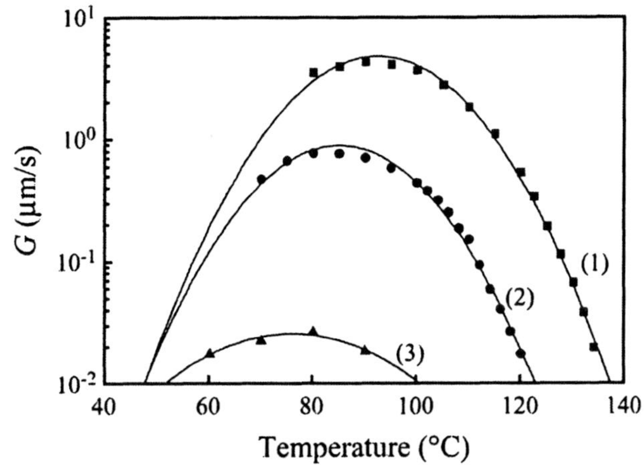

Figure 6. Dependence of the crystal growth rate G on the crystallization temperature T_c of P(3HB-*co*-3HH) samples containing 0 mol% (1), 4.6 mol% (2), and 11.9 mol% (3) of 3HH units.

In this study the universal empirical values of $U^* = 1500$ cal/mol and $T_\infty = T_g - 30$K[30] was used for the kinetic analysis of P(3HB-*co*-3HH) samples. In addition, the value of 194 °C, determined by the Gibbs-Thomson equation[31] has been used for the equilibrium melting temperature T_m^0 of P(3HB-*co*-3HH) copolyesters with different compositions in consideration of the same crystal structure of copolyesters.

Figure 7 represents the relations between $\mathrm{Ln}G + U^*/R(T_c - T_\infty)$ and $1/T_c(\Delta T)f$ for P(3HB) homopolymer and P(3HB-*co*-3HH) copolyester containing 4.6 mol% of 3HH units. It is found that the experimental data for each sample are fitted well by two discontinuous straight lines, indicating that there are two regimes in the crystallization temperature regions, i.e. regime III in a relatively lower temperature region, and regime II in a relatively higher temperature region. The intersection point of two discontinuous lines is the transition between the two regimes. The transition temperatures of $T_{III\rightarrow II}$ are around 124 °C and 104 °C for the homopolymer and the copolymer sample, respectively. Similarly to the maximum crystal growth rate G (Figure 6), the transition temperature $T_{III\rightarrow II}$ also shifts to the lower temperature region with increasing the 3HH contents for P(3HB-*co*-3HH) copolyesters.

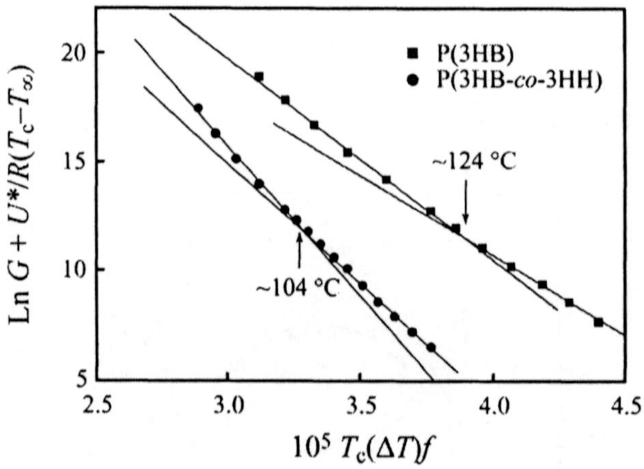

Figure 7. Plots of $\text{Ln}G+U^*/R(T_c-T_\infty)$ *vs*. $1/T_c(\Delta T)f$ for P(3HB) and P(3HB-co-3HH) containing 4.6 mol% of 3HH units according to the secondary nucleation theory.

Figure 8 shows the lamellar morphology of P(3HB) and P(3HB-*co*-3HH) samples after melt-crystallization at different temperatures. In the case of P(3HB), at 130 °C (Fig. 8b) corresponding to the regime II, lamellar crystals grow in a relatively regular shape with ~500 nm width, while only fibril-like or featureless crystals are observed at 90 °C (Fig. 8a) corresponding to the regime III. Similar results can be observed for the copolymer sample. The morphological results in Figure 8 provide a direct and visible evidence of the existence of two regimes, as indicated by kinetic analysis. According to the secondary nucleation theory, the relative magnitudes of nucleation rate i and of the lateral spreading rate g on the front of a growing crystal are different under different regimes. In regime II, the crystal growth rates are slow and below 0.3 μm/sec for the two samples, as seen in Figure 6; therefore, polymer chains have enough time to pack well into crystals. In the case of regime II, nucleation and lateral spread have similar rates, so the growing crystals may keep the growth habit of single crystals. Therefore, lath-like lamellar crystals for two samples can grow in regime II (Fig. 8b and 8d), with shape similar to that of P(3HB) single crystals. On the contrary, in regime III, the much fast nucleation rate *i* compared to the lateral spreading rate *g* brings about the growth of crystals with many defects on the surface or lateral sides. These defects may partially melt and then rapidly recrystallize to form the crystals with higher perfection. It is supported by the melting behaviour of the P(3HB-*co*-3HH) sample containing 4.6 mol% 3HH units that has the regime transition temperature $T_{III \to II}$ around 104 °C. As shown in Figure 3, the melting peaks corresponding to the recrystallization process are clearly observed in regime III at crystallization temperatures up to 100 °C, whereas they disappear in regime II at crystallization temperatures above 100 °C. These results suggest that the recrystallization process is related to the crystal growth mechanism.

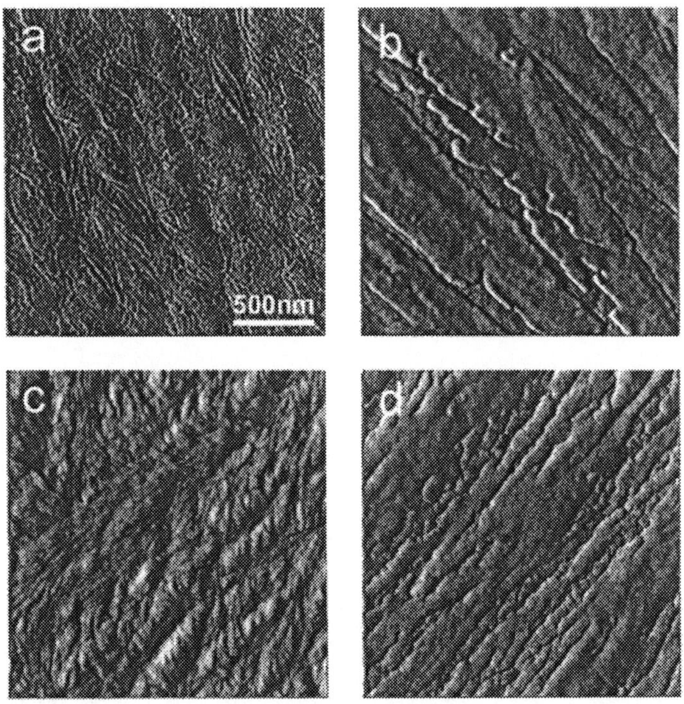

Figure 8. AFM deflection images of P(3HB-co-3HH) thin films after melt-crystallization at different temperatures. P(3HB) at 90 °C (a) and 130 °C (b), P(3HB-*co*-3HH) containing 4.6 mol% 3HH units at 60 °C (c) and 110 °C (d).

4. CONCLUSIONS

Both WAXD and solid-state ^{13}C NMR results confirm that the randomly distributed 3HH units in P(3HB-*co*-3HB) copolyesters are excluded from the

P(3HB) crystalline phase, thus reducing the crystallinity and the crystallization rate of copolyesters. The presence of 3HH units also results in the distribution of 3HB sequences with different length that is responsible of the complicated melting behaviour with multiple peaks. The existence of two types of crystals in P(3HB-*co*-3HH) has been suggested on the basis of the melting behaviour. The crystals with melting peak T_{m1} are composed of 3HB sequences with a length longer than the critical length $\zeta(T_c)$; the growth of these crystals is affected by both temperature and chemical composition. On the other hand, the crystals with melting temperature T_{s1} are thought to be composed of shorter 3HB sequences and their crystallization is not affected by the composition. The results of the annealing treatment seem to indicate that T_{s1} crystals are located outside the T_{m1} crystals.

ACKNOWLEDGEMENTS

This research was supported by the SORST (Solution Oriented Research for Science and Technology) grant from Japan Science and Technology Corporation (JST).

REFERENCES

1. Doi, Y., 1990, *Microbial Polyesters*, VCH Publishers.
2. Sudesh, K., Abe, H., and Doi, Y., 2000, Synthesis, structure and properties of polyhydroxyalkanoates: biological polyesters. *Prog. Polym. Sci.* **25**: 1503-1555.
3. Anderson, A.J., and Dawes, E.A., 1990, Occurrence, metabolism, metabolic role, and industrial uses of bacterial polyhydroxyalkanoates. *Microbiol. Rev.* **54**: 450-472.
4. Muller, H.M., and Seebach, D., 1993, Poly(hydroxy fatty acid esters): a fifth class of physiologically significant organic biopolymers? *Angew. Chem., Inter. Ed. Engl.* **32**: 477-502.
5. Steinbuchel, A., and Valentin, H.E., 1995, Diversity of bacterial polyhydroxyalkanoic acids. *FEMS Microbiol. Lett.* **128**: 219-228.
6. Holmes, P.A., 1988, Biologically produced (R)-3-hydroxyalkanoate polymers and copolymers. In *Developments in Crystalline Polymers-2* (D.C. Bassett, ed.), Elsevier, Amsterdam, pp 1-65.
7. Organ, S.J., and Barham, P.J., 1991, Nucleation, growth and morphology of poly(hydroxybutyrate) and its copolymers. *J. Mater. Sci.* **26**: 1368-1374.
8. Owen, A.J., Heinzel, J., Skrbic, Z., and Divjakovic, V., 1992, Crystallization and melting behavior of poly(hydroxybutyrate) (PHB) and PHB/hydroxyvalerate copolymer. *Polymer* **33**: 1563-1567.
9. Rule, R.J., and Liggat, J., 1995, Time-resolved synchrotron small angle x-ray scattering studies of poly(3-hydroxybutyrate) and poly(3-hydroxybutyrate-co-3-hydroxyvalerate) polymers. *Polymer* **36**: 3831-3840.

10. Spyros, A., and Marchessault, R.H., 1996, Segmental dynamics in poly(3-hydroxybutyrate-co-4-hydroxybutyrate)s above the glass transition temperature: ^{13}C nuclear magnetic relaxation in the amorphous phase. *Macromolecules* **29**: 2479-2486.
11. VanderHart, D.L., Orts, W.L., and Marchessault, R.H., 1995, ^{13}C NMR determination of the degree of cocrystallization in random copolymers of poly(β-hydroxybutyrate-co-β-hydroxyvalerate). *Macromolecules* **28**: 6394-6400.
12. Yoshie, N., Sakurai, M., Inoue, Y., and Chujo, R., 1992, Cocrystallization of isothermally crystallized poly(3-hydroxybutyrate-co-3-hydroxyvalerate). *Macromolecules* **25**: 2046-2048.
13. Mitomo, H., Barham, P.J., and Keller, A., 1987, Crystallization and morphology of poly(β-hydroxybutyrate) and its copolymer. *Polym. J.* **19**: 1241-1253.
14. Nobes, G.A.R., Marchessault, R.H., Briese, B.H., and Jendrossek, D., 1998, Microscopic visualization of the enzymatic degradation of poly(3HB-co-3HV) and poly(3HV) single crystals by PHA depolymerases from Pseudomonas lemoignei. *J. Environ. Polym. Degrad.* **6**: 99-107.
15. Orts, W.J., Marchessault, R.H., and Bluhm, T.L., 1991, Thermodynamics of the melting point depression in poly(β-hydroxybutyrate-co-β-hydroxyvalerate) copolymers. *Macromolecules* **24**: 6435-6438.
16. Bloembergen, S., Holden, D.A., Hamer, G.K., Bluhm, T.L., and Marchessault, R.H., 1986, Studies of composition and crystallinity of bacterial poly(β-hydroxybutyrate-co-β-hydroxyvalerate). *Macromolecules* **19**: 2865-2871.
17. Kamiya, N., Sakurai, M., Inoue, Y., Chujo, R., and Doi, Y., 1991, Study of cocrystallization of poly(3-hydroxybutyrate-co-3-hydroxyvalerate) by solid-state high-resolution carbon-13 NMR spectroscopy and differential scanning calorimetry. *Macromolecules* **24**: 2178-2182.
18. Kobayashi, G., Shiotani, T., Shima, Y., and Doi, Y., 1994, Biosynthesis and characterization of poly(3-hydroxybutyrate-co-3-hydroxyhexanoate) from oils and fats by *Aeromonas* sp. OL-338 and *Aeromonas* sp. FA-440. In *Biodegradable Plastics and Polymers* (Y. Doi, K. Fukuda, Eds.), Elsevier, Amsterdam, pp. 410-416.
19. Shimamura, E., Kasuya, K., Kobayashi, G., Shiotani, T., Shima, Y., and Doi, Y., 1994, Physical properties and biodegradability of microbial Poly(3-hydroxybutyrate-co-3-hydroxyhexanoate). *Macromolecules* **27**: 878-880.
20. Doi, Y., Kitamura, S., and Abe, H., 1995, Microbial synthesis and characterization of poly(3-hydroxybutyrate-co-3-hydroxyhexanoate). *Macromolecules* **28**: 4822-4828.
21. Vonk, C.G., 1973, Computerization of Ruland's x-ray method for determination of the crystallinity in polymers. *J. Appl. Crystallogr.* **6**:148-152.
22. Torchia, D.A., 1978, The measurement of proton-enhanced carbon-13 T1 values by a method which suppresses artefacts. *J. Magn. Reson.* **30**: 613-616.
23. Gan, Z.H., Abe, H., and Doi, Y., 2000, Biodegradable poly(ethylene succinate) (PES). 2. Crystal morphology of melt-crystallized ultrathin film and its change after enzymatic degradation. *Biomacromolecules* **1**: 713-720.
24. Liu, T., Yan, S., Bonnet, M., Lieberwirth, I., Rogausch, K.D., and Petermann, J., 2000, DSC and TEM investigations on multiple melting phenomena in isotactic polystyrene. *J. Mater. Sci.* **35**: 5047-5055. See the references therein.
25. Liu, T., and Petremann, J., 2001, Multiple melting behavior in isothermally cold-crystallized isotactic polystyrene. *Polymer* **42**: 6453-6461.
26. Alizadeh, A., Richardson, L., Xu, J., McCartney, S., Marand, H., Cheung, Y.W., and Chum, S., 1999, Influence of structural and topological constraints on the crystallization and melting behavior of polymers. 1. Ethylene/1-octene copolymers. *Macromolecules* **32**: 6221-6235.

27. Verma, R.K., and Hsiao, B.S., 1996, Some new insights into the crystallization and melting mechanism in semicrystalline semistiff polymers. *Trends Polym. Sci.* **4**: 312-319.
28. Flory, P.J., 1955, Theory of crystallization in copolymers. *Trans. Faraday Soc.* **51**: 848-857.
29. Abe, H., Kikkawa, Y., Aoki, H., Akehata, T., Iwata, T., and Doi, Y., 1999, Crystallization behavior and thermal properties of melt-crystallized poly[(R)-3-hydroxybutyric acid-co-6-hydroxyhexanoic acid] films. *Inter. J. Biol. Macromol.* **25**: 177-183.
30. Hoffman, J.D., Davis, G.T., and Lauritzen, J.I., Jr, 1976, The rate of crystallization of linear polymers with chain folding. In *Treatise on Solid State Chemistry, Vol.3: Crystalline and Noncrystalline Solids* (N.B. Hannay, Ed.), Plenum Press, New York, Chapter 7, 497-614.
31. Abe, H., Doi, Y., Aoki, H., and Akehata, T., 1998, Solid-state structures and enzymatic degradabilities for melt-crystallized films of copolymers of (R)-3-hydroxybutyric acid with different hydroxyalkanoic acids. *Macromolecules* **31**: 1791-1797.

Biobased Polymeric Materials for Agriculture Applications

Biobased Polymeric Materials for Agriculture Applications

EMO CHIELLINI, FEDERICA CHIELLINI, PATRIZIA CINELLI, and VASSILKA IVANOVA ILIEVA

Department of Chemistry and Industrial Chemistry, University of Pisa, Via Risorgimento 35, 56126 Pisa, Italy.

1. INTRODUCTION

Owing to their low production cost, good physical properties and lightweight, plastic objects have slowly substituted glass, paper and metals in several fields of application including agriculture. Polymeric materials started to be applied in agricultural practices from the sixty's, mostly in replacing glass as greenhouses and tunnels covering. Thus plastics made possible the introduction of mulching films, a novel agriculture technology not applied before the production of plastic films. The used polymeric materials consisted mainly of polyethylene and poly(vinyl chloride)[1].

Although plastics in agriculture comprise less than 2% of total plastic usage in Europe and about 4% in the USA, larger amounts are used in Mediterranean countries (Spain 8%, Israel 12%) and in China (20%) where agriculture is much more intensive[2]. Particularly the new technology based on polyethylene had a strong impact on the growth of soft fruits and vegetables[3,4].

In most of agricultural practices implying applications of polymeric materials, recovery of plastics is not economically feasible, viable, controllable or attractive, and the plastics remain often as litter. The characteristics of strength and durability designed into plastics in order to meet end-use requirements, coupled with the problems associated with their post-consume disposal, played a significant role in offering new alternative options to plastics recycle and landfilling practice. Particular efforts were devoted to options based on the recovery of plastic waste as raw material or energy source and whenever applicable (biodegradable materials) via a biorecyclying[5].

Several technologies have been recognized as viable options along with environmental degradation, such as recovery of plastic materials, including mechanical recycling, recycling to monomeric components followed by polymerization to the same or new polymeric materials, recycling to hydrocarbon feedstocks by pyrolysis, direct incineration and composting. Each of these disposal technologies is holding a role in plastic waste management[6]. The choice among them depends on several factors, including available infrastructure, processing facilities, collection and reclamation of waste material, cost of new polymers, property requirements, and specific service responses.

Utilization of plastics in agriculture in the form of mulch films, greenhouse components, irrigation tubes and general-purpose containers continue to generate plastic waste in large quantities. Currently, any systematic collection of plastic waste for recycling and/or disposal is rather expensive and is limited only to certain communities. Particularly, when plastics are contaminated with soil, foods or chemicals their recycling is rather difficult[7].

In the European Community, directives have been so far issued on wastes[8], on dangerous wastes[9], on waste from packaging applications[10] in which waste has been classified depending on its origin and its potential risk for human beings and word-life. This new classification has introduced a new category of "special wastes". Waste deriving from agricultural and agrochemical activities has been placed in this category with the need of post-consume reclamation and further treatment in controlled infrastructures, thus leading to substantial increase of disposal costs, that in some cases may be even higher than the cost of the virgin material itself.

In the last decade, research was aimed at developing polymeric materials for applications in which they offer unique advantages over the competitive alternatives. In that respect, interest for biodegradable plastics used in agriculture has grown, as costs may be reduced by using photo-thermal degradable or biodegradable polymers, thus avoiding the intense labour demanding and costly steps required by waste collection and sorting practice for making the recovery of the free energy content of the plastic waste technologically and eventually economically feasible.

With this point in mind, degradable polymers based on polyethylene started to be investigated since the 1970s[11-20] and brought to the development of degradable materials (Table 1) as recently reviewed by Scott and Wiles[21].

The biodegradability and utilization of agricultural polymeric materials is a topic rising in importance in the last years[22]. Industries have started to develop several products based on environmentally degradable polymeric materials to be applied in agricultural practices such as mulching films, green houses sheets, laminates, containers, seedling pots, and for applications such as soil structurization and controlled release systems of chemicals such as fertilizers, herbicides, growth stimulants and pesticides.

Table 1. Degradable materials based on polyolefins[21]

Category	Polymeric Material	Commercial name
Photolytic Polymers	Poly(ethylene-*co*-carbon monoxide)	E-CO,
	Ethylene-vinyl ketone copolymers	Ecolyte,
Oxodegradable polymers	Polyethylene, prooxidants	Plastor , TDPA, (EPI)
	Polyethylene/Starch Blends	PE-Starch Coloroll St Lawrence Starch

The market of biodegradable polymers is at this moment focusing on products in which biodegradability provides beneficial effects (e.g. waste-disposal, recycling) and a number of biodegradable materials is already marketed or close to market introduction and customer acceptance, as reported in Tables 2 and 3[23].

Table 2. Companies involved in the manufacturing of agricultural products based on biodegradable polymers in Europe[23]

Company	Trademark	Application	Materials
Bayer	BAK 1095®	Films and sheets	Polyester amides
Biotec	Bioplast®	Films and sheets	TPStarch [a]
	Bioflex®	Plant pots	Cellulose acetate
	Biopur®	Trays for cultivating plants	PCL-Starch
BSL [b]	Sconacell	Films, flower pots	Starch acetate
Idroplast	AgriBag®	Chemicals distribution	PVA [c]
	SoilBag®	Plant root wraps	PVA [c]
Novamont	MaterBi™	Mulch films, nursey pots	PCL-Starch
Solvay Sa	CAPA®,600	CR of fertilizer	PCL

[a] TPStarch = Thermoplastic Starch, [b] BSL = Buna Sow Leuna Olenfinverbund, [c] PVA = Poly(vinyl alcohol).

Much effort has been focused in recent years to develop eco-compatible polymeric materials by starting from renewable resources as an alternative to petroleum-based synthetic polymers[24]. Indeed, the current utilization of natural resources cannot be sustained forever. Most of the fuel utilized in our societies comes from fossil fuel, such as oil that, other than being subjected to price fluctuations, eventually will be depleted[25]. Moreover, the rising atmospheric carbon dioxide levels from combustion of fossil fuels are thought to be increasing global temperature that, in turn, may cause droughts, crop losses, and storm damage[26].

Table 3. Companies involved in the manufacturing of agricultural products based on biodegradable polymers in USA, Canada and Asia[23]

Company [a]	Tradename	Application	Composition [b]
Bioplastic, Inc.	Envar™	Mulch films	PCL –starch
Cargill Dow Chemicals	EcoPLA	Mulch films, green houses pots	PLA
Chronopol Inc.	Heplon™	Plant growth stimulant, films	PLA
DuPont	Biomax®	Mulch containers, plant pots	Polyester Resins
Eastman	Eastar Bio	Seed mats, root covers	Copolyesters of diacids and glycols
Kemira Agro Oy	Ns [c]	Controlled release	ns
Marshall Plastic Film, Inc	ns	Agricultural films	Degra-Novon[c]
Metabolix, Inc.	ns	Lawn and leaf bags, films	PHA
Monsanto	Biopol®	Mulch films, plant pots	PHA
NATCO	E-Z Turf™	Biodegradable seeded grass	Wheat-starch
	Sta-Wet™	Mats for instant lawns	super absorbent
Nova Chem Ltd.	Nutri Save®	Coating for fruit and vegetables	Carboxymethyl derivatives of chitosan
	Phero Release®	Delivery of pheromones for pest control	Carboxymethyl derivatives of chitosan
Novon	Degra-Novon®	Mulch films	Modified starch, other chemicals
	Aqua-Novon®	Sanitary applications	Modified PVA
	Poly-Novon®	Mulch films	Starch
Poly Expert	ns	Mulch films	Novon starch based polymer
PPT, Inc.	EnviroPlastic®	Mulch films	PCL alloys
	EnviroPlastic®	Controlled release of fertilizers	Urethane acrylics, styrene, vegetable oils
TPS, Inc.	Vinex™	Packaging for chemicals	PVA
Union Carbide Corp	TONE®	Mulch films, pots, controlled release	PCL

[a] NATCO = Natural Absorbent Technology Co., PPT, Inc. = Planet Polymer Technologies, Inc.; TPS, Inc. = Texas Polymer Services, Inc. [b] PCL = polycaprolactone, PHA = polyhydroxyalkanoates, PLA = poly(lactic acid), PVA = poly(vinyl alcohol), ns = not specified.

The Kyoto Protocol that has been signed by the major industrialized countries and emerging countries is paying specific attention to the reduction of green house gases (GHG)[27]. Fuel shortage and the waste accumulation in the environment are generating a worldwide interest in alternative resources and particularly for the use of renewable resources both as an energy source[28] and as raw materials for the production of polymers and plastics[29]. There is increasing pressure for a wider utilization of biomass feed-stocks for specialty items. The total biomass produced on earth is estimated as approximately 170 billion tons, of which a very small portion, less than 4 %, is used[30].

Table 4. Fossil and renewable resources[30]

Biomass	Consumption [a]	Reserves [b]
Renewable biomass [c]	6.0	170
Mineral oil	3.2	135
Natural gas	1,900 [d]	140,000 [d]
Coal	3.4	850

[a] Billions metric tons/year if not otherwise stated. [b] Billions metric tons if not otherwise stated
[c] By photosynthetic process [d] Cubic meters.

Concerns about climate change and the preservation of natural resources represents a worldwide driving force to reduce the consumption of fossil fuel feedstock that is currently stimulating academic and industrial researchers as well as decision makers[31-32]. In industrial production sustainability must be achieved, but keeping in mind that business will fail unless a minimum margin profit is guaranteed. A normative strategy has been proposed for resource choice and recycling to meet the criteria of sustainability, and the "green" chemistry concepts and principles have been drawn as suitable effective tools[33,34].

Thus biodegradability is an advantageous property in those cases where it is implicitly demanded by the application; that is where recycling is inhibited by a fairly high management cost for disposal, or may be technically impossible. Moreover the availability of raw materials from renewable resources should not interfere with food production, whose demand is expected to be almost doubled in the next 40-50 years a front of a 50% population increase[21,28]. In the meantime, the request for goods will almost quadruplicate the present demand within the same time frame[25].

The use of environmentally degradable polymeric materials based on agricultural wastes, crop surplus, and by-products allows for cheap materials from renewable resources to be used either as raw components or after modification. The worldwide increased agricultural productivity has promoted the utilization of agriculture products in the design and formulation of environmentally acceptable plastic items.

The usefulness of a variety of agricultural co-products as cheap source of polymers from renewable resources, for the production of commodity plastics substitutes, is under consideration of researchers in academies and industries[35].

The term “polymers from renewable resources” refers to natural products that are intrinsically polymeric or can be converted to polymeric materials by conventional or enzymatic synthetic procedures[36].

Natural polymers, or biopolymers, are produced in nature by living organisms, and by plants through biosynthetic processes that involve carbon dioxide consumption[37]. Natural polymers are ultimately degraded and consumed in nature in a continuous recycling of resources. Arguments in favour of “natural” polymers are: biodegradability, renewability, recyclability, non-waste producing, neutrality on green house effect, and functionality. However, in some cases natural polymers such as rubber, lignin and humus present a slow rate of biodegradation that will not satisfy the rapid mineralization criteria currently advocated by standards committees for synthetic polymers[22]. Since they are produced in nature there is no major concern about it.

Polymers based on renewable resources from the agriculture feedstock include among the others: polysaccharides, such as cellulose, starch, lignin and vegetable proteins, natural rubbers, and microbial polyesters, such as polyhydroxyalkanoates (PHAs).

Polymers derived from renewable resources can be broadly classified according to the method of production. A first category encompasses polymers directly extracted/removed from plants such as carbohydrates, aromatic plant products, polyisoprenes, and proteins.

Thus under that heading one can include *natural polymers,* used as direct feedstock for plastic production, as well as *artificial polymers* as those obtained by chemical modification of preformed natural polymers or by polymerization of monomers deriving from renewables[38], such as cellulose esters and ether or polylactate from starch.

The last category includes polymers produced by native microorganisms or genetically transformed bacteria. The best-known example of this category is constituted by polyhydroxyalkanoates (PHAs), mainly polyhydroxybutyrate (PHB) and copolymers of hydroxybutyrate and hydroxyvalerate (PHBV).

2. POLYMERS PRODUCTION FROM AGRICULTURAL FEEDSTOCKS

Carbohydrates such as starch and cellulose find application in agriculture practices either as raw materials, in the form of fillers and components in composite films and sheets, or after suitable chemical modification. Starch is the major form of carbohydrate storage in green plants. It is the main component of most seeds, tubers, and roots and it is produced commercially from corn, wheat, rice, tapioca, potato, sago, cassava, and other sources. Starch price ranges in 0.35-0.50 €/kg[39]. Most commercial starch is produced from corn that is comparatively cheap and abundant throughout the world. Wheat, tapioca, and potato starch are produced on smaller scale and at higher prices. Starch can be easily converted to glucose from which a variety of cyclic and acyclic polyols, aldehydes, ketones, acids, esters, and ether can be obtained[40]. Starch has been reacted with synthetic polymer by graft polymerization. In this technique a free radical is initiated on the starch backbone and then allowed to react with polymerizable vinyl or acrylic monomers. Initiation can be chemically or radioactively induced. These systems have been reviewed by various authors[41-45]. Several starch graft polymers have been proposed as thickeners for aqueous systems, flocculants, clarification aids for wastewaters, retention aids in paper manufacturing, and many other uses.

Cellulose is the main component of higher plant cell walls. Cellulose for industrial uses comes from wood and scores of minor sources such as bagasse, the stalks of sugar cane after the extraction of the juice by press technology. Cellulose is a relatively cheap raw material available at a price of 0.5-1 €/kg before derivatization[42]. After industrial treatment, with and without chemical derivatization, cellulose is made into diverse products; for example cellulose acetate can be synthesized by the reaction of acetic anhydride with cotton linters or wood pulp, and cellulose esters from sugar cane bagasse have also been proposed[46]. The glycosidic bonds in cellulose are strong and this polymer is stable under a wide variety of reaction conditions. It is an insoluble, highly crystalline polymer. Industrially important chemical modifications of this polymer generally involve reaction with free hydroxyl groups in 2, 3, and 6 position[47]. These reactive sites undergo most of the characteristic reactions of alcohols; their etherification and esterification are of particular importance for cellulose. The chemical modification of cellulose from the melt or in solution facilitates its processing under conditions used for thermoplastic polymeric materials. Numerous derivatives are commercially available, such as cellulose acetate, ethyl cellulose, hydroxyethyl cellulose, and hydroxypropyl cellulose. Hydroxyalkyl cellulose and carboxymethyl celluloses have found applications as matrices for drug delivery and as soil conditioners[48]. Cellulose is soluble only in unusual and complex systems[49]. When dissolved, cellulose molecules are still fairly extended, but exist as random coils with relatively large end-to-end distances. Commercially, dissolving pulps, which have lower molecular weights, are used along with strong alkali and derivatization. Cellulose subjected to high temperature and pressure during the steam explosion process can be dissolved in strong base. For film production cellulose is dissolved in an aggressive toxic mixture of sodium hydroxide and carbon disulphide ("Xanthation") and then recast into sulphuric acid to give cellophane films. This procedure that was largely applied in the past for the production of regenerated cellulose will probably be banned in the future because of its negative environmental impact. A more recent way to impart solubility and melt processabilty to cellulose, and other hydroxy polymers, has been identified in trimethylsilylation using different silylating agents. Recently Mormann and Demeter reported a method for cellulose silylation with hexamethyldisilazane in liquid ammonia[50]. Ammonia is known to activate cellulose by intercalation into the lattice breaking up the inter- and intramolecular hydrogen bonds. In the process reported by Mormann and Demeter ammonia is the only by-products generated from hexamethyldisilazane upon conversion of hydroxy groups into trimethylsilyloxy groups, and can be removed together with the ammonia used as reaction medium. Saccharin was used as a catalyst at concentration of 0.5 mol % saccharin/mol of hydroxy groups. A ratio of trimethylsilyl groups to OH of 3.4 was found to be suitable for complete silylation. With a similar process Mormann and Spitzer[51] have reported the silylation of OH-containing polymers, such as cellulose and poly(vinyl alcohol) by reactive extrusion (Figure 1).

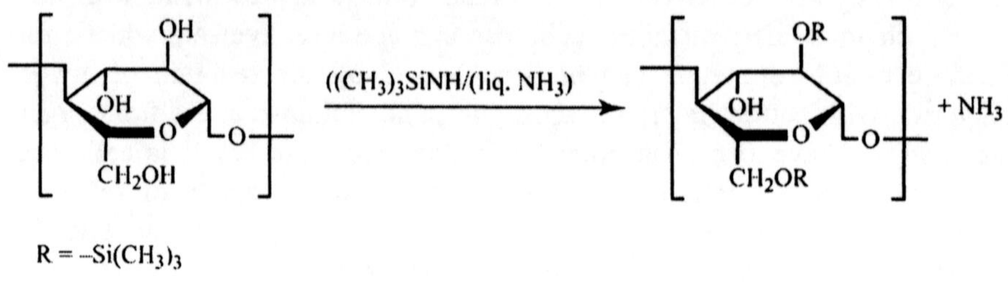

Figure 1. Cellulose silylation with hexamethyldisilazane in liquid ammonia[51].

A possible application of silylated cellulose is in the field of regeneration of cellulose after spinning or molding. Silylation of cellulose avoids the problems connected with the huge amounts of salts, wastewater, and toxic reagents like carbon disulfide.

Aliphatic polyesters, produced from agricultural feedstocks such as PLA and PHAs, have excellent mechanical properties and biodegradability and are well suited to agricultural applications but their relatively high cost represents a major drawback. Polyesters have been produced by biotechnology from refined raw materials (e.g. glucose and sodium propionate). Poly(lactic acid) is a biodegradable aliphatic polyester attainable by polycondensation of lactic acid, a monomeric precursor that can be obtained from renewable resources. In the fermentation process, sugar feedstocks, such as dextrose (glucose) are obtained either directly from sources such as sugar beets or sugar cane, or through the conversion of starch from corn, corn steep liquor, potato peels, wheat, rice and other starch source. Lactic acid in the L-form is actually produced by fermentation from nearly any renewable resource. Yields of lactic acid are greater than 90%, and in the batch production lactic acid can be produced at the rate of 2 g per litre per hour[52]. Polyhydroxyalkanoates (PHAs) can be produced from saccharides, alcohols, and low molecular weight fatty acids.

Epoxydized vegetable oils have found industrial application as cross-linkers in environmentally sound solvent-free powder coatings and can serve as building blocks for the preparation of binders based on renewable feedstock exhibiting good drying properties[53]. Warwel *et al.* applied catalytic methods of olefin chemistry to achieve polymer building blocks and polymers, such as functionalized polyolefins, polyesters, polyethers, polyamides as well as sugar-based surfactants (Figure 2)[30].

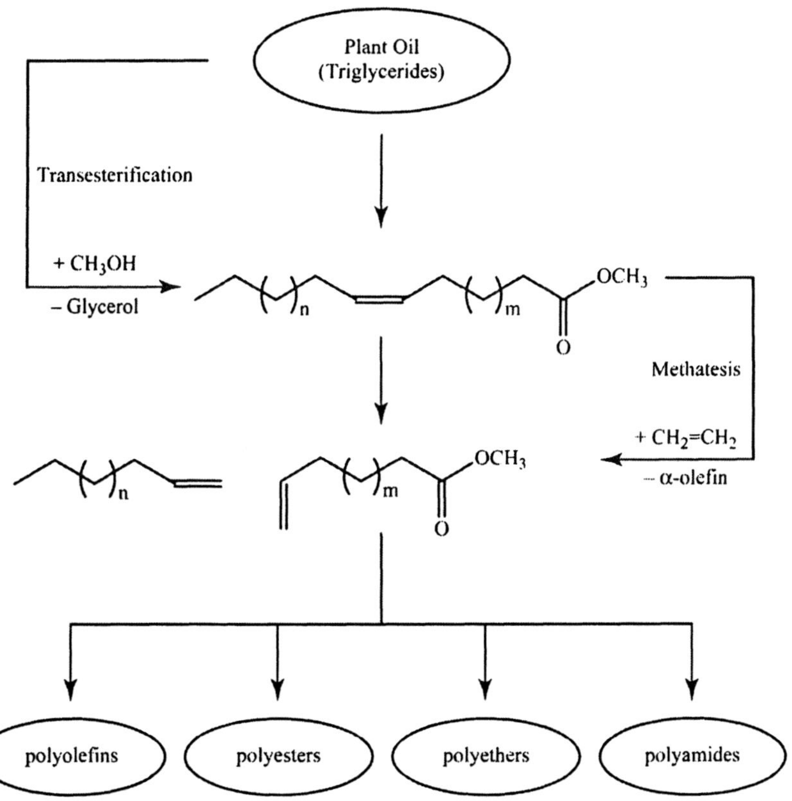

Figure 2. Schematic representation of the production of polymeric materials from plant oils[53].

The fundamental approach consisted in the polymer synthesis based on unsaturated fatty acid methyl esters, which are available by industrially applied transesterification of fats and oils with methanol. First, unsaturated fatty acid methyl esters obtained from plant oils were converted to terminally unsaturated esters and α-olefins by metathesis with ethylene using heterogeneous rhenium or homogeneous ruthenium catalysts. These esters were directly copolymerized with ethylene by an insertion-type palladium-catalyzed polymerization to functionalized polyolefins. Polyesters were synthesized by metathesis dimerization of ω-unsaturated esters and subsequent polycondensation of the produced internally unsaturated dicarboxyl esters or by acid transesterification with petrochemical diols and additional acyclic diene metathesis polymerization. ω-Epoxidized fatty acid methyl esters are obtained by a new method of chemo-enzymatic epoxidation. They are converted into polyethers displaying a comb-like structure by using alumoxanes as catalyst. The same epoxy derivatives can be converted into sugar surfactants by nucleophilic ring-opening reaction with amino carbohydrates.

Polyesters could as well be produced by a much cheaper way starting from agricultural wastes (e.g. molasses, maltose, glycerol phase from biodiesel production, whey, bagasse) as long as these materials have a known composition and are available in appropriate quantities[54,55]. New production technology are under development to reduce polyester final cost.

2.1 Genomics and Metabolic Engineering: an Alternative Route to the Production of Bio-Polymers

The increasing attention to renewable resources for the synthesis of chemicals by "green" processes is the driving force for the development of new and alternative synthetic routes arising from the knowledge of functional genomics and metabolic engineering[56].

Agricultural raw materials are a rich, renewable source of carbohydrates that could provide the basis for the production of many products for agricultural applications. One of the most efficient routes for the employment of these resources is through biotechnological methods, provided that suitable pathways are engineered. An important example of this approach is represented by the long-time efforts focused on metabolic engineering for the microbial production of 1,3-propanediol, an important intermediate in the synthesis of polyesters and polyurethanes. Recently, a number of studies are developing biological synthetic routes to 1,2-propanediol that is chemically derived from propylene[57,58].

Application of molecular genetics knowledge from a range of species may results in sufficient understanding of biosynthetic genes to allow isolation and modification of potentially valuable candidates to be transformed into higher-yielding productions systems[59]. The production of biodegradable thermoplastic PHAs is an example of the application of genomics and metabolic engineering tools. Microbial PHAs polyesters are attracting increasing attention from scientific and industrial communities because of several properties including biodegradability, biocompatibility and piezoelectricity[60]. PHA biosynthesis is a complex process controlled by several enzymes catalyzing different metabolic pathways[61]. A few bacteria have been found to produce blend of polyhydroxybutyrate (PHB) and medium-chain-length PHAs when grown on specifically selected carbon sources, such as fatty acids[62,63]. Besides the biotechnological approaches that imply the use of microbial systems, attention is being focused on the potential of plants for the production of renewable resources. The agricultural production of new biological products from carbon dioxide, and driven by sunlight, might therefore in the long term be preferable to approaches involving microorganisms and bioreactors. As an example, the copolymer poly(3-hydroxybutyrate-*co*-3-hydroxy-valerate) produced via fermentation requires high cost for energy input, thus resulting not economically competitive. In the recent years researcher focused on the production of genetically engineered plants have to produce such copolymer by introducing four bacterial genes and driving the biosynthesis from endogenous carbon flow.

The copolymer was successfully produced in transgenic *Arabidopsis* and rapeseed, but the level of expression was fairly low and in the range of 3%. The improved control of expression and regulation of the added metabolic pathway segment required for commercial production in crops is a main task that can be developed with the aid of biotechnology[64].

3. APPLICATION IN AGRICULTURE

All main classes of polymeric materials, i.e. plastics, elastomers, foams, fibres, coatings, and water-soluble polymers are presently utilized in applications including mulching films, green-houses sheets, soil conditioning, controlled release of pesticides, seed coating, plant protection, gel planting, water transport, and packaging[1]. In this overview we will focus on polymeric materials derived from agriculture feedstocks for application in mulching, hydromulching, seed coating and controlled release sectors.

The main agroindustrial applications of polymeric materials of synthetic and natural origin, deriving respectively from fossil fuel and agricultural feedstocks are collected in Table 5.

Table 5. Agricultural applications of polymeric materials from agriculture feedstocks

Application	Polymeric Material
Mulching and Sheets	Photodegradable polyethylene and starch
	Poly(ethylene-*co*-acrylic acid) and starch
	Poly(vinyl alcohol) and starch
	Poly(lactic acid) and starch
	Poly(ε-caprolactone) and starch
	Polyesters and starch
	Poly(vinyl alcohol) and lignocellulosics
	Kraft paper and polymerised vegetable oils
	Pectin and starch
Soil Amendments	Poly(amino acids)
	Polyesters
	Poly(vinyl alcohol)
	Poly(vinyl alcohol) and lignocellulosics
	Carboxymethylcellulose
	Hydrolysed starch-*g*-polyacrylonitrile
	Galactomannan (Guar bean)
	Cellulose and starch xanthates
	Cellulose microfibrills
Seed Coating	Carboxymethyl, hydroxyethyl, and hydroxymethyl cellulose
	Hydrolyzed starch-*g*-polyacrylonitrile
Controlled Release	Starch
	Cellulose
	Alginic acid
	Lignin

3.1 Mulching

In the past mulch practice has been performed by the use of natural materials such as straw and leaves to provide an insulating layer around the roots of vegetables and soft fruits. At present, the use of plastic sheets or films in mulching is the largest single application of plastics in agriculture. Mulch controls radiation, soil temperature and humidity, weed growth, insect infestation, soil compaction, and the degree of carbon dioxide retention. In some cases weed control has been reported because of solar heating plastic films mulches (solarization). Not only mulch-grown crops mature faster, but also yields are increased and in most cases the product is easier to harvest and cleaner with evident returns on the production costs[21].

The use of black mulching films with elimination of weeds and avoidance of soil compaction eliminates the need for cultivation thus preventing from root damage and plants stunting or killing. Fertilizer and water requirements are also reduced; the use of plastic mulches results in 50% saving of irrigation water and as much as 30% saving in nitrogenous fertilisers even in temperate climates[12]. Low-density polyethylene, poly(vinyl chloride), polybutylene, and copolymers of ethylene with vinyl acetate have been generally used for mulching. In 1998 the worldwide annual consumption of polyethylene mulch films alone was around half million tons[21]. The fact that plastics do not degrade as fast as the previously used natural materials may sound as an advantage because it ensures a coherent protective barrier between the roots of the plants and the environment throughout the growing life of the plant. If left in place, however, conventional plastic films can cause problems during the next year harvesting or cultivating operations. Many soft fruit crops are now harvested automatically with a procedure that lives the stems and leaves on the ground. The presence of plastic fragments mixed with the crop residue may clog the engines of harvesting machineries and makes automatic collection hard to be performed. Removal and disposal are costly and inconvenient. Attempts to promote collection systems, recycling technology and applications for the recycled material deriving from mulching films have shown a series of difficulties. Transportation of the long film strips, compaction, and washing were found to be the most critical and labour intensive steps in the process, because of film deterioration and high level (30-40% by weight) of soil contamination[7]. Moreover, many landfills reject mulch films because of pesticide residues for which they must be treated as hazardous waste. Furthermore, nowadays the thickness of mulching films can be as low as 8-10 μm that makes them too fragile to be easily and efficiently collected from the field after cropping.

Interest in the development of biodegradable or photodegradable films with controllable short service lifetimes has grown. Degradable mulches should break down to small brittle pieces, to pass through harvesting machinery without difficulty and should not interfere with subsequent planting operations. The induction time therefore must be variable, predictable, and reproducible[66]. Crop yields could be considerably reduced whether the film degrades before the end of the growing season. In addition, toxic residues are unacceptable; processing stability must not be affected by films components; storage must not modify the mechanical and physical product properties. In order to improve degradability, polymers derived from agricultural feedstock have started to be introduced in materials for mulching application, in particular starch has been used both as an additive in photodegradable films and as component or filler in blends and composites with synthetic polymers. Photodegradable films, such as poly(1-butene) have been proposed for mulching[67]. Interest in the development of new photo-degradable films has recently mounted in mainland China and Taiwan[68-70]. As reported before, in these countries the plastic consumption for agricultural practice is very high (20%)[21]. In these materials a polyolefin is blended with modified starch as coupling agent, a photodegrading agent, an oxidation accelerant, a self-oxidant, and a degradation-controlling agent. The point is that after degradation and disintegration of the films, debris of the hydrocarbon polymers tends to accumulate in the soil.

In other approaches, degradable mulching films were prepared by blending synthetic polymers with natural fillers such as starch. Films based on starch blended with: polyethylene[71,72], poly(vinyl alcohol)[73], poly(ethylene-*co*-acrylic acid)[74-76], and poly(vinyl chloride) (PVC)[77] were developed. In some of these materials only the starch component was degraded while the continuous matrix represented by the synthetic polymer accumulated in the environment. Especially for starch-polyethylene films, the fragments resulting from film deterioration may require decades to completely biodegrade. Moreover, the effect of long-term soil exposure to polyhydrocarbon debris is largely unknown[78]. For these reasons the interest focused toward blends based on starch and biodegradable synthetic polymers, such as poly(vinyl alcohol)[79-82], poly(ε-caprolactone) (Materbi)[83], poly(lactic acid) (PLA)[84-86], and other synthetic polyesters[87,88]. More recently, systems based on polyethylene added of thermal oxidant promoters (TDPA-PE) that help the polyhydrocarbon matrix to first oxidize and thermally disintegrate to fragment that have been shown to biodegrade in soil and nature compost have been introduced[89].

The advantages connected with the use of material from renewable resources as filler in blends with synthetic degradable polymers have promoted the interest for a wide series of polymers from non-food crops or over production, such as lignocellulosic materials, pectin, proteins, and oils from vegetal sources.

Pectin[90,91], sugar cane bagasse and fruit juice extraction by-products such as apple and orange wastes[92-94], and soy protein[95] have been blended with poly(vinyl alcohol).

The use of biodegradable thermoplastic polymers from renewable resources as the continuous matrix is attracting recent research activity[96-99]. Several patents dealing with blends of natural polymers with polyesters from renewable resources for agricultural films production have been filed and assigned[100-102]. Films and laminates produced with natural polymers have been also object of research activity. Thus films fabricated from chitosan and pectin[103], starch and pectin[104], soy protein and starch[105,106] have been claimed. In our laboratories hybrid composite films have been developed based on natural and synthetic degradable polymers blended with waste materials from agriculture practices. Materials such as sugar cane bagasse have been blended with natural polymers such as gelatin waste from pharmaceutical industry. The prepared composites presented interesting mechanical properties[107,108] and degradation times[109-110] for agriculture applications. Moreover, animal and vegetal protein based materials possess an intrinsic agronomic value because of their fairly high nitrogen content

(10-12%). Mulching practice based on the use of recycled materials has also attracted interest as an alternative to the use of plastic films[111]. Therefore, organic mulches such as paper, leaves, straw are sometimes used by the farmers. Kraft paper coated by polymerised vegetable oils has been recently proposed as biodegradable mulch[112,113].

3.2 Liquid Mulch and Seed Coatings

The capacity of some polymers to hold water allows for their spraying and blowing alone or in slurries with other mulching materials and nutrients for soil conditioning or seed coating. Few soils possess the optimum physical and chemical characteristics for maximum productivity without the addition of some type of amendments. Poor soil physical conditions induce inadequate aeration, restricted water filtration, unfavourable water and nutrient retention, and sometimes crust formation on the soil surface[114]. Most of these amendments are chemicals but several studies have been conducted to evaluate the influence of both mineral and organic polymers on soil physical properties[115]. Polymers may be present as tackyfiers to help holding the mulch in place once applied. In some cases, a type of thatch is formed that protects seeds and soil against erosion. Hydrophilic polymers, such as polyacrylamide (PAAm), poly(vinyl alcohol) (PVA), carboxymethyl cellulose, and hydrolysed starch-*g*-polyacrylonitrile copolymers (HSPAN) have been proposed as soil conditioners in techniques called hydro-mulching[116,117]. Thus semi-dry and liquid mulches have a large range of application, chiefly as soil structuring agents or binders for seedling on friable and inclined fields[118-121]. The stabilization of friable soils has a great importance in order to avoid land slip. Soil erosion threatens water quality and agricultural productivity because of the loss of valuable top soil and the runoff of chemicals. Improvements in soil conservation have been achieved by addition of polymeric materials to the inflowing water[122-124]. In the past polymers were applied as dry granular material, while most of the newer polymers may be applied at low concentration in irrigation water. The successful use of PAAm in irrigation water raises the interest for the use of other polymers with similar properties. However, concerns have arisen about the widespread use of PAAm in open agricultural environment since the monomer, acrylamide, is a neurotoxin[125]. Even the use of PAAm almost devoid of monomer ($< 0.05\%$) does not eliminate the concern that the monomer can occur as a degradation product by early removal of the amine group from the polymer backbone[125]. In agricultural applications of polymeric materials the interest for the final fate of applied polymers and eventually ecotoxicity of its degradation products is raising[126].

Polymers from agriculture feedstock and their derivatives have been widely investigated as an alternative to synthetic polymers, or eventually to be used in blends with synthetics. Polysaccharides, such as starch-hydrolysed polyacrylonitrile graft compounds (HSPAN) have been used as soil conditioners. This material swells in the presence of water, forming a lattice that is able to absorb water and other polar charged molecules[127]. The influence of a gel-forming conditioner containing 24.5% humic acids and 3.8% polysaccharides on water penetration of soil columns has been investigated[128]. Rates of emergence of tomato, cotton (*Gossypium hirsutum* L.), and lettuce (*lactuca sativa* L.) were increased by the use of different combinations of PAAm and polysaccharide soil conditioners[129,130]. Tomato seedling showed maximum response with a mixture of 31.2 kg/ha PAAm and 6.2 kg/ha of polysaccharide[131].

One popular form of galactomannan derives from guar bean (*Cyamposis tetragonoloba* L. Taub.)[132]. Guar products can be non-ionic, anionic, and cationic[133] and have a molecular weight of 200,000 to 2 million in comparison with PAAm, which range between 10 to 15 million[115]. Cationic polysaccharide guar derivatives have been applied with sprinkler water and better maintenance of water infiltration by derivatives having higher charge density was evidenced. Anyway, the polymer was relatively ineffective in subsequent applications with untreated sprinkler due to impact energy. Spraying concentrated polymer solutions on the soil surface was not effective in preventing crust formation following rain events except in the case of lower charge density guar polysaccharides sprayed in $CaCl_2$ solution. Cationic guar polymer resulted effective in increasing flocculation in soils with sodium saturation ratio of 1 to 15[134].

Cellulose and starch xanthates are promising alternatives to PAAm, because they have been previously applied as flocculating agents[135,136] and soil stabilizers[137-139]. Also cellulose microfibrills have been proposed as alternative to PAAm. Cellulose microfibrils are obtained during acid hydrolysis of pure cellulose, and represent the basic crystalline unit of the cellulose fiber[125]. They gain a charge on their outer surface during acid hydrolysis, allowing them to disperse rapidly in water. Because of microfibril large size, affinity to soil aggregates, affinity to soil via surface charge, and stability in aqueous suspensions they appear as suitable soil structuring agents.

As already mentioned, hydrophilic polymers can be used to form thatch that protects seeds and soil against erosion. In particular, poly(vinyl alcohol) has shown to effectively maintain soil structure[140-145]. Liquid mulch formulations have been investigated in our laboratories by premixing powdered poly(vinyl alcohol) with starch, or lignocellulosic natural fillers such as wheat flour, sugar cane bagasse, by-products of wood industries[146-149]. These formulations are applied directly to the soil, by spraying with conventional apparatus in order to confer a structuring and colouring effect to the soil.

In seed coating, usually a hydrophilic polymer is coated directly onto the seed surface. After planting, the polymer absorbs water and thereby increases the rate of germination as well as the percentage of germinated seeds. However, depending on the application, the type of polymeric coatings can be varied to delay germination, inhibit root growth, control pests, fertilize, and bind the seed to the soil. Agar, water-soluble cellulose ethers, such as carboxymethyl, hydroxyethyl, and hydroxymethyl cellulose, and hydrolyzed starch-*g*-polyacrylonitrile copolymers (HSPAN) have been studied to a great extent in seed coating. HSPAN coatings have been applied to a variety of seeds, including soybean, cotton, corn, sorghum, sugar beet, and a number of vegetables[150].

3.3 Controlled Release

The effect on the environment deriving from the use of fertilizers and pesticides is an issue of global concern. Controlled release (CR) is a method by which biologically active chemicals are made available to a target species at a specified rate and for a predetermined time. The polymer serves primarily to control the rate of delivery, mobility, and period of effectiveness of the chemical component.

The principal advantage of CR formulations is that fewer chemicals are used for a given time, thus lowering the impact on non-target species and limiting leaching, volatilisation, and degradation. The macromolecular nature of polymers is the key to limiting loss of chemicals by these processes. CR systems can be divided into two broad categories. In the first one, the active agent is dissolved, dispersed, or encapsulated by the polymeric matrix. Release generally occurs by diffusion controlled processes or by the biological or chemical breakdown of the matrix. In the second category, polymers contain the active agent as part of the macromolecular backbone or pendant side chain. Release takes place by biological or chemical cleavage of the bond between the bioactive agents and the polymer. Physical systems that incorporate agricultural chemicals include microcapsules, blends, dispersions, laminates, hollow fibres, and membranes. Kinetic models for release have been developed for each type of device[151,152]. Starch, cellulose, alginic acid, and lignin are among the natural polymers used in CR systems[153-159]. Although they possess functionality for derivatization, they have the one significant disadvantage of being insoluble in standard solvents suitable for encapsulation, dispersion, and formulation. Systems have been developed that overcome the solvent problem by *in situ* encapsulation, whereby gelatinised starch containing a chosen pesticide is crosslinked by adding calcium chloride[154], boric acid[156], or by xanthation followed by oxidation[157]. The pesticide, as a result, is entrapped within the granular particles formed.

4. CONCLUSIONS

The modern agricultural technology is ever more demanding for agrochemicals and materials and manufacts that are eco-compatible and attainable at a reasonably competitive price. In that respect, the introduction of polymeric materials from renewable resources in a large variety of agricultural applications appears to be a viable solution provided these materials may be derived from cheap raw materials and eventually from agroindustrial waste or a suitable combination of natural resources and fossil fuel.

The extremely active academical and industrial research focused on the use of biodegradable polymers for agricultural applications has induced the introduction of several products to the market. These products based on biodegradable materials are supposed to be applied as mulching films, green houses sheets, laminates, and containers in agricultural practices and for application such as soil structurization and controlled release of chemicals such as fertilizers and herbicides.

The number of biodegradable items for agricultural application to enter the market is going to increase in the early future.

REFERENCES

1. Mc. Cormick, C. L., 1984, Agricultural Applications. In *Encyclopedia of Polymer Science and Engineering* John Wiley & Sons, New York, 2nd Ed. **1:** 611-622.
2. Scott, G., 1997, Abiotic control of polymer biodegradation, *Trends Polym. Sci.* **5**: 361-368.
3. Gilead, D., 1995, Photodegradable Plastics in Agriculture. In *Degradable Polymers: Principles and Applications* (G. Scott and D. Gilead, eds), Chapman & Hall, London, pp. 186-199.
4. Fabbri, A., 1995, The Role of Degradable Polymers in Agricultural Systems. In *Degradable Polymers: Principles and Applications* (G. Scott and D. Gilead, eds.), Chapman & Hall, London, pp. 200-215.
5. Doane, W.M., 1994, Biodegradable plastics. *J. Polym. Mater.* **11**: 229-237.
6. Feil, H., 1998, Sense and Non-Sense about Polymers. *Macromol. Symp.* **127**: 7-11.
7. Llop, C., and Perez, A., 1992, Technology Available for Recycling Agricultural Mulch Film. *Makromol. Chem., Macromol. Symp.* **57**: 115-121.
8. Council Directive 91/156/EEC of 18 March 1991 amending Directive 75/442/EEC on waste, http://europa.eu.int/

9. Council Directive 91/689/EEC 75/442/EEC of 12 December 1991 on hazardous waste, http://europa.eu.int/
10. European Parliament and Council Directive 94/62/EC of 20 December 1994 on packaging and packaging waste, http://europa.eu.int/
11. Scott, G., 1999, Environmental Impact of Polymers. In *Polymers and the Environment* (G. Scott, ed.), Royal Soc. Chem., Cambridge, UK, pp. 19-37.
12. Guillet, E.J., 1973, *Polymers and Ecological Problems*. Plenum Press, New York.
13. Scott, G., 1976, Time Controlled Stabilization of Polyolefins. *J. Polym. Sci. Polym. Symp.* **57**: 357- 374.
14. Omichi, H., 1983, In *Degradation and Stabilization of Polyolefins* (S.N. Allen, ed.), Applied Science Publisher, London, 4.
15. Griffin, G.J.L., 1994, Particulate Starch Based Products. In *Chem. Technol. Biodegrad. Polym.*, (G.J.L. Griffin ed.), Blackie, Glasgow, UK, pp. 18-47.
16. Scott, G., 1994, Environmental Biodegradation of Hydrocarbon Polymers: Initiation and Control. In *Biodegradable Plastic and Polymers* (Y. Doi and K. Fukuda, eds.), Elsevier Science BV, Amsterdam, pp. 79-81.
17. Scott, G., 2000, "Green" Polymers. *Polym. Degr. Stab.* **68**: 1-7.
18. Scott, G., 1995, Photo-Biodegradable Plastics. In *Degradable Polymers: Principles and Applications* (Chapman & Hall, eds.), London, **9**: 169-185.
19. Guillet, J., 1974, Plastics, Energy and Ecology - a Harmonius Triad. *Plast. Eng.* August: 47-56.
20. Wiles, D.M., Cermak, B.E., Gho, J.G., and Hare, C.W., 1998, Controlled - Lifetime, Environmentally Degradable Plastics Based on Conventional Polyethylenes. In Official Book of Papers, *International Plastics Additives and Modifiers Conference, London, Nov. 9-10, 1998*, Rapra Technology, Shrewsbury, UK, Paper 14/1-Paper 14/6.
21. Scott, G., and Wiles, D.M., 2001, Programmed-Life Plastics from Polyolefins: A New Look at Sustainability. *Biomacromolecules* **2**: 615-622.
22. Various authors, 1990, *Agricultural and Synthetic Polymers. Biodegradability and Utilization* (J.E. Grass, and G. Swift, eds.) ACS Symp. Ser., Washington DC, vol. 433.
23. Chiellini, E., Cinelli, P., D'Antone, S., Ilieva, V.I., 2002, Environmentally Degradable Polymeric Materials in Agriculture Applications - an Overview. *Polymery* **47**: 21-27.
24. Chiellini, E., Chiellini, F., and Cinelli, P., 2003, Polymers from Renewable Resources. In *Degradable Polymers: Principles and Applications,* 2nd Ed (G. Scott, ed.), Kluwer, UK, pp. 163-233.
25. Eissen, M., Metzger, J.O., Schmidt, E., and Schneidewind, U., 2002, 10 Years after Rio—Concepts on the Contribution of Chemistry to a Sustainable Development. *Angew. Chem. Int. Ed.* **41:** 414-436.
26. Shogren, R.L. and Bagley, E.B., 1999, Natural Polymer as Advanced Materials: Some Research Needs and Directions. *ACS Symp. Ser.* **723**: 2-11.
27. http://unfccc.int/resource/convkp.html
28. El Bassam, N., 2001, Renewable energy for rural communities. *Renew. Energ.* **24**: 401-408.
29. US government ordinance on biobased materials
30. Warwel, S., Brüse, F., Demes, C., Kunz, M., and Klaas, M.R., 2001, Polymers and Surfactants on the Basis of Renewable Resources. *Chemosphere* **43**: 39-48.
31. Ayong Le Kama, A.D., 2001, Sustainable Growth, Renewable Resources and Pollution. *J. Econ. Dynam. Control* **25**: 1911–1918.
32. Reijnders, L., 2000, A Normative Strategy for Sustainable Resource Choice and Recycling. *Resour. Conserv. Recycl.* **28**: 121-133.

33. Van Wyk, J.P.H., 2001, Biotechnology and the Utilization of Biowaste as a Resource for Bioproduct Development. *Trends Biotechnol.* **19**: 172-177.
34. Heyde, M., 1998, Ecological Consideration on the Use Production of Biosynthetic and Synthetic Biodegradable Polymers. *Polym. Degr. Stab.* **59**: 3–6.
35. Various authors, 1994, *Polymers from Agricultural Coproducts* (M.L. Fishman, R. Friedman, and S.J. Huang, eds.) ACS Symp. Ser. **575**, Washington DC, pp. 1-247.
36. Sperling, L.H., and Carraher, C.E., 1988, Polymers from Renewable Resources In *Encyclopedia of Polymer Science and Engineering* (H.F. Mark, N.M. Bikales, C.G. Overberger, and G. Menges, eds.), Wiley, New York, **12**: 658-690.
37. MacGregor, E. A., and Greenwood, C .T., 1980, *Polymers in Nature.* John Wiley & Sons, New York, USA.
38. Various authors, 2001, *Chemicals and Materials from Renewable Resources* (J.J. Bozell, ed.) ACS Symp. Ser. **784**, ACS, Washington D.C., USA.
39. Petersen, K., Nielsen, P.V., Bertelsen, G., Lawther, M., Olsen, M.B., Nilsson, N.H., and Mortensen, G., 1999, Potential of Biobased Materials for Food Packaging. *Trends Food Sci. Technol.* **10**: 52-68.
40. Otey, F.H., and, Doane, W.M., 1984, Chemicals from Starch. In *Starch Chemistry and Technology*- 2nd ed. (R.L. Whistler, J.M. Bemiller, and E.F. Paschall, eds*.),* Academic Press, New York, pp. 389-416.
41. Fanta, G.F., and Bagley, E.B., 1977, Starch, Graft Copolymers. In *Encyclopedia of Polymer Science and Technology-Suppl.* Wiley-Interscience, New York. **2**: 665-699.
42. Fanta, G.F., and Doane, W.M., 1986, Grafted Starches, Modified Starches: Properties and Uses. In *Grafted Starches, Modified Starches: Properties and Uses* (O.B. Wurzburg, ed.), CRC Press. Boca Raton, pp. 149-178.
43. Weaver, M.O., Bagley, E.B., Fanta, G.F. and Doane, W.M., 1976, Method of Reducing Water Content of Emulsions, Suspensions, and Dispersions with Highly Absorbent Starch-Containing Polymeric Compositions. *U.S. Pat. 3935099.*
44. Weaver, M.O., Bagley, E.B., Fanta, G.F. and Doane, W.M., 1976, Highly-Absorbent Starch-Containing Polymeric Compositions. *U.S. Pat. 3997484.*
45. Swanson, C.L., Shogren, R.L., Fanta, G.F. and Imam, S.H. (1993) Starch-Plastic Materials- Preparation, Physical Properties, and Biodegradability (A review of Recent USDA Research), *J. Environ. Polym. Degrad.* **1,** 155-166.
46. Elion, G.R., 1993, Synthesis of Plastic from Recycled Paper and Sugar Cane. *U.S. Pat. 5244945.*
47. Simon, J., Müller, H.P., Koch, R., and Müller V., 1998, Thermoplastic and Biodegradable Polymers of Cellulose. *Polym. Degr. Stab.* **59**: 107-115.
48. Vert, M., 2001, Biopolymers and Artificial Biopolymers in Biomedical Applications, an Overview. In *Biorelated Polymers, Sustainable Polymer Science and Technology* (E. Chiellini, H. Gil, G. Braunegg, J. Buchert, P. Gatenholm, and M. Van der Zee, eds.), Kluwer Academic/Plenum Publishers, pp. 63-79.
49. French, A.D., Bertoniere, N.R., Battista, O.A., Cuculo, J.A., and Gray, D.G., 1993, Cellulose. In *Kirk-Othmer Encyclopedia of Chemical Technology, 4th Ed*., John Wiley & Sons, New York, **5**: 476-496.
50. Mormann, W., and Demeter, J., 1999, Silylation of Cellulose with Hexamethyldisilizane in Liquid Ammonia-First Examples of Compltely Trimethylsilylated Cellulose. *Macromolecules* **32**: 1706-1710.
51. Mormann, W. and Spitzer, D., 2001, Silylation of OH-Polymers by Reactive Extrusion, *Macromol. Symp.* **176**: 279-287.
52. Stevens, E.S., 2002, Biopolymers. In *Green Plastics*, Princenton University Press, Princenton, New Jersey, USA, pp. 83-103.

53. Buisman,, G.J.H., Overeem, A., Cuperus, F.P., 1999, Synthesis of Epoxidized Novel fatty Acids for Use in Pain Applications. In: Knothe, G., Derksen, J.T.P. (Eds), *Recent Developments in the Synthesis of Fatty Acid Derivatives*. AOCS Press, Champaign, IL, USA, pp. 128-140.
54. Braunegg, G., Genser, K., Bona, R., Haage, G., Schellauf, F., and Winkler, E., 1999, Production of PHAs from Agricultural Waste Material. *Macromol. Symp.* **144**: 375-383.
55. Nonato, R.V., Mantelatto, P.E., and Rossell, C.E.V., 2001, Integrated Production of Biodegradable Plastic, Sugar and Ethanol. *Appl. Microbiol. Biotechnol.* **57**: 1-5.
56. Stafford, D.E., and Stephanopoulos, G., 2001, Metabolic Engineering as an Integrating Platform for Strain Development. *Curr. Opin. Microbiol.* **4**: 336-340.
57. Altaras, N.E., and Cameron, D.C., 1999, Metabolic Engineering of a 1,2-Propanediol Pathway in *Escherichia Coli*. *Appl. Environ. Microbiol.* **65**: 1180-1185.
58. Follstad, B.D., Balkarcel, R.R., Stephanopoulos, G., and Wang, D.I.C., 1999, Metabolic Flux Analysis of Hybridoma Continuous Culture Steady State multiplicity. *Biotechnol. Bioeng.* **63**: 675- 683.
59. McLaren, J.S., 2000, Future Renewable Resource Needs: Will Genomics Help?. *J. Chem. Technol. Biotechnol.* **75**: 927-932.
60. Wu, Q., Sun, S.Q., Yu, P., and Chen, G.Q., 2000, Environmental Dependence of Microbial Synthesis of Polyhydroxyalkanoates. *Acta Polym. Sin.* **6**: 751-756.
61. Sudesh, K., Abe, H., and Doi, Y., 2000, Synthesis, Structure and Properties of Polyhydroxyalkanoates: Biological Polyesters. *Prog. Polym. Sci.* **25**: 1503-1555.
62 Yao, J., Zhang, G., Wu, Q., Chen, G.Q., and Zhang, R.Q., 1999, Production of Polyhydroxyalkanoates by *Pseudomonas Nitroreducens*. *Anton. Leeuw. Int. J. G.* **75**: 245-349.
63. Hong, K., Chen, G.Q., Yu, P.H., Zhang, G., Liu, Y., and Chua, H., 2000, Effect of C: N Molar Ratio on Monomer Composition of Polyhydroxyalkanoates Produced by *Pseudomonas Mendocina* 0806 and *Pseudomonas Pseudoalcaligenes* YSI. *Appl. Biochem. Biotechnol.* **84-86**: 971-980.
64. Slater, S., Mitsky, T.A., Houmie, K.L., Hao, M., Reiser, S.E., Taylor, N.B., Tran, M., Valentin, H.E., Rodriguez, D.J., Stone, D.A., Padgette, S.R., Kishore, G., and Gruys, K.J., 1999, Metabolic Engineering of *Arabidopsis* and *Brassica* for Poly(3-hydroxybutyrate-co-3-hydroxyvalerate) Copolymer Production. *Nat. Biotechnol.* **17**: 1011-1016.
65. Jensen, M.H., 1998, 27th National Agricultural Plastics Congress.
66. Gilead, D., 1985, Plastics that self destruct. *Chemtech.* **15**: 299-301.
67. Potts, J.E., 1984, Plastics, Environmentally Degradable. In *Kirk-Othmer Encyclopedia of Chemical Technology,* 3rd ed., John Wiley & Sons, NY, Suppl., pp. 626-668.
68. Zhao, X, 1998, Base Material Capable of Light Degradation and Biological Degradation. *CN 1193026 A 16 Sep 1998.*
69. Lu ,Y, and Yang, C.D., 1999, Polymeric Degradable Resin Composition and its Preparation Process. *CN 1239725 A 29 Dec 1999.*
70. Xiong, G., 1999, Optical- and Bio-degradable Plastics. *CN 1235174 A 17 Nov 1999.*
71. Griffin, G.J.L., 1977, Biodegradable Synthetic Resin Sheet Material Containing Starch and a Fatty Material. *U. S. Pat. 4016117.*
72. Otey, F.H., Mark, A.M., Mehltretter, C.L., and Russell, C.R., 1974, Starch-Based Film for Degradable Agricultural Mulch. *Ind. Eng. Chem., Prod. Res. Develop.* **13**: 90-92.
73. Lahalih, S.M., Akashah, S.A, and Al-Hajjar, F.H, 1987, Development of Degradable Slow Release Multinutritional Agricultural Mulch Film. *Ind. Eng. Chem. Res.* **26**: 2366-2372.

74. Otey, F.H, Westhoff R.P, and Russell, C.R., 1977, Biodegradable Films from Starch and Ethylene-Acrylic Acid Copolymer. *Ind. Eng. Chem. Prod. Res. Dev.,* **16**: 305-308.
75. Otey, F.H., Westhoff, R.P., and Doane, W.M., 1980, Starch-Based Blown Films. *Ind. Eng. Chem. Res.,* **19**: 592-595.
76. Otey, F.H., Westhoff, R.P., and Doane W.M., 1987, Starch-Based Blown Films. *Ind. Eng. Chem. Res.,* **26**: 1659-1663.
77. Westhoff, R.P., Otey, F.H., Mehitretter, C.L., and Russell, C.R., 1974, Starch-Filled Polyvinyl Chloride Plastics. Preparation and Evaluation. *Ind. Eng. Chem. Prod. Res. Develop.* **13**: 123-125.
78. Albertsson, A.C., and Karlsson, S., 1988, The Three Stages in Degradation of Polymers - Polyethylene as a Model Substance. *J. Appl. Polym. Sci.* **35**: 1289-1302.
79. Lawton, J.L., and Fanta, G.F., 1994, Glycerol-Plasticized Films Prepared from Starch-Poly(vinyl alcohol) Mixtures: Effect of Poly(ethylene-co-Acrylic Acid). *Carbohydr. Polym.* **23**: 275-280.
80. Chen, L., Imam, S.H., Gordon, S.H., and Green, R.V., 1997, Starch-Polyvinyl Alcohol Crosslinked Film - Performance and Biodegradation. *J. Environ. Polym. Degr.* **5**: 111-117.
81. Ishigaki, T., Kawagoshi, Y., Ike, M., and Fujita, M., 1999, Biodegradation of a Polyvinyl Alcohol - Starch Blend Plastic Film. *World J. Microb. Biot.* **15**: 321-327.
82. Mao, L., Imam, S.H., Gordon, S.H., Cinelli, P., and Chiellini, E., 2000, Extruded Cornstarch—Polyvinyl Alcohol Blends: Mechanical Properties, Morphology, and Biodegradability. *J. Polym. Environ.* **8**: 205-211.
83. http://www.materbi.it/welcomeSi.html
84. http://agproducts.unl.edu/pla.htm
85. Bogaert, J.C., and Coszach, Ph., 2000, Poly(lactic acid): A Potential Solution to Plastic Waste Dilemma. *Macromol. Symp.* **153**: 287-303.
86. http://www.cargilldow.com
87. Petersen, K., Nielsen, P.V., and Olsen, M.B., 2001, Physical and Mechanical Properties of Biobased Materials Starch, Polylactate and Polyhydroxybutyrate. *Starch/Starke* **53**: 356-361.
88. Martin, O., Schwach, E., Avérous, L., and Couturier, Y., 2001, Properties of Biodegradable Multilayer Films Based on Plasticized Wheat Starch. *Starch/Starke* **53**: 372-380.
89. Chiellini, E., Corti, A., Swift, G., Biodegradation of Thermally-Oxidized Fragmented LDPE Samples, *Polym. Degr. Stab.* in press.
90. Coffin, D.R., and Fishman, M.L., 1994, Physical and Mechanical Properties of Highly Plasticized Pectin/Starch Films. *J. Appl. Polym. Sci.* **54**: 1311-1320.
91. Coffin, D.R., Fishman, M.L., and Ly, T.V., 1996, Thermomechanical Properties of Blends of Pectin and Poly(vinyl alcohol). *J. Appl. Polym. Sci.* **61**: 71-79.
92. Chiellini, E., Cinelli, P., Imam, S.H., and Mao, L., 2001, Composite Films Based on Biorelated Agro-Industrial Waste and Poly(vinyl alcohol). Preparation and Mechanical Properties Characterization. *Biomacromolecules* **2**: 1029-1037.
93. Chiellini, E., Cinelli, P., Imam, S.H., and Mao, L. 2001, Composite Films Based on Poly(vinyl alcohol) and Lignocellulosic Fibres: Preparation and Characterization. In *Biorelated Polymers-Sustainable Polymer Science and Technology* (E. Chiellini, H. Gil, G. Braunegg, J. Buchert, P. Gatenholm, and M. Van der Zee, eds.), Kluwer Academic/Plenum Publishers, pp. 87-100.
94. Cinelli, P., Chiellini, E., Imam, S.H., and Gordon, S.H., 2003, Characteristics and Degradation of Hybrid Composite Films Prepared from PVA, Starch and Lignocellulosics. *Macromol. Symp.* in press.

95. Zhang, Y., Ghasemzadeh, S., Kotliar, A.M., Kumar, S., Presnell, S., and Williams, L.D., 1999, Fibers from Soybean Protein and Poly(vinyl alcohol). *J. Appl. Polym. Sci.* **71**: 11-19.
96. Kajikawa, Y., Ito, M., Shimizu, K., Sakane, M., and Tanigawa, M., 2000, Aliphatic Copolymer, Production Process, Aliphatic Polyester Resin Composition, Various Uses, Coating Composition, and Agricultural or Horticultural Particulate Composition comprising Degradable Coating Film. *PCT Int. Appl. WO 0029460.*
97. Ishikawa, M., Nakata, K., Shimizu K., Daito, T., Nishimura, K., and Murakami, T., 2000, Biodegradable Polyester Resin Composition, Biodisintegrable Resin Composition, and Molded Objects of These . *PCT Int. Appl. WO 9963001.*
98. Kim, D.H., Chung, H.S., Jun, J.N., Lee, J.W., and Lee, S.W., 2000, Polyester Resin and a Process for Preparing the same. *PCT Int. Appl. WO 0011063.*
99. Kitahara, Y., Watanabe, T., Aihara, H., Ajioka, M., Nakata, T., Obuchi, S., Suzuki, K., and Yoshida, Y., 1999, Polylactic Acid Composition and Film thereof. *PCT Int. Appl. WO 9945067 A1 10 Sep 1999.*
100. Darmawan, D., Makuuchi, K., Mitomi, H., Yoshii, F., and Murakami, T., 1999, Lactone-Containing Resin Composition, Molded Object thereof, and Film. *PCT Int. Appl. WO 9946332.*
101. Iannace, S., Nocilla, G., and Nicolais, L., 2000, Processing and Properties of Biodegradable Composites for Agricultural Applications. Institute of Composite Material Technology ECCM-8, *8th Eur. Conf. Compos. Mater.: Sci., Technol. Appl.* (V.I. Crivelli, ed.), Woodhead, Cambridge, UK, **1**: 631-637.
102. White, J.E., Doane, W.M., Mang, M.N., and Xu, W., 2000, Method of Marking Biodegradable Polymer Compositions. *PCT Int. Appl. WO 200011064.*
103. Hoagland, P. D., 1999, Biodegradable Laminated Films Fabricated from Pectin and Chitosan. *U.S. Pat. 5919574.*
104. Coffin, D.R., and Fishman, M.L., 1995, Films Fabricated from Mixtures of Pectin and Starch. *U.S. Pat. 5451673.*
105. Otaigbe, J.U., 1998, Processability and Properties of Biodegradable Plastics Made from Agricultural Biopolymers. *Annual Tech. Conf. - Soc. Plast. Eng., 56th*, **2**: 1777-1781.
106. Otaigbe, J.U, Goel, H., Babcock, T., and Jane, J, 1999, Processability and Properties of Biodegradable Plastics Made from Agricultural Biopolymers. *J. Elastom. Plast.* **31**: 56-71.
107. Chiellini, E., Cinelli, P., Fernandes, E.G., Kenawy, E.R., and Lazzeri, A., 2001, Composite Materials Based on Gelatin and Fillers from Renewable Resources. In *Biorelated Polymers-Sustainable Polymer Science and Technology* (E. Chiellini, H. Gil, G. Braunegg, J. Buchert, P. Gatenholm, and M. Van der Zee eds.), Kluwer Academic/Plenum Publishers, pp.101-114.
108. Chiellini, E., Cinelli, P., Fernandes, E.G., Kenawy, E.R., Lazzeri, A., 2001, Gelatin-Based Blends and Composites. Morphological and Thermal Mechanical Characterization. *Biomacromolecules* **2**: 806-811.
109. Chiellini, E., Cinelli, P., Corti, A., Kenawy, E.R., Fernandes, E.G., and Solaro, R., 2000, Environmentally Sound Blends and Composites Based on Water-Soluble Polymer Matrices. *Macromol. Symp.* **152**: 83-94.
110. Chiellini, E., Cinelli, P., Corti, A., and Kenawy, E.R., 2001, Composite Films Based on Waste Gelatin: Thermal-Mechanical Properties and Biodegradation Testing. *Polym. Degr. Stab.* **73**: 549-555.
111. Anderson, D.F., Garisto, M.-A., Bourrut, J.-C., Schonbeck, M.W., Jaye, R., Wurzberger, A., and De Gregorio, R., 1995, Evaluation of a Paper Mulch Made from

Recycled Materials as an Alternative to Plastic Film Mulch for Vegetables. *J. Sustain. Agric.* **7**: 39-61.

112. Shogren, R. L., 1999, Preparation and Characterization of a Biodegradable Mulch: Paper Coated with Polymerized Vegetable Oils. *J. Appl. Polym. Sci.* **73**: 2159-2167.
113. Shogren, R. L., 2000, Biodegradable Mulches from Renewable Resources. *J. Sustain. Agric.* **16**: 33-47.
114. Wallace, A., Wallace, G.A., and Abouzamzam, A.M., 1986a, Effects of Soil Conditioners on Water Relationships in Soils. *Soil Sci.* **141**: 346-352.
115. Sartain, J.B., 1995, Effects of Clay and Polymer Amendments on the Physical and Chemical Properties of Soil. *J. Turfgrass Management* **1**: 1-18.
116. Lentz, R.D., and Sojka, R.E., 1994, Field Results Using Polyacrylamide to Manage Furrow Erosion and Infiltration. *Soil Sci.* **158**: 274-282.
117. Fanta, G.F., and Bagley, E.B., 1977, Starch, Graft Copolymers. In *Encyclopedia of Polymer Science and Technology,* Wiley-Interscience, NY, Suppl. 2: 665-699.
118. Kay, B.L., Evans, R.A., and Young, J.A., 1977, Effects of Hydroseeding on Bermudagrass. *Agron. J.* **69**: 554-557.
119. Young, J.A., Kay, B.L., and Evans, R.A., 1977, Accelerating the Germination of Common Bermudagrass for Hydroseeding. *Agron. J.* **69**: 115-119.
120. Mellon, M.G., 1989, Hydro-Sprigging Revolution. *The Landscape Architet and Specifier News* **5**: 40-43.
121. Mellon, M.G., 1990, Hydro-Sprigging is Quietly Revolutionizing Planting of Warm Season Grasses. *Southern Golf-Landscape and Resort Management.* May/June.
122. Meadows, G.W., 1956, Process for Conditioning Soil with Polysaccharide Xanthates. *U.S. Pat. 2761247.*
123. Menefee, E., and Hautala, E., 1978, Soil Stabilization by Cellulose Xanthate. *Nature* **275**: 530-532.
124. Mitchell, A.R., 1986, Polyacrylamide Application in Irrigation Water to Increase Infiltration. *Soil Sci.* **141**: 353-358.
125. Orts, W.J., Sojka, R.E., and Glenn, G.M., 2000, Biopolymer additives to reduce erosion-induced soil losses during irrigation. *Ind. Crop. Prod.* **11**: 19-29.
126. Fritz, J., Sandhofer, M., Stacher, C., and Braun, R., 2002, Strategies for Detecting Undesired Ecotoxicological Effects of Biodegradable Materials in Agricultural Application. *7th World Conf. on Biodegradable Polymers and Plastics*, Tirrenia (Pisa, Italy), 4-8 June, pp. 59.
127. Taylor, K.C., and Halfacre, R.G., 1986, The Effect of Hydrophilic Polymer on Media Water Retention and Nutrient Availability to Ligustrum Lucidum. *HortScience* **21**: 1159-1161.
128. Mustafa, M.A., Al-Omran, A.M., Shalaby, A.S., and Al-Darby, A.M., 1988, Horizontal Infiltration of Water in Soil Columns as Affected by a Gel-Forming Conditioner. *Soil Sci.* **145**: 330-336.
129. Wallace, A., and Wallace, G.A., 1986a, Effects of Soil Conditioners on Emergence and Growth of Tomato, Cotton, and Lettuce Seedlings. *Soil Sci.* **141**: 313-320.
130. Wallace, A., and Wallace, G.A., 1986e, Effect of Polymeric Soil Conditioners on Emergence of Tomato Seedlings. *Soil Sci,* **141**: 321-323.
131. Wallace, A., Abouzamzam, A.M., and Cha, J.W., 1986, Interactions between a Polyacrylamide and a Polysaccharide as Soil Conditioners when Applied Simultaneously. *Soil Sci* **141**: 374-376.
132. Wallace A., 1986, A Polysaccharide (Guar) as a Soil Conditioners. *Soil Sci.* **141**: 371-374.

133. Helalia, A.M., and Letey, J., 1988. Polymer Type and Water Quality Effects on Soil Dispersion. *Soil Sci. Soc. Am. J.* **52**: 243-246.
134. Aly, S.M., and Letey, J., 1990, Physical-Properties of Sodium-Treated Soil as Affected by Two Polymers. *Soil Sci. Soc. Am. J.* **54**: 501-504.
135. Maher, G.G., 1981, Flocculation of Coals with Water-Soluble Starch Xanthates. *U.S. Pat.* 4253970.
136. Coltrinari, E., 1994, Selective Recovery of Heavy Metals Using Xanthates. *U.S. Pat.* 5320759.
137. Meadows, G.W., Process for Conditioning Soil with Polysaccharide Xanthates. *U.S. Pat.* 2761247.
138. Meadows, G.W., Cement and Method of Stabilizing Soil. *U.S. Pat. 2884334.*
139. Menefee, E., and Hautala, E., 1978, Soil Stabilization by Cellulose Xanthate. *Nature* **275**: 530-532.
140. Stefanson, R.C., 1973, Polyvinyl Alcohol as a Stabilizer of Surface Soils. *Soil Sci.* **115**: 420-428.
141. Stefanson, R.C., 1974,Soil Stabilization by Polyvinyl Alcohol and its Effect on the Growth of Wheat. *Aust. J. Soil Res.* **12**: 59-62.
142. Stefanson, R.C., 1975, The Fate of Polyvinyl Alcohol in Soils. *Soil Sci.* **119**: 426-430.
143. Oades, J.M., 1976, Prevention of Crust Formation in Soils by Poly(vinyl alcohol). *Aust. J. Soil Res.* **12**: 139-148.
144. Painuli, D.K., and Pagliai, M., 1990, Effect of Polyvinyl Alcohol, Dextran and Humic Acid on some Physical Properties of a Clay and Loam Soil. I. Cracking and Aggregate Stability. *Agrochimica* **34** (1-2): 117-130.
145. Painuli, D.K., Pagliai, M., La Marca, M., and Lucamante, G., 1990, Effect of Polyvinyl Alcohol, Dextran and Humic Acid on some Physical Properties of a Clay and Loam Soil. II. Hydraulic Conductivity and Porosity. *Agrochimica* **34**(1-2): 131-146.
146. Kenawy, El-.R., Cinelli, P., Corti, A., Miertus, S., and Chiellini, E., 1999, Biodegradable Composite Films Based on Waste Gelatin. *Macromol. Symp.* **144**: 351-364.
147. Cinelli, P., 1999, Ph.D Thesis Formulation and Characterization of Environmentally Compatible Polymeric Materials for Agriculture Applications, University of Pisa.
148. Bargiacchi, E., Chiellini, E., Cinelli, P., Corti, A., Magni, S., and Miele, S., 2002, Liquid or Semifluid self Sustaining Mulch for Protecting Crops. *U.S. Pat. 02129544.*
149. Chiellini, E., Cinelli, P., D'Antone, S., Ilieva, V.I., Magni, S., Miele, S., and Pampana, S., 2003, Hybrid Liquid Mulch Based on Poly(Vinyl Alcohol) and Agroindustrial Waste. Interaction with the Soil. *Macromol. Symp.* in press.
150. Fanta, G.F., and Bagley, E.B., 1977, Starch, Graft Copolymers, In *Encyclopaedia of Polymer Science and Technology*, Wiley-Interscience, NY, Suppl. 2, pp. 665-699.
151. Paul, D.R., 1976, Controlled Release Polymeric Formulations. *ACS Symp. Ser.* **33**: 2-14.
152. McCormick, C.L., Anderson, K.W., and Hutchinson, B.H., 1982, Controlled Activity Polymers with Pendently Bound Herbicides. *J. Macromol. Sci.- Rev. Macromol. Chem. Phys.* **22**: 57-87.
153. Stout, E.L., and Doane, W.M., 1979, Pilot-Plant Process for Starch Xanthide Encapsulated Pesticides. *J. Appl. Polym. Sci.* **24**: 153-159.
154. Shasha, B.S., Trimnell, D., and Otey, F.H., 1981, Encapsulation of Pesticides in a Starch-Calcium Adduct. *J. Polym. Sci., Polym. Chem.* **19**: 1891-1899.
155. Shasha, B.S., Trimnell, D., and Otey, F.H., 1981, Shelf Life Indicators for Encapsulated Diazinon. *J. Agric. Food Chem.* **29**: 1099-1100.
156. Trimnell, D., Shasha, B.S., Wing, R.E., and Otey, F.H., 1982, Pesticide Encapsulation Using a Starch-Borate Complex as Wall Material. *J. Appl. Polym. Sci.* **27**: 3919-3928.

157. Riley, R.T., 1983, Starch-Xanthate-Encapsulated Pesticides: a Preliminary Toxicological Evaluation. *J. Agri. Food Chem.* **31**: 202-206.
158. McCormick, C.L., and Lichatowhich, D.K., 1979, Homogeneous Solution Reactions of Cellulose, Chitin, and Other Polysaccharides to Produce Controlled-Activity Pesticide Systems. *J. Polym. Sci.* **17**: 479-484.
159. Chandra, R. and Rustgi, R., 1998, Biodegradable Polymers. *Prog. Polym. Sci.* **23**: 1273-1335.

PART 3

BIOMEDICAL APPLICATIONS

Hydrophilic/Hydrophobic Copolymers: Fluorinated Hydrogels as Biomaterials

SAMUEL J. HUANG*, DAWN A. SMITH*, and JEFFREY T. KOBERSTEIN#

** Institute of Materials Science, University of Connecticut, Storrs, Connecticut 06269-3136, USA, # Department of Chemical Engineering and Applied Chemistry, Columbia University, 500 West 120 Street, New York, New York , USA*

1. INTRODUCTION

Bio-organs are generally multi-component systems containing both hydrophilic and hydrophobic parts. Several hydrophobic biodegradable polymers have been successfully used as biomedical materials. They become hydrophilic as they are degraded by hydrolytic and oxidative processes and their biocompatibility improves with the degradation. It is commonly observed that polymer systems with hydrophilic surfaces which become 'soft' when swollen with water have better biocompatibility with respect to blood proteins interaction than similar systems with hydrophobic and 'hard' surfaces which tend to cause denaturing of the absorbed proteins. We have studied the design, synthesis, characterization, testing, and application of multi-component hydrophilic-hydrophobic biodegradable polymer systems as biomedical materials[1-5]. These include copolyesters containing hydroxy groups derived from tartaric acid as drug release systems[6,7]; hydrogen-bond containing poly(enol-ketone)s and poly(amide-enamine) as drug release matrices[8]; and semi-interpenetrating and interpenetrating networks of polycaprolactone and poly(2-hydroxyethyl methacrylate) as tendon implants[9-11]. It become clear to us that it is important to control the adhesion/interaction between polymer surfaces and blood proteins. Our aims are two –fold: minimizing the non-specific interactions and maximizing the

Biodegradable Polymers and Plastics, Edited by Chiellini and Solaro
Kluwer Academic/Plenum Publishers, New York, 2003

desirable interactions. Among our approaches is the synthesis of hydrophilic-hydrophobic multi-component polymer system with well-defined structures. We reported here some of our recent results on fluorine containing hydrogel systems.

2. MATERIALS, SYNTHESIS, AND CHARACTERIZATION

Four monomers were chosen for the preparation of multi-component hydrogels: 2-hydroxyethyl methacrylate (HEMA), N-vinyl-2-pyrrolidinone (NVP), and poly(ethylene glycol) methacrylate (PEGMA) with average Mn of 360 were from Aldrich Chemical Company and purified by distillation; 2-(N-ethylperfluorooctanesulfonamido)ethyl acrylate (FOSA) from 3 M and purified by recrystallisation from methanol.

Azobis*iso*butylnitrile (AIBN), used as initiator, was purified by recrystallisation from ethanol. Chemical and/or spectroscopic grades solvents were used.

Instruments used form characterization include: Nicolet Magna-IR 560 Fourier transform infrared spectrometer, Perkin Elmer Physical Electronics PHI 5300 X-ray photoelectrom spectrometer, Brucker DMX-500 spectrometer, Water 150-C ALC/GC, Perkin Elmer DSC 7 Differential Scanning Calorimeter, Beckman Glucose Analyzer, and Wallace 1282 Compugamma Universal Gamma Counter.

Hydrogels were prepared in Teflon dishes by both thermal polymerization with AIBN as initiator at 70 °C under nitrogen for 9 hrs and photo-polymerization with KIP-150 oligomeric hydroxyketone from Sartomer as initiator and 10 W mercury lamp as 365 nm light source for 15 minutes followed by placement in a oven at 85 °C for 25 minutes. Gels were soaked and washed several times with 50/50 acetone/water followed by distilled water.

Swelling of gels were carried by soaking in acetone and dried in vacuum at 40 °C for 24 hrs and then soaked in de-ionized water until constant weight was achieved.

The adsorption of the proteins, human albumin, human fibrinogen, and immonoglobulin G (IgG) was determined using radioactive technique. IgG labelled with ^{125}I bonded to tyrosine in the protein. The hydrated gels were soaked in protein solutions for various periods of time, rinsed with copious amounts of buffer saline solution and then read with a gamma counter.

Glucose permeability was determined the amount of glucose diffused through gel covered polycarbonate membrane in 0.9% NaCl at 20 minute intervals detected by a Beckman Glucose Analyzer.

3. RESULTS AND DISCUSSION

Polymer surfaces play an important role in many polymer applications especially biomaterials. The design and synthesis of desired surface properties of biomedical polymers are among our approaches to biomedical polymers. Biocompatibility was defined in 1987[9] as "the ability of a material to perform with an appropriate host response in a specific application". Bulk and surface properties of biomaterials used for implant devices directly influence the tissue interface dynamics from initial implantation until explantation. The most important influence on how proteins, cells, and the organism respond to a material is the surface structure, in both chemical and morphological terms.

Every implanted material, synthetic or natural, suffer some degree of rejection from the body. Simply the act of implantation causes injury and an unavoidable chain of events summed up in term of acute inflammation[10]. During acute inflammation, there is a race to the surface of the materials. Small proteins and peptides arrive first, followed by larger proteins, and then cells. The conformation and composition of the protein layer on the surface of the implant mediate the cell recognition and the resulting inflammatory response. There are more than 200 different proteins in blood and plasma. These proteins will absorb in many conformations and orientations dependent on the surface of the implant materials. If the surface of the materials causes the proteins to absorb non-specifically, often followed by conformational changes, then the cells that arrive to the surface will not recognize the surface as nature never uses non-specifically absorbed proteins. The cells will see the surface as foreign: and what the body sees as foreign, it will either try to degrade or wall off. In many cases of implants, this "walling off" or a fibrous capsule will compromise the function of the implant.

There are two main approaches to design biocompatible surfaces. The first approach is to create a "non-interacting" surface where undesirable bio-interactions are reduced. Low fouling coatings resistant to protein adsorption and cell adhesion and antithrombogenic surfaces have been studied. The second approach is to design a bioactive surface, which enhances desirable bio-interaction by immobilization of biomolecules with desirable interactions (bio-recognitions).

Non-interacting surfaces are designed to adsorb a minimum amount of proteins on their surfaces. These materials can be divided into two general categories: low energy surfaces such as fluorinated surfaces or extremely hydrophilic and mobile surfaces such as hydrogels which result in less conformational changes in absorbed proteins than hard surfaces. In all these cases the lack of non-specific interactions results in "stealth" materials.

Hydrogels base on HEMA (PHEMA) were chosen for our study because of their prevalent uses as biomaterials. Earlier we increased the physical strength of PHEMA by combinations of PHEMA with polycaprolactone (PCL), a hydrophobic biodegradable polyester, in various forms such as copolymers, semi-interpenetrating networks and interpenetrating networks[11-13]. Both thermal and photo- initiated polymerizations result in multi-component hydrogels. Thermal process provides gels of various volumes and shapes whereas photo-process provides thin films. Structures of gels were verified with FTIR and XPS.

HEMA and NVP do not copolymerize together very well as the reactivity ratios are $r_1 = 4{,}81 \pm 0.133$ and $r_2 = -0.019 \pm 0.046$[14,15] calculated according to the method Kelen and Tüdös[16]. The negative r_2 has no physical meaning and should be considered to be close to zero. Since the yield of polymerization is quite high it can assumed that HEMA reacts first and NVP later giving blocky structures. The presence of NVP increases the extent of water swelling suggesting HEMA-NVP gels are more "hydrophilic" than HEMA gels. On the other hand the incorporation of small amount of FOSA, as confirmed by XPS spectra, improves the strength of gels. It is likely due to phase separation with hydrophobic FOSA regions serving as physical crosslinks for swollen gels. Swelling and water content of swollen gels are close to that reported for HEMA with similar structure[17]. As expected increasing PEGMA in feed increases the amount of covalent crosslinking and thus reduces the extent of water swelling. Results on thermal and photo-polymerized gels are listed in Table 1 and Table 2, respectively.

Table 1. Thermal Polymerization of Hydrogels with AIBN as Initiator

Hydrogel	Monomer feed (wt%)				Average Hydration
	HEMA	FOSA	NVP	PEGMA	(%)
1	100	-	-	-	73
2	94.5	5.5	-	-	64
3	62	6	32	-	97
4	35	6	59	-	244
5	56	6	28	10	110
6	40	6	24	30	140

Table 2. Synthesis of Hydrogels by Photo-initiation with KIP (3wt%)

Hydrogel	Monomer feed (wt%)				Yield	Hydration	Water
	HEMA	FOSA	NVP	EGDMA	(wt%)	(%)	(%)
A	94	6.0	-	0.5	80.2	76±6	43.1±2
B	100	-	-	0.5	88.5	77±6	43.4±2
C	32	6.0	61.5	0.5	88.4	220±12	68.7±1
D1	34	-	65.5	0.5			
D2	34	-	65.0	1.0			
D3	34	-	63.0	5.0	92.5	93±9	48.1±2

The glucose permeability of the multi-component gels is shown in Figure 1. The rate of glucose diffusion is roughly proportional to the hydrophobicity and/or crosslinking of the gels. Gels with certain compositions should find potential applications as anti-fouling coating for bio-sensors.

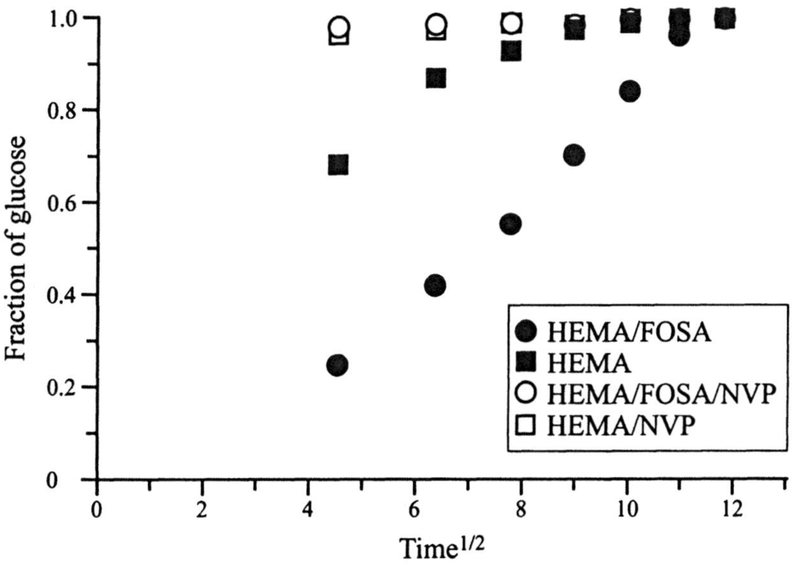

Figure 1. Glucose permeability of hydrogels.

Figure 2 shows results of plasma protein immonoglobulin (IgG) adsorption on thermally polymerized gels. High feed of PEGMA (30%) had the greatest IgG adsorption at 4 days whereas lower PEGMA feeds show little effect on PHEMA–IgG adsorption. The exact reason is not clear at this point although the high content of relative short PEGMA chains might have resulted in "hydrophobic" polymerized methacrylate domains and thus increased the IgG adsorption.

Figures 3 and 4 show the fibrinogern absortion. Incorporation of either the "hydrophilic" NVP or the "hydrophobic" FOSA decreases the fibrinogen adsorption on PHEMA with the tri-component gel containing HEMA, NVP and FOSA having the lowest adsorption.

The adsorption of albumin is show in Figures 5 and 6. Again the tri-component gel C with HEMA, NVP and FOSA having the lowest adsorption.

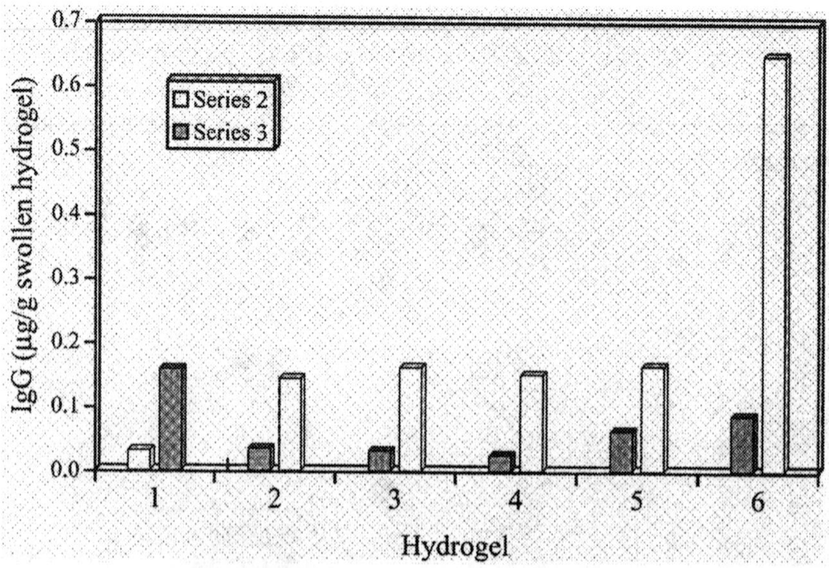

Figure 2. IgG adsorption of hydrogels 1-6 at 4 hrs (series 2) and at 1 day (series 3).

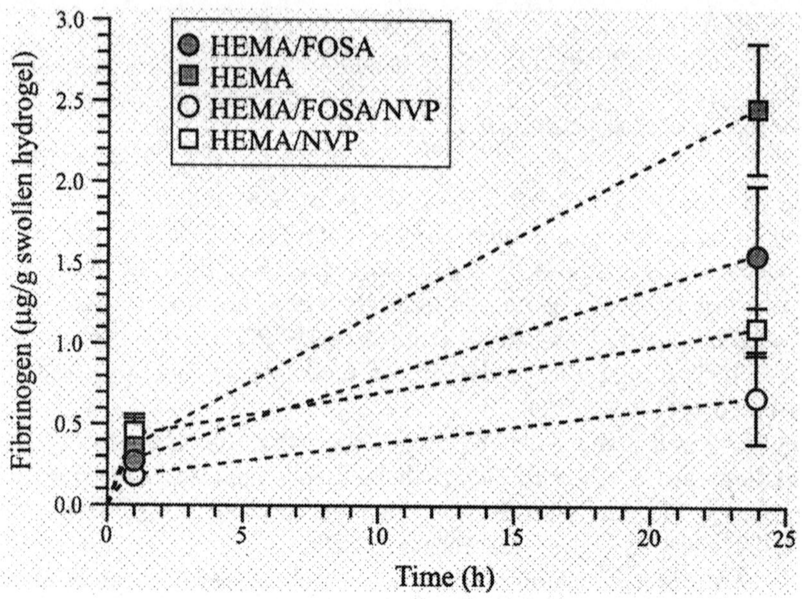

Figure 3. Fibrinogen adsorption at 1 hr and 24 hrs for hydrogels A, B, C, and D3

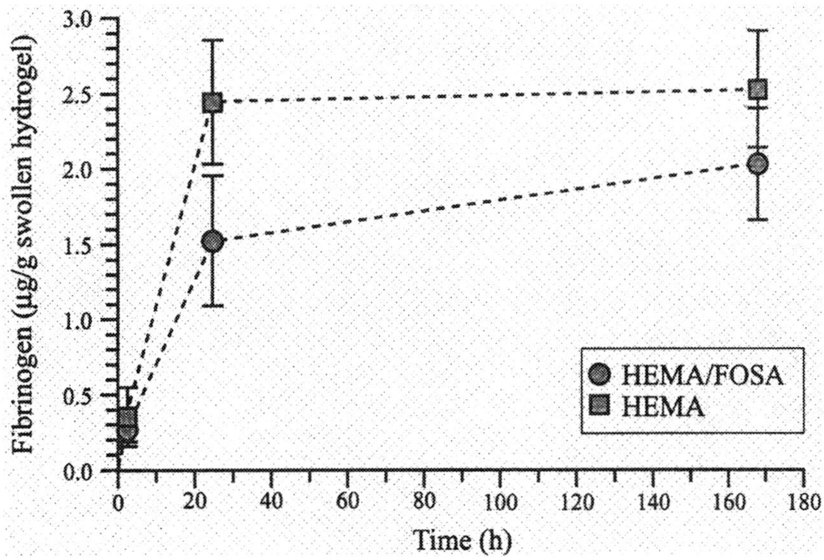

Figure 4. Fibrinogen adsorption at 1 hr, 24 hrs, and 1 week for hydrogels A and B.

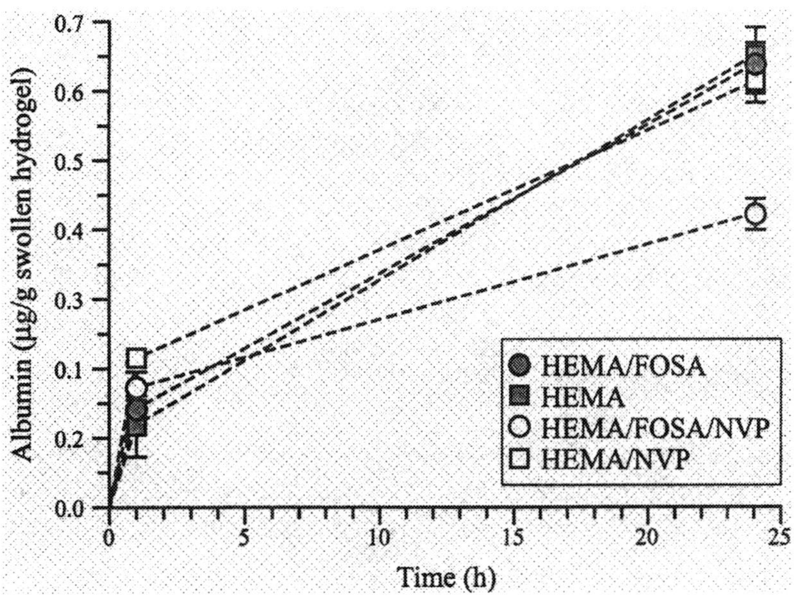

Figure 5. Albumin adsorption of hydrogels A, B, C, and D3 at 1 hr and 1 day.

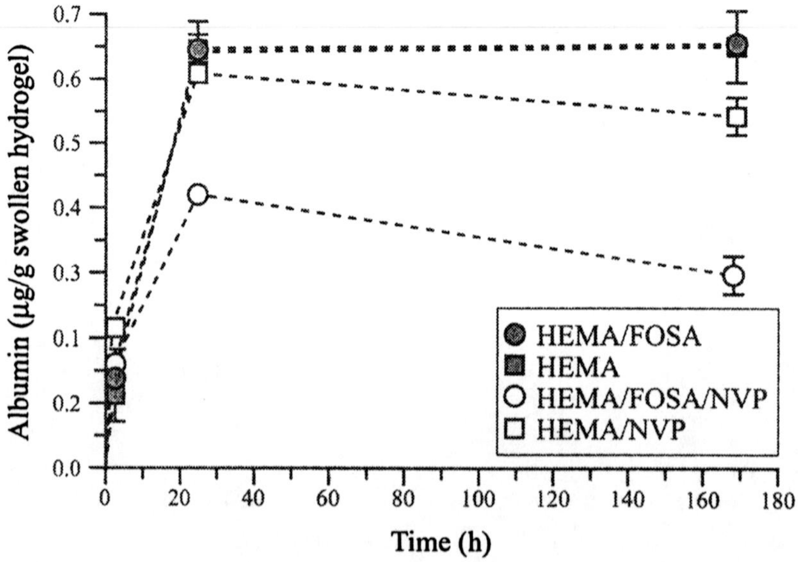

Figure 6. Albumin adsorption of hydrogels A, B, C, and D3 at 1 hr, 1 day, and 1 week.

4. CONCLUSIONS

Multi-component gels have been prepared from HEMA, NVP, FOSA, and PEGMA with various structures and properties. Blood and plasma proteins adsorption on PHEMA based gels can be reduced by the incorporation of the "hydrophilic" NVP or the "hydrophobic" FOSA. Indications have been obtained that the presence of both results in gels with the least protein adsorption.

ACKNOWLEDGEMENT

Financial supports form NSF and NIH are gratefully acknowledged. We thank Ms. Karan Pasquale and Ms. Victoria Wagner of the University of Connecticut Health Center for assistance in protein adsorption experiments.

REFERENCES

1. Huang, S. J., 1985, Biodegradable Polymers. In *Encyclopedia of Polymer Science and Engineering* (H. Mark, N. Bikales, C. G. Overberger, and G. Menges, eds.), John Wiley & Sons, New York, Vol. 2, pp. 220-243.

2. Huang, S. J., 1989, Biodegradable Polymers. In *Comprehensive Polymer Science*, (G. Eastmond, A. Ledwith, S. Russo, and P. Sigwalt, eds.), Pergamon Press, Oxford, Vol. 6, Chap. 21, pp. 597-607.
3. Huang, S. J., Bitritto, M., Leong, K. W., Pavilisko, J., Roby, M., and Knox, J. R., 1978, The Effects of Some Structural Variations on the Biodegradability of Step-growth Polymers. In *Stabilization and Degradation of Polymers*, Am. Chem. Soc., Chap. 17, pp. 205-214.
4. Huang, S. J., 1994, Biodegradation of Polymers. In *The Encyclopedia of Advanced Materials* (D. Bloor, R. J. Brook, M. C. Flemings, and S. Mahajan, eds.), Pergamon, Oxford, pp. 238-249.
5. Huang, S. J., Bell, J. P., Knox, J. R., Atwood, H., Bansleben, D., Betrotto, M., Borghard, W., Chapin, T., Leong, K. W., Natarjan, K., Nepumuceno, J., Roby, M., Sobosjai, J., and Shoemaker, N., 1977, Design, Synthesis and Degradation of polymers Susceptible to Hydrolysis by Proteolytic Enzymes. In *Proceedings of 3rd International Biodegradation Symposium* (J. M. Sharley and A.M. Kaplan, eds.), Applied Science Publishers, England, pp. 731-741.
6. DiBenedetto, L. J., and Huang, S. J., 1989, Biodegradable Hydroxylated Polymers as Controlled Release Agents. *Polym. Prepr.* **30**(1): 453.
7. DiBenedetto, L. J., and Huang, S. J., 1994, Poly(alkylene tartrate)s as Controlled Release Agents. *Polym. Degrad. Stab.* **45**: 249-257.
8. Huang, S. J., Ho, L.-H., Hong, E., and Kitchen, O., 1994, Hydrophilic-hydrophobic Biodegradable Polymers: Release Characteristics of Hydrogen-bonded Ring Containing Polymer Matrices. *Biomaterials* **15**: 1243-1247.
9. Williams, D. F., 1987, Definitions in Biomaterials. *Proc.a Consensus Conference of the European Society for Biomaterials*, Chester, England, March 3-5, 1986, Elsevier, New York, Vol. 4.
10. Ratner, B. D., Hoffman, A. S., Schoen, F. J., and Lemons, J. E., eds, 1996, *Biomaterials Science-An Introduction to Materials in Medicine*, Academic Press, San Diego.
11. Davis, P. A., Nicolais, L., Ambrosio, L., and Huang, S. J., 1988, Poly(2-hydroxyethyl methacrylate)/polycaprolactone Semi-interpenetrating Polymer Networks. *J. Bioact. Compat. Polym.* **3**: 205-218.
12. Davis, P. A., Huang, S. J., Nicolais, L., and Ambrosio, L., 1991, A Biodegradable Composite Artificial Tendon. *J. Mater. Sci.: Mater. Med.* **3**: 359-364.
13. Eschbach, F. O., and Huang, S. J., 1994, Hydrophilic-hydrophobic interpenetrating polymer networks and semi-interpenetrating polymer networks. In *Interpenetrating Network* (D. Klempner, L. H. Sperling, and L. A. Utrake, eds.), Am. Chem. Soc., pp. 205-219.
14. Brandrup, J., Immergut, E. H., 1989, *Polymer Handbook*, 3rd Ed., Wiley and Sons, New York.
15. Reddy, B. S. R., Arshady, R., and George, M. H., 1985, Copolymerization of N-vinyl-2-pyrrolidone with 2,4,5-trichlorophenyl acrylate and with 2-hydroxyethyl methacrylate. Reactivity ratios and molecular weights. *Eur. Polym. J.* **21** (6): 511-515.
16. Kelen, T., and Tüdös, F., 1975, Analysis of the linear methods for determining copolymerization reactivity ratios. I. New improved linear graphic method. *J. Macromol. Sci.-Chem.* **A9 (1)**: 1-7.
17. Künzler, J. F., and McGee, J. A., 1995, Contact lens materials. *Chem. Ind. London* **16**: 651-655.

Characteristics and Applications of Star-Shaped Architecture for PLA and PGCL

YOUNG HA KIM[*], SOO HYUN KIM[*], SEOUNG SOON IM[#], SOO HONG LEE[*], JI WON PACK[*], and JUN WUK PARK[#]
[*]*Biomaterials Research Center, Korea Institute of Science and Technology, P.O.Box 131, Cheongryang, Seoul, 130-650, Korea;* [#]*Dept. of Textile & Polymer Engineering, Hanyang University, 17 Haengdang-Dong, Seongdong-Gu, Seoul, 133-791, Korea*

1. INTRODUCTION

Several biodegradable polymers including polyglycolide (PGA), polylactide (PLA), and their copolymers are now widely used for biomedical applications in surgery, such as surgical suture, drug delivery systems, and internal bone fixation.

For these applications it is sometimes desirable to modify the physical properties and, especially, the degradation rate by copolymerization with other monomers, such as ε-caprolactone (CL) or trimethylene carbonate. Such an incorporation of chemically different comonomers induces a decrease of the crystallinity of copolymers to accelerate their degradation. For further structural modification, there are many studies on block copolymers[1,2] or star-shaped polymers[3,4], where multifunctional hydroxyl compounds were used as initiators for the ring-opening polymerization of these monomers. Branched or star-shaped polymers would be interesting as the physical properties and the degradation rate should be different from ordinary linear ones. In addition, the multi-functional end groups are very useful to investigate their effects on properties, such as degradation rate, or to incorporate biofunctional compounds to be coupled to the end groups.

The preferred route for the preparation of high molecular weight (MW) PGA or PLA is the bulk polymerization of glycolide or lactide in the presence of a suitable catalyst, such as stannous octoate (Sn-oct)[5], which was approved by FDA as a food additive. The mechanism of ring-opening polymerization of glycolide or lactide initiated with Sn-oct was reported somewhat controvertibly in several papers whether a coordinated insertion mechanism[6] or cationic one[7]. On the other hand, it is generally accepted that glycolide or lactide polymerization can be initiated by alcohols via a cationic ring-opening mechanism[8] where the alcohols act as molecular weight control agents.

In this study, lactide was polymerized by Sn-oct in the presence of polyfunctional alcohols such as glycerol or pentaerythritol. The resulting PLA had multi-armed chains to yield a star-shaped architecture. The microstructure, thermal properties, and degradation behaviour were studied to compare the effect of the different architecture, linear PLA with star-shaped one. In addition, it was very beneficial to study the effect of various chain end groups on the thermal and hydrolytic stability of this multi-armed PLA. Therefore, OH terminal groups of PLA were converted into Cl, NH_2 or COOH groups.

The star-shaped architecture was also applied to poly(glycolide-*co*-ε-caprolactone) (PGCL) for developing a flexible monofilament suture. The microstructure, thermal properties, and degradation rate of linear and star-shaped PGCL was studied in terms of the different chain structure. Additionally, the mechanical properties of drawn fibres made of linear and star-shaped PGCL were compared.

2. EXPERIMENTAL METHODS

Polymerization: Polymerization of L-lactide or glycolide/ε-caprolactone was described in detail elsewhere[9,10]. Briefly, freshly recrystallized monomers and various amounts of alcohols such as 1-dodecanol, glycerol or pentaerythritol (PET) were placed into the ampoule and Sn-oct in toluene was added. The ampoule was sealed in vacuum after 3 times nitrogen purging then heated up to the polymerization temperature. After the reaction was completed, the products were dissolved in chloroform, recovered by precipitation in methanol, and dried in a vacuum oven at 60 °C overnight. For comparison, the polymerization was also carried out with Sn-oct in the absence of polyfunctional alcohols.

Modification of end groups: The OH end groups of linear, three-armed, and four-armed OH-PLA were converted to Cl, NH_2, or COOH as reported in detail elsewhere[11]. Briefly, OH-PLA was reacted with thionyl chloride in toluene at 100 °C for 4 hrs by using pyridine as catalyst. Pyridine hydrochloride was filtered off and the residue was dissolved in chloroform and precipitated in methanol to yield Cl-PLA. Cl-PLA was reacted with sodium azide in N,N-dimethylformamide (DMF) at 80 °C for 2 hrs. DMF was then evaporated under vacuum, and the polymer, N_3-PLA, was recovered from chloroform solution by using methanol as a non-solvent. N_3-PLA was hydrogenated on Pd/C for 6 hrs in chloroform. After filtering the catalyst off, NH_2-PLA was isolated by precipitation into methanol.

OH-PLA was reacted with succinic anhydride in 1,4-dioxane overnight at room temperature in the presence of 4-dimethylaminopyridine (DMAP) and triethylamine (TEA). Dioxane was removed under vacuum, and the residue was dissolved in chloroform and then precipitated in methanol to get COOH-PLA. All polymer structures were confirmed by NMR; their molecular weight did not change after the modification of end groups.

Measurements: The structure of polymers was analyzed by a Brucker 200 MHz (AM-200) or 600 MHz (Varian Unity Plus 600) ^{1}H-NMR apparatus in $CDCl_3$ or hexafluoroacetone solution, and by an Alpha Centauri FT-IR (Mattson Instruments) apparatus. PLAs were also characterized by light scattering using 1,1,1,3,3,3-hexafluoro-2-propanol (HFIP) as solvent[9]. The intrinsic viscosity of the polymers was determined in chloroform at 25 °C. Size Exclusion Chromatography (SEC) measurements were carried out at 35 °C using a Waters ALC/GPC 150C equipped with μ-styragel columns and calibrated with polystyrene standards which covered a MW range of 1,000-1,000,000. Chloroform was used as eluent at 1.0 ml/min flow rate. Thermal analysis data were obtained by a Du Pont TA2000 DSC (Differential Scanning Calorimetry) at 10 °C/min heating rate in nitrogen atmosphere. The polymer crystallinity was determined by Wide-angle X-ray diffraction (WAXD) (MAX Science Co.) using Ni-filtered CuKα radiation at 45 kV and 250 mA at 2 °/min scanning rate.

3. STAR-SHAPED POLYLACTIDE

3.1 Synthesis

In this study, L-lactide was polymerized with Sn-oct in the presence of pentaerythritol (PET). The four primary hydroxyl groups are expected to initiate the polymerization of lactide yielding a four-armed structure as shown in Figure 1. Whether all four hydroxyl groups of PET are reacted to form a four-armed polymer can be investigated by analyzing the unreacted hydroxyl groups at the polymer chain ends. If the molecular weight of the resulting PLA is not too large, this analysis can be performed by NMR.

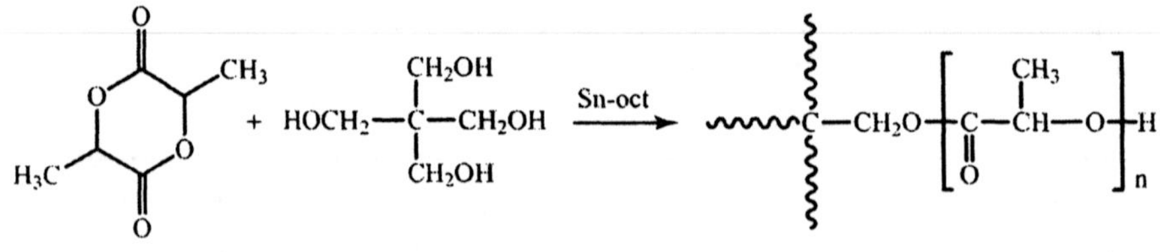

Figure 1. Scheme of the synthesis of four-armed star-shaped PLA[9].

Therefore, low molecular weight PLA oligomers were prepared in bulk with large amounts of PET and Sn-oct. The prepared branched PLA oligomers are viscous liquids or white solids with molecular weights below 5000 and melting points (T_m) far below that of ordinary PLA. However, the linear PLA polymerized with Sn-oct only is of completely different nature and is characterized by high MW and T_m. Figure 2 shows the NMR spectrum and the peak assignments of PLA oligomer prepared with [L]/[PET] molar ratio of 4.

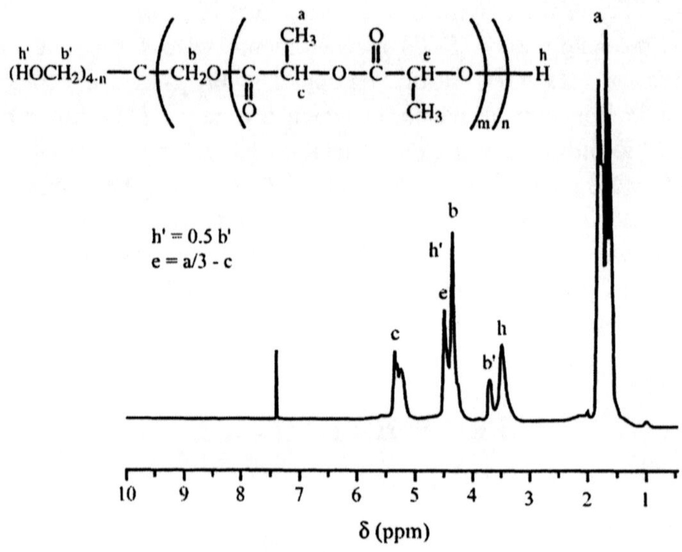

Figure 2. ^{1}H-NMR of the four-armed PLA oligomer[9].

From the conversion of PET reacted with lactide the number of chain arms could be evaluated, as shown in Figure 3. Under the assumptions that each PET molecules takes part in the lactide initiation and all lactide monomers reacted (this assumption seems to be valid, as discussed below), when the [L]/[PET] ratio is 4, theoretically all PET hydroxyl groups should react to form four-armed polymers (the dotted line in Figure 3). However, some unreacted PET hydroxyl groups were observed when [L]/[PET] was 4.

As the [L]/[PET] ratio was increased enough, the PET alcoholic groups completely reacted to form four-arm star-shaped molecules[9].

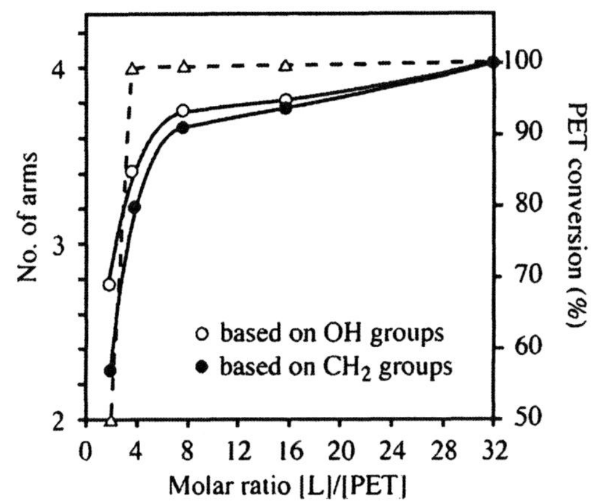

Figure 3. Number of arms of PLA prepared by using different [L]/[PET] molar ratios[9].

3.2 Characterization of star-shaped PLA

The molecular weight of branched and linear PLAs was evaluated by viscometry in $CHCl_3$ solutions and by light scattering in hexafluoro*iso*propyl alcohol solutions. Figure 4 shows a linear dependence of the degree of polymerizations (DP) of PLA on the [L]/[PET] ratio; DP coincides with the theoretical values (dotted line) calculated under the assumption that all PET hydroxyl groups participate in lactide initiation and all monomer is consumed[9]. This result strongly supports the living character of the lactide polymerization where PET might be a true initiator for the polymerization and Sn-oct might act just as a catalyst, in accordance with what was reported by many authors[12]. A. Kowalski *et al.* recently reported that tin alkoxides obtained by alcoholysis of Sn-oct are the initiating species[15]. The addition of lactide seems to proceed by cationic[7] or modified coordination mechanism[13,15].

High molecular weight PLA was also prepared by using [L]/[Sn-oct] molar ratios from 1,000 to 50,000. At low PET concentration, the molecular weight of star PLAs first increased with increasing [L]/[PET] molar ratio and then decreased to exhibit bell-shaped curves for both linear and star PLAs, witnessing the effect of transesterification in the later stages of polymerization. However, star-shaped PLAs were found to achieve higher molecular weights than linear ones, as there are four polymerizing centres in each star-architectured molecule to yield a higher molecular weight at lower conversion.

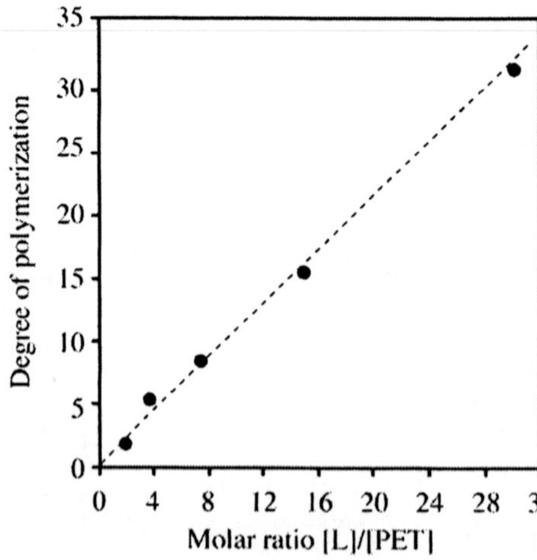

Figure 4. Dependence of the degree of polymerization of PLA on [L]/[PET] molar ratio[9].

Figure 5 shows the dependence of the intrinsic viscosity of linear ($[\eta]_L$) and star ($[\eta]_S$) PLAs in chloroform solution on the polymer molecular weight, as determined by light scattering in HFIP. It is evident that branched PLAs have smaller intrinsic viscosity than linear ones at the equivalent molecular weight. This is a clear demonstration of the branched architecture of PLA polymerized with PET, since branched polymers are known to have smaller hydrodynamic volumes than their linear counterparts.

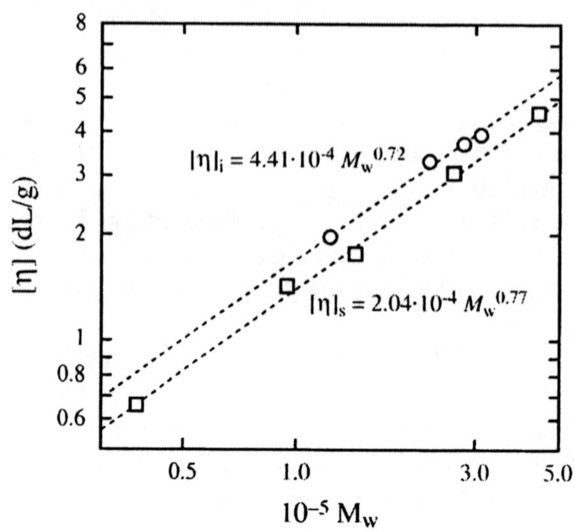

Figure 5. Mark-Houwink relationships for high molecular weight PLAs[10].

Mark-Houwink parameters can be evaluated from the plots in Figure 5 by least square analysis[10]:

$$[\eta]_L = 4.41\cdot10^{-4}\, M_w^{0.72}$$
$$[\eta]_S = 2.04\cdot10^{-4}\, M_w^{0.77}$$

The parameters obtained for linear PLA are slightly different from literature data[14] but still in reasonable agreement.

In addition, the second virial coefficient (A_2) can also be used as a measure of the branched architecture. The A_2 values of star PLAs measured by light scattering were smaller than those of linear ones, which is again consistent with the theoretical prediction and confirms a branched architecture of the PLAs[10].

4. DEGRADATION OF END GROUP MODIFIED POLYLACTIDES

4.1 Modification of End Groups of Star-shaped PLA

Aliphatic polyesters such as PGA, PLA, and their copolymers are degraded in the body by hydrolysis of ester groups. Therefore, the chemical structure, especially hydrophilicity determines the degradation rate so that PGA is the fastest, PLA slower, and PCL the slowest degrading polymer. It is generally accepted that degradation occurs preferentially in the amorphous regions of the polymer. The incorporation of co-monomers generally decreases the crystallinity and therefore accelerates biodegradation. In addition to the chemical structure and the presence of co-monomers, there are many other factors, such as molecular weight, crystallinity, surface area, end groups, additives, medium pH and temperature that may affect the degradation rate. Among those, the influence of hydroxyl or carboxyl end groups is often emphasized for the weak thermal stabilities of these aliphatic polyesters especially at thermal processing. The multi-armed structure of star-shaped PLA is very beneficial to investigate the effect of end groups on the degradation. In this study, three-armed and four-armed OH-PLAs were obtained by Sn-oct catalyzed polymerization of lactide with glycerol and PET, respectively.

Then, the hydroxyl end groups were converted to chlorine (Cl-PLA), amine (NH_2-PLA), and carboxyl (COOH-PLA) groups by reaction with thionyl chloride or sodium azide and then hydrogen or succinic anhydride, respectively, as shown in Figure 6 (see also section 1.1). The structures of polymers with different end groups were confirmed by NMR; as expected, the polymer molecular weight did not change after end group modification[11].

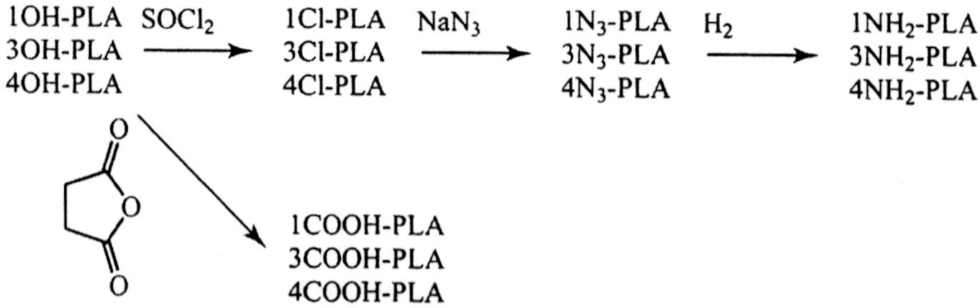

Figure 6. Schematic representation of the modification of end groups of multi-armed PLAs[11].

4.2 Thermal Property of PLA with Modified End Groups

Data relevant to the DSC analysis of PLAs with different end groups are shown in Table 1 and Figure 7. Four-armed and especially three-armed PLAs have lower T_m than linear PLA. 4Cl-PLA, 4NH_2-PLA, and in particular 4COOH-PLA have low T_m and ΔH_m (heat of melting) and therefore low crystallinity as compared with other PLAs. 4COOH-PLA exhibits the lowest crystallinity (37%). However, the cold crystallization temperatures are rather increased for star-shaped PLA. Lower melting points and higher cold crystallization temperatures are typical characteristics of star-shaped PLAs[16].

Figure 8 shows the TGA thermograms of four-armed functionalized PLAs in nitrogen atmosphere. The onset of the decomposition of 4OH-PLA occurs at about 210 °C, indicating a relatively poor thermal stability. This can be attributed to the thermally unstable nature of OH-PLA, which is easily decomposed to form the cyclic monomer. On the other hand, 4Cl-PLA possesses the highest thermal stability. Linear and three-armed PLAs show the same behaviour. The thermal degradation rate decreases in the order of 4OH-PLA > 4COOH-PLA > 4NH_2-PLA ≥ 4Cl-PLA. This order is not related with the polymer crystallinity as 4OH-PLA has the highest crystallinity (51 %) while 4COOH-PLA the lowest one (37 %)[11].

Table 1. Thermal properties and crystallinity (X_c) of PLAs with different end groups[11]

PLA Sample	T_g (°C)	T_c	T_m (°C)	ΔH_m (J/g)	X_c (%)
1OH	55	92	167	54	58
3OH	54	101	157	41	44
4OH	58	98	162	47	51
4Cl	58	128	163	40	42
4NH_2	55	123	162	40	43
4COOH	54	125	159	35	37

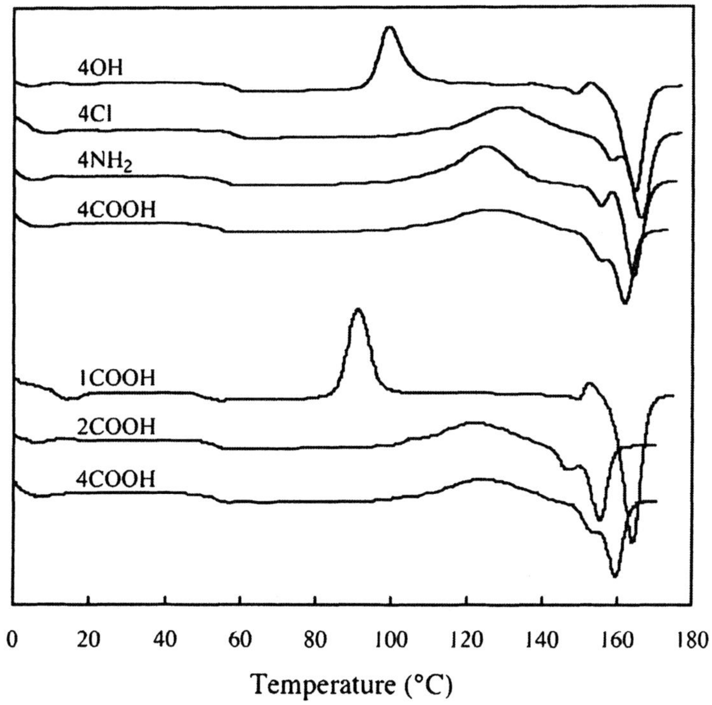

Figure 7. DSC thermograms of PLAs with different end groups[11].

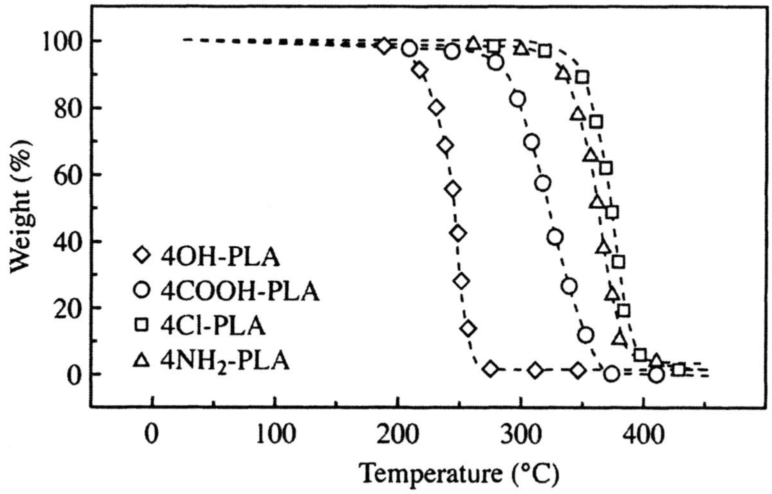

Figure 8. TGA thermograms of PLAs with different end groups[11].

4.3 Hydrolytic Degradation of PLA with Modified End Groups

The possible factors contributing to hydrolytic degradation of PLAs are the crystallinity, the nature of the end group and the number of terminal groups. Figure 9 illustrates the molecular weight reduction and the mass loss, respectively, which occurred by maintaining the investigated samples in pH 7.4 buffer solution at 80 °C.

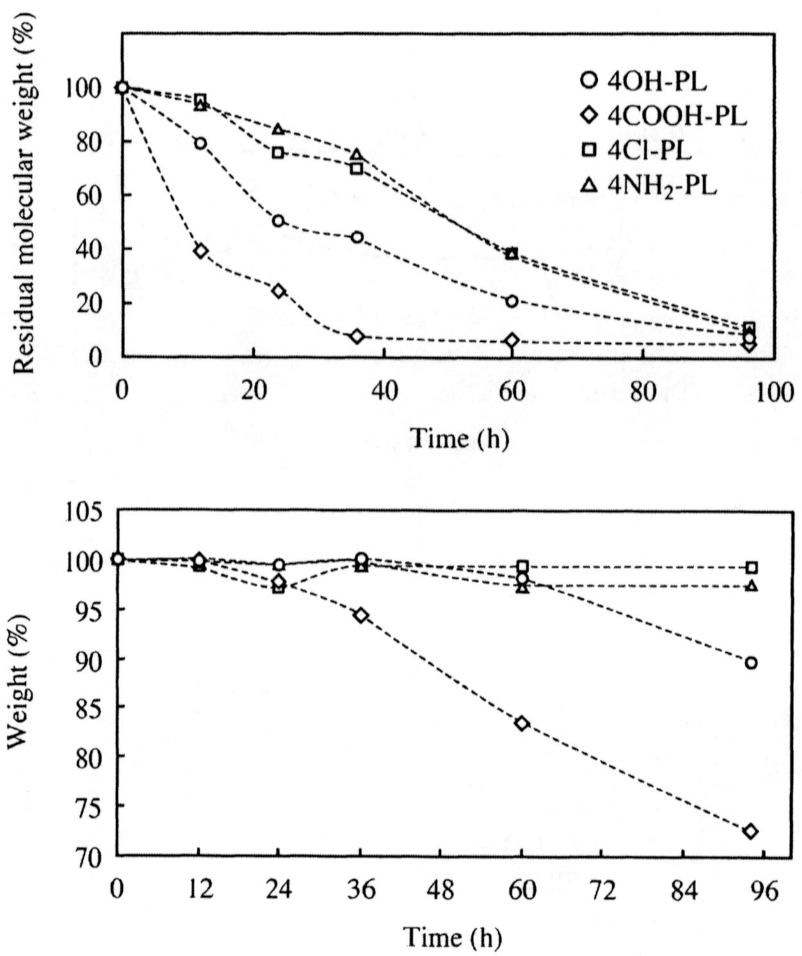

Figure 9. Hydrolytic degradation of PLA with different end groups[11]

The hydrolysis rate decreased in the order of 4COOH-PLA >> 4OH-PLA > 4Cl-PLA ≥ $4NH_2$-PLA. COOH-PLA was hydrolyzed faster than the other PLAs as its crystallinity is the lowest and the acidic end groups have a role as hydrolysis catalyst of the backbone ester bonds. It is well accepted that hydrolysis proceeds preferentially in the amorphous regions of low crystallinity. However, NH_2-PLA and Cl-PLA were more resistant to hydrolysis than OH-PLA due to the nonpolar character of NH_2 and Cl groups, although their crystallinity was lower than that of OH-PLA. In the case of NH_2-PLA, the NH_2 groups should complex the acidic by-products of hydrolysis and delay the hydrolysis process. Therefore, both the polymer crystallinity and the nature of end groups are important factors affecting PLA degradation[11].

5. STAR-SHAPED PGCL COPOLYMER

5.1 The Rationale

Surgical sutures are one of the most important applications of biodegradable aliphatic polyesters. PGA was the first product to replace catguts on the market. For biodegradable surgical sutures, it is important to have strong strength, proper degradation rate to maintain the strength for 2-6 weeks, and knot stability. PGA or PLGA (lactide content < 10 %) are too strong to make stable knots so that they should be braided with multi-filaments. Such a braided structure has disadvantages, such as hard to control degradation rate and ease of infection. Therefore, several flexible monofilament sutures made of polydioxanone or various copolymers of glycolide, lactide or ε-caprolactone (CL) have been developed.

In this study, star-shaped PGCL copolymers were prepared in order to investigate the relationships between their structure and their thermal and mechanical properties, in view of a possible application as flexible monofilament sutures. The concept of the preparation of star-shaped structures was extended to PGCL as shown in Figure 10.

5.2 Microstructure of Linear PGCL

First of all, various mixtures of glycolide (G) and ε-caprolactone (CL) were copolymerised with only Sn-oct to give linear PGCLs in high yields above 95%, as shown in Table 2. All copolymers were characterized by relatively high molecular weights, as indicated by their intrinsic viscosities, and their final compositions were not much different from those in the feed.

The PGA homopolymer exhibited T_g at 39.7 °C, T_m at 218 °C, and high crystallinity. PCL showed T_g at -58 °C, T_m at 58 °C, and was flexible.

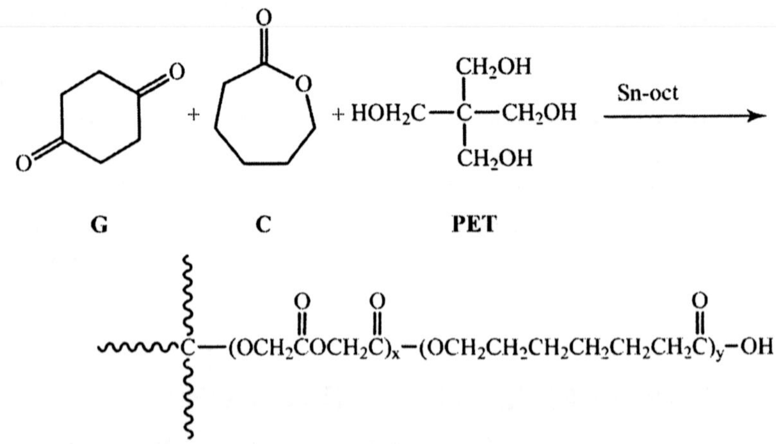

Figure 10. Scheme of star-shaped PGCL copolymerization.

The prepared copolymers were analyzed by 600 MHz ^{1}H-NMR and 50 MHz ^{13}C NMR to determine their monomer compositions and sequence distribution. Kricheldorf *et al.* reported the detailed NMR analysis of PGCL copolymers synthesized at 100 °C by a series of acidic and complexing initiators, except Sn-oct. They could not assign the peaks of the 270 MHz ^{1}H-NMR spectrum, but successfully attributed all signals present in the 22.63 MHz ^{13}C NMR spectrum[18].

Table 2. Microstructure and thermal properties of linear PGCL polymerized at 170 °C for 20 h, [M]/[Sn-oct] = 7,500/1[17]

Polymer [G]/[CL]	N_G/N_{CL} Unit ratio	T_g (°C)	T_m (°C)	ΔH_m (J/g)	X_c (%)	Average block length l_G	l_{CL}
100 / 0	-	39.7	218	111	52	-	0
88/12	7.3	-3.4	217	74	36	8.4	1.2
74/26	2.9	-13.7	215	37	16	3.5	1.2
65/35	1.8	-15.4	-	-	0	2.7	1.4
15/85	0.2	-53.2	43.9	46	28	1.2	5.9
0/100	0	-58.2	58.0	59	41	0	-

In Figure 11, the 600 MHz ^{1}H-NMR spectrum and the assignments of each peak of the 83/17 PGCL copolymer in the 4.5-5.3 region are illustrated. In this paper, glycolyl structural units are represented as G, although glycolide contains 2 G units. First of all, two protons present in G units and several protons in hydroxycaproyl (CL) units are observed enough apart each other to allow determining the monomer compositions. Furthermore, the H_G signal relevant to G units is split into four peaks (H_G, H_G‘, H_G“, and

H_G‘”), corresponding to four different monomer distribution triads (GGG, GGC, CGG, and CGC), respectively (see Figure 11)[18]. The proton peaks (not shown here) of CL units were split according to CC, CG or GC monomer distribution dyads.

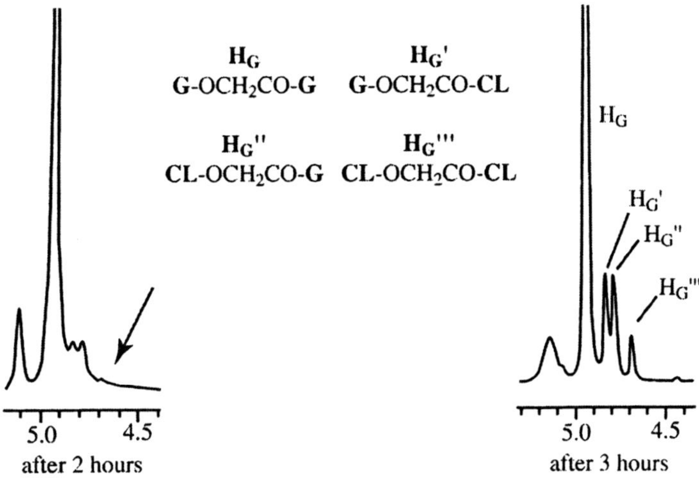

Figure 11. 600 MHz ^{1}H-NMR of linear PGCL in the 4.5-5.3 ppm region[17].

It is noticeable to observe the CGC peak (H_G‘”), because this triad cannot be incorporated in the polymer backbone by polymerization, as the glycolide monomer contains two G units. Therefore, this sequence should result from transesterification reactions occurring during the polymerization time.

The intensities of the CGC triad increased with increasing CL content. As the CGC peak was observed since 2 hrs after the polymerization start, the transesterification might occur even in the early step. In addition, the intensities of the H_G‘ and H_G“ peaks, attributed to the CGG and GGC triads, respectively represent the rates of the two crossover polymerization reactions, which should have the same probabilities[17]. The intensity of each peak (area) represents the probability of the formation of the corresponding dyad or triad, so that information on the copolymerization behaviour and the resulting polymer microstructure can be extracted. The calculated average block lengths of G units (l_G) and CL units (l_{CL}) are shown in Table 2. The N_G/N_{CL} and l_G/l_{CL} ratios, which in principle should be identical, are actually very close. It was further observed that both the copolymer G content and L_G decreased as the polymerization temperature was increased, in other words, the blocky character decreased and the randomness increased. This should be attributed to the larger enhancement of CL reactivity as compared with that of the glycolide monomer and/or the incidence of the transesterification was increased.

The reactivity ratios for PGCL copolymerization at 170 °C were calculated as $r_G = 6.84$ and $r_{CL} = 0.13$ to indicate the preferential reactivity of G over CL and the formation of a blocky structure[17].

The microstructure of linear PGCLs, especially monomer block lengths will affect crystallinity and thermal property of the copolymers. The T_g and T_m of the copolymers of various compositions are presented in Table 2 and Figure 12. PGA homopolymer exhibits T_g at 39.7 °C and T_m at 218 °C whereas T_g and T_m of PCL are –58.2 °C and 58.0 °C, respectively, indicating that at room temperature PGA is a stiff material whereas PCL is a soft one.

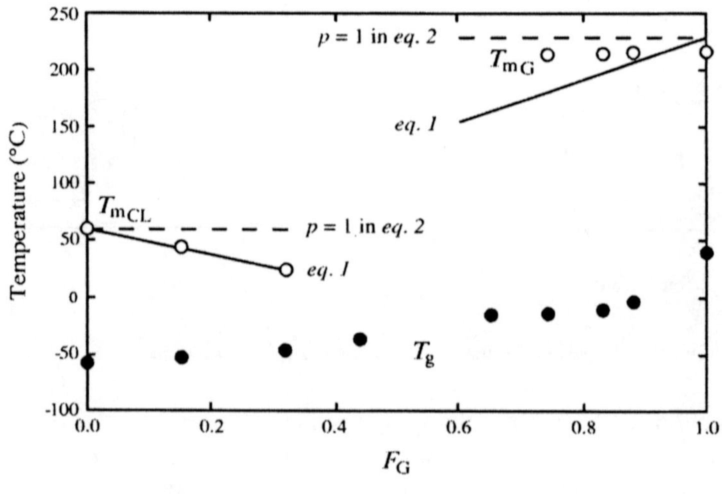

Figure 12. Thermal properties of linear PGCL on various mole fractions F_G[17].

The T_g of the copolymers decreased or increased from these values proportionally according to their compositions to exhibit an almost linear dependence as shown in Figure 12. On the contrary, the T_m of the copolymer crystalline regions appeared only at high molar [G] or [CL] contents, as expected. The dependence of T_m on the chemical composition of the copolymers having G content lower than 30 % fitted Flory's equation (eq. 1) that is usually applied to random copolymers:

$$1/T_m - 1/T_m^\circ = -R/\Delta H_m^\circ \ln X_i \quad (1)$$

where X_i indicates the mole fraction of the i-th component[17].

However, for copolymers having G content higher than 70 %, T_m did not follow equation 1 but rather equation 2 with $p = 1$:

$$1/T_m - 1/T_m^\circ = -R/\Delta H_m^\circ \ln p \quad (2)$$

where *p* denotes the sequence propagation probability. This indicates that G unit has a very high preference for addition in the copolymerisation steps.

Additionally, it is very interesting to note that the melting transition can be observed up to the 75/25 [G]/[CL] composition. At this composition, the N_G/N_{CL} ratio decreases to 2.9 and l_G to 3.5.

Even understanding that these N_G /N_{CL} or L_G values are statistically averaged ones and therefore some chains containing larger values can form crystals, such an observation is very unusual in common copolymers. Therefore, the copolymerization of G and CL should have a preference for the block formation in spite of an appreciable possibility of transesterification. The copolymers should be very heterogeneous systems containing crystalline domains, which are composed of relatively long PGA or PCL chain segments formed maybe in the initial polymerization period. The presence of crystallites was confirmed by WAXD. The PGA homopolymer exhibited diffraction peaks at $2\theta = 22.46°$ and 29.04°, whereas PCL at $2\theta = 21.58°$ and 23.88°. This confirmed once more that PGL and PCL homopolymers are highly crystalline materials with 52.2 and 41.4% crystallinity, respectively. The copolymer PGCL (75/25) demonstrated clearly a reflection pattern corresponding to the presence of PGA crystallites with 16.5% crystallinity[17].

5.3 Microstructure and Degradation of Star-shaped PGCL

Data relevant to four-armed PGCL copolymers prepared with Sn-oct and PET are summarized in Table 3. The T_g and T_m of linear PGCL are at −14.9 °C and 223 °C, respectively, but the star-shaped samples have higher T_g and lower T_m, which ranged between 212- and 218 °C. Both T_m and the heat of melting (ΔH_m) decreased with decreasing the monomers/PET molar ratio. In addition, l_G and l_{CL} sequence lengths also decreased with increasing the PET concentration as the polymerization might have been enhanced and therefore the randomness was increased.

The hydrolysis kinetics of drawn films of linear and star-shaped PGCL copolymers was investigated at different temperatures and draw ratios (DR). The recorded weight loss profiles are reported in Figures 13 and 14, respectively. It was found that all drawn films degraded slower than the control, and the hydrolysis was delayed proportionally to the draw ratio increase for linear PGCL, but not so regularly for star-shaped one. As compared with linear copolymers, star-shaped PGCL films degraded slower in the initial stage and faster in the final step. This is one of the characteristics of the multi-armed structure, which might be useful for their application as surgical sutures.

Table 3. Microstructure and thermal property of star-shaped PGCL copolymers prepared at 170 °C for 20 h, [monomers]/[Sn-oct] = 7,500/1, [G]/[CL] = 82/18

[Monomers]/[PET] Molar ratio	Polymer [G]/[CL]	T_g (°C)	T_m (°C)	ΔH_m (J/g)	Average block length l_G	l_{CL}
100	81/19	-7.2	212	32.8	5.3	1.3
1,000	81/19	-9.0	217	26.5	5.7	1.1
2,000	82/18	-12.3	216	32.0	5.9	1.1
5,000	83/17	-14.2	218	48.0	6.8	1.1
10,000	82/18	-10.9	218	45.9	6.6	1.3
no PET	**83/17**	**-14.9**	**223**	**56.4**	**7.2**	**1.3**

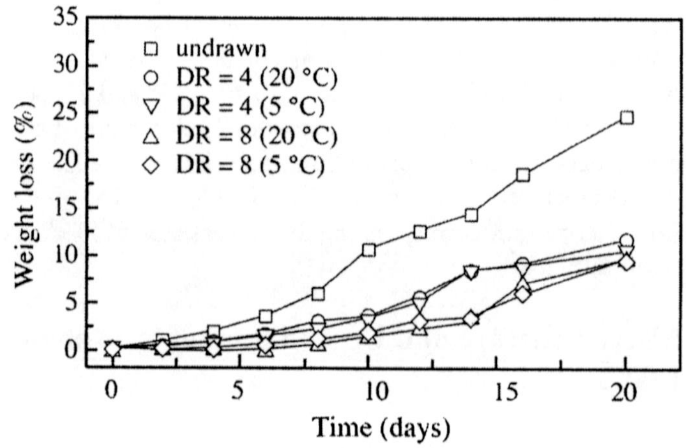

Figure 13. Hydrolytic weight loss of linear PGCL films at different temperatures and draw ratios (DR). Copolymer samples were prepared at 190 °C for 20 h, [M]/[Sn-oct] = 7,500/1, [G]/[CL] = 82/18.

Fibres made of star-shaped PGCL were more soft and flexible and showed lower modulus and larger elongation as compared with linear PGCL. However, the star-shaped PGCL fibres demonstrated enough strength and knot strength. In addition, the fibres made of star-shaped PGCL exhibited a higher degradation rate than those from the linear sample (Figure 15).

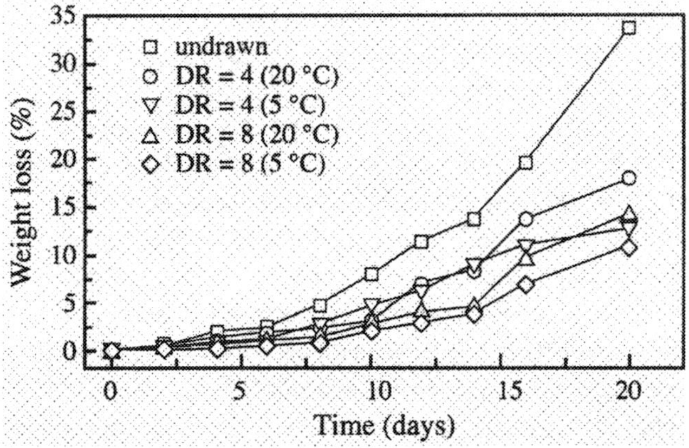

Figure 14. Hydrolytic weight loss of star-shaped PGCL films at different temperatures and draw ratios (DR). Copolymer samples were prepared at 190 °C for 20 h, [M]/[Sn-oct] = 7,500/1, [G]/[CL] = 82/18.

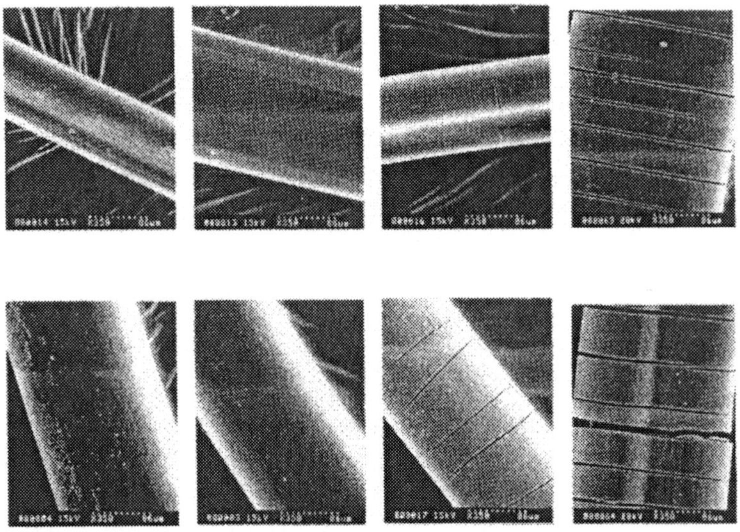

Figure 15. SEM micrograph (350X) of linear (top) and star-shaped (bottom) PGCL fibres after hydrolytic degradation for 2, 3, 4, and 6 weeks, respectively.

6. CONCLUSIONS

Lactide was polymerized by stannous octoate in the presence of polyfunctional alcohols such as glycerol and pentaerythritol. The resulting polylactide (PLA) had multi-armed chains to yield a star-shaped architecture. The T_m and degree of crystallinity of star-shaped PLA were lower than those of linear PLA, whereas the hydrolysis rate was enhanced very likely because of the larger number of end groups.

It was very beneficial to study the effect of various chain end groups on the thermal and hydrolytic stability of multi-armed PLA. Therefore, the OH terminal groups of multi-armed PLA were converted into Cl, NH_2, and COOH moieties. The melting points and crystallinity of the variously terminated PLA decreased in the order OH > Cl $\geq$ NH_2 > COOH. The thermal stability was rather poor for OH terminated sample and increased in the order OH << COOH < $NH_2 \leq$ Cl, whilst the hydrolytic stability was least for COOH end group and increased in the order COOH << OH < Cl $\leq NH_2$. These end group effects were more prominent upon increasing the number of chain arms.

The star-shaped architecture was also applied to PGCL for developing a flexible monofilament suture. In the linear copolymerization of PGCL at 170 °C, r_G and r_{CL} reactivity ratios were calculated from the monomer sequences as 6.84 and 0.13, respectively. This indicates a blocky character of glycolide units, which was confirmed by thermal and crystallographic analysis. Star-shaped PGCL, prepared in the presence of pentaerythritol exhibited lower melting point, crystallinity, and chain orientation than linear PGCL due to the different architecture. Drawn fibres made of star-shaped PGCL are more soft and flexible and show lower modulus and larger elongation than those made of linear PGCL. However, star-shaped PGCL fibres demonstrated enough strength and knot strength and higher degradation rate.

ACKNOWLEDGEMENTS

The authors thank Prof. T. Chang (Pohang Institute of Science and Technology) for the light scattering works. This work was supported by both the Korean Ministry of Science and Technology and Ministry of Welfare and Health.

REFERENCES

1. Schindler, A., Hibionada, Y. M., and Pitt, C. G., 1982, Aliphatic polyesters. III. Molecular weight and molecular weight distribution in alcohol-initiated polymerizations of ε-caprolactone. *J. Polym. Sci., Polym. Chem.* **20**: 319-326.
2. Cohn, D., and Younes, H., Biodegradable PEO/PLA block copolymers. 1988, *J. Biomed. Mater. Res.* **22**: 993-1009.
3. Brich, Z., and Kissel, T., 1985, Polyol esters, their preparation and use in depot forms of pharmacologically active agents. *Brit. Pat.* 2145422.

4. Bruin, P., Veenstra, G. J., Nijenhuis, A. J., and Pennings, A. J., 1988, Design and synthesis of biodegradable poly(ester-urethane) elastomer networks composed of non-toxic building blocks. *Makromol. Chem., Rapid Commun.* **9**: 589-594.
5. Gilding, D. K., and Reed, A. M., 1979, Biodegradable polymers for use in surgery. Polyglycolic/poly(lactic acid) homo- and copolymers. *Polymer* **20**: 1459-1464.
6. Leenslag, J. W., and Pennings, A. J., 1987, Synthesis of high-molecular-weight poly(L-lactide) initiated with tin 2-ethylhexanoate. *Makromol. Chem.* **188**: 1809-1814.
7. Nijenhuis, A. J., Grijpma, D. W., and Pennings, A. J., 1992, Lewis acid catalyzed polymerization of L-lactide. Kinetics and mechanism of the bulk polymerization. *Macromolecules* **25**: 6419-6424.
8. Frazza, E. J., and Schmitt, E. E., 1971, A new absorbable suture. *J. Biomed. Mater. Res.* **1**: 43-58.
9. Kim, S. H., Han, Y-K., Kim, Y. H., and Hong, S. I., 1992, Multifunctional initiation of lactide polymerization by stannous octoate/pentaerythritol. *Markromol. Chem.* **193**: 1623-1631.
10. Kim, S. H., Han, Y-K., Ahn, K-D., Kim, Y. H., and Chang, T., 1993, Preparation of star-shaped polylactide with pentaerythritol and stannous octoate. *Markromol. Chem.* **194**: 3229-3236.
11. Lee, S-H., Kim, S. H., Han, Y-K., and Kim, Y. H. , 2001, Synthesis and Degradation of End-Group-Functionalized Polylactide. *J. Polym. Sci., Polym. Chem.* **39**: 973-985.
12. Dubois, P., Jacobs, C., Jerome, R., and Teyssie, P., 1991, Macromolecular engineering of polylactones and polylactides. 4. Mechanism and kinetics of lactide homopolymerization by aluminum isopropoxide. *Macromolecules* **24**: 2266-2270.
13. Kricheldorf, H. R, Kreiser-Saunders, I., and Boettcher, C., 1995, Polylactones: Sn(II)octoate-initiated polymerization of L-lactide: a mechanistic study. *Polymer* **36**: 1253-1259.
14. Schindler, A., and Harper, D., 1979, Polylactide. II. Viscosity-molecular weight relationships and unperturbed chain dimensions. *J. Polym. Sci., Polym. Chem.* **17**: 2593-2599.
15. Kowalski, A., Duda, A., and Penczek, S., 2000, Kinetics and Mechanism of Cyclic Esters Polymerization Initiated with Tin(II) Octoate. 3. Polymerization of L,L-Dilactide. *Macromolecules* **33**: 7359-7370.
16. Kasperczyk, J., and Bero, M., 1993, Coordination polymerization of lactides. 4. The role of transesterification in the copolymerization of L,L-lactide and ε-caprolactone. *Makromol. Chem.* **194**: 913-925.
17. Pack, J. W., Kim, S. H., Kim, Y. H., Cho, I-W., and Park, S. Y., 2002, Microstructure analysis and thermal property of copolymers made of glycolide and ε-caprolactone by stannous octoate. *J. Polym. Sci. Polym. Chem.* 40: 544-554.
18. Kricheldorf, H. R, Mang, T., and Jonte, J. M., 1984, Polylactones. 1. Copolymerizations of glycolide and ε-caprolactone. *Macromolecules* **17**: 2173-2181.

Hydrogels And Hydrophilic Partially Degradable Bone Cements Based On Biodegradable Blends Incorporating Starch

LUCIANO F. BOESEL[1,2], JOÃO F. MANO[1,2], CARLOS ELVIRA[3], JÚLIO SAN ROMÁN[3], and RUI L. REIS [1,2]

[1]Dept. of Polymer Eng., Univ. of Minho, Campus de Azurém, 4800-058 - Guimarães, Portugal: [2]3B's Research Group, Univ. of Minho, Campus de Gualtar, 4710-057 - Braga, Portugal: [3]Institute of Science and Technology of Polymers, CSIC, c/Juan de la Cierva, 3, 28006 - Madrid, Spain

1. INTRODUCTION

Bone cements are, traditionally, injectable systems based on acrylic polymers. They are constituted by a solid and a liquid component that harden after mixing due to the polymerization of acrylic monomers in the liquid. Just some minutes after mixing, the paste attains high viscosity and must then be injected into the bone cavity, where the final stages of polymerization will take place. The solid is in most cases constituted by poly(methyl methacrylate) (PMMA) powder (or a copolymer of MMA with others monomers), benzoyl peroxide (BPO, the initiator of the polymerization), and a radio-opacifier, while the liquid is formed by MMA monomer (in some cases with *n*-butyl methacrylate, BuMA) and dimethyl-*p*-toluidine (DMT, the activator of the initiator).

Although bone cements are on the market since the 60's, only small variations in their compositions have been implemented, as the systems have performed reasonably well their function. This included the efficient transfer of body weight and service loads from the prosthesis to the bone and the increase of the load carrying capacity of the prosthesis – bone cement – bone system[1]. However, although at least 90% of the cemented arthroplasties are still in function after 15 years of implantation[2], the currently used formulations present yet several problems[1,3]. This includes: thermal necrosis of the bone and impaired local blood circulation due to the high temperature of the exothermic cement polymerization, chemical necrosis of the bone due to the release of unreacted MMA, shrinkage during polymerization, weakness of the cement (and of the interfaces) when compared with the prosthesis and the bone, interaction of the cement particles with the surrounding tissues (causing inflammatory responses and osteolysis of the bone), and stress shielding due to improper transfer of load causing changes of distribution of bone mass. The most frequent long-term complication in arthroplasties is aseptic loosening of the joint prosthesis, which is related to interfacial or bond failure, bone remodelling and mechanical failure of the bone cement.

The use of cementless arthroplasties, that are a former alternative to the drawbacks of bone cement, also presented several problems like high tight pain, failure of the bone-implant interface, and osteolysis of the bone[1,3,4]. Therefore, research efforts towards solving the main bone cement disadvantages were and are encouraged[5-7]. For that reason, several new formulations with enhanced mixing methods or based on reinforced cements, among other alternatives, have been developed. Strategies also comprised the increase of the cement-bone interfacial interaction by the incorporation of bioactive glasses in the formulations[5,6] or the substitution of PMMA by poly(ethyl methacrylate) (PEMA) and the use of BuMA instead of MMA for increasing the ductility of the cement[7]. However, such systems are still based on hydrophobic acrylic monomers that could still show some drawbacks.

Consequently, the substitution of MMA with an hydrogel forming monomer in the bone cement formulation would change its character to an hydrophilic one, making it possible to improve several of the cement properties or even providing it with new characteristics. Hydrogels can be obtained if cross-linked structures are formed, from the incorporation of multi-functional monomers or by association through secondary bonds (hydrogen bonding, strong van der Waals forces,...)[8]. Due to their high water uptake, they are able to swell high amounts of biological fluids, turning out to be very biocompatible systems. Their affinity towards water can be tailored by copolymerization with less hydrophilic (or hydrophobic) monomers, making them suitable to different applications.

For example, Netti et al.[9] have shown that the coating of a prosthesis with a hydrogel, (poly(hydroxyethyl methacrylate) - PHEMA) could act as the bone cement. After immersing the coated prosthesis in water (or after implantation in the body) the hydrogel absorbed water, generating hydrostatic pressure (due to the constrained volume) that could stabilize the implant.

This principle can be applied to bone cements. After polymerization of the (hydrophilic) monomer inside the femoral cavity and insertion of the prosthesis, the hydrogel would start absorbing water, due to the same constraints referred to above, fixing the implant in place. Since the bone cement must fulfil some minimum requirements in terms of mechanical properties, the combination of hydrophilic and hydrophobic (MMA) monomers will allow for the tailoring of water uptake and, consequently, the mechanical properties of the system.

Although these systems would have lower mechanical properties, their hydrophilicity would allow great improvements of other important parameters. For instance, Taguchi et al.[10,11] have shown that apatite (the same mineral present in bone) can be induced to grow not only on the surface but also inside a hydrogel by appropriately soaking it in Ca and P solutions. It was also shown that the amount of apatite formed increases with the increment of the swelling ratio. So one can anticipate that *in vivo* new bone would grow also inside the cement, making the incorporation of bioactive glass as reinforcing filler and the stress transfer more efficient and improving the interfacial strength cement-bone. The swelling associated with the water uptake would also compensate for the shrinkage during polymerization, press-fitting the implant and improving the strength of the whole system.

Since there are acrylic monomers hydrophilic enough to form hydrogels (e.g., acrylic acid (AA), HEMA,...) they could be incorporated into the bone cement formulation without changing the polymerization system, using the same initiator and activator. On the other hand, different acrylic monomers could be selected to tailor the extent of water uptake of the system or even to impart some additional behaviour, like pH- or temperature sensitivity.

For instance, hydrogels based on AA have been used due to their pH sensitive behaviour to release drugs in neutral pH environments[12]. The extent of swelling increases as the pH increases due to the ionization of the –COOH groups and the consequent electrostatic repulsion. By using neutral comonomers (such as HEMA, or MMA), the swelling, pH responsiveness and mechanical properties of the hydrogel can be adjusted for the intended application[12]. When grafted in other polymers, AA can provide high water uptake properties, while maintaining the bulk properties of the original polymer. In the last few decades, graft polymerization of AA and other vinyl monomers, such as MMA or acrylonitrile, onto starch has been a subject of academic and industrial interest since it is one of the most effective methods to incorporate desirable properties in starch without sacrificing its biodegradable nature[13,14]. Some researchers reported that poly(acrylic acid) or poly(methacrylic acid) - PAA and PMAA - give more stability to starch when grafted onto it, owing to the hydrogen bonding between their carboxylic groups and the –OH groups of starch. Clausen and Schnürch[15] used these interactions to produce PMAA-starch compositions for controlled drug delivery in neutral conditions, since at acid pH the carboxylic groups would not be ionized and could then protect the starch through hydrogen bonding. By altering the ratio PAA/starch, Athawale and Lele[14] could reach water uptake values of up to 250 g/g, confirming the very good swelling properties of these systems.

Based on these findings, our group started to study the possibility of incorporating an hydrogel forming monomer (AA) in the liquid component of the typical bone cement formulation, partially substituting MMA in order to adjust the mechanical and the swelling properties of the system. Alternatively, AA or mixtures of AA with other acrylic monomers, such as acrylamide (AAm) or bisacrylamide (B-AAm) replaced totally the MMA. As these later systems would present much lower mechanical properties and much higher degree of swelling, they were intended only for designing systems for controlled drug delivery.

The solid component, PMMA, was replaced by biodegradable blends of corn starch with different synthetic polymers, materials that have already shown potential for several biomedical applications, such as scaffolds for tissue engineering[16-18], systems for controlled drug delivery[19] and for bone replacement and regeneration[20,21]. The studied blends were two commercial ones, namely SEVA-C (corn starch/ethylene-*co*-vinyl alcohol) and SCA (corn starch/cellulose acetate), supplied by Novamont, Italy. The advantage of using biodegradable polymers, especially when combined with bioactive glasses, is the possibility of allowing tissue (in this case, bone tissue) growth into the volume previously occupied by the polymer being degraded. Additionally (regarding drug delivery systems), they would confer a second release mechanism to the system, that is a degradation-dependent release.

2. PREPARATION OF THE FORMULATIONS

The hydrogels and bone cement formulations presented in this work can be manipulated, prepared and applied using the classical technique of the acrylic self-curing bone cements and acrylic dental composites and sealants. This means that no change of surgical procedure is proposed or needed, what is very important from a practical clinical point of view. As a consequence of the polymerization reaction, a semi-interpenetrating polymer network with eventual graft copolymer chains of poly(AA-*co*-AAm) or poly(AA-*co*-MMA) onto the starch-based polysaccharides is formed by transfer reactions of the growing radicals on the side substituents of the pyranosyl cycles. In addition the incorporation of bisacrylamide with AA in some hydrogel formulations will give raise to chemically crosslinked hydrogels with improved mechanical properties. The composition of the prepared formulations is shown in Table 1.

Table 1. Composition of all prepared hydrogel and bone cement formulations

Formulation	Solid phase (S)	Liquid phase (L)	S/L ratio
Hydrogel	SEVA-C	Thermoplastic AA/AAm	
		65/35 or 80/20 or 90/10	60/40
		65/35 or 80/20 or 90/10	50/50
		65/35 or 80/20 or 90/10	40/60
		Crosslinked AA/B-AAm	
		97/03 or 99/01	60/40
		97/03 or 99/01	50/50
		97/03 or 99/01	40/60
	SCA	AA	50/50
			30/70
			15/85
Bone cement	SCA	MMA/AA	
		18/82	60/40
		74/26 or 85/15	59/41
		41/59 or 74/26	54/46
		35/65	50/50
		45/55	40/60
		41/59 or 74/26	32/68
		54/46	30/70
	SCA+(HA or sHA) [a]	MMA/AA	
		84/16	55/45
		76/24	50/50
		69/31	45/55

[a] HA = non-sintered hydroxyapatite, sHA = sintered hydroxyapatite.

The characterization of SCA thermoplastic hydrogels and SEVA-C crosslinked hydrogels (with B-AAm) by FTIR spectroscopy identified the most characteristic signals[22,23] of the carboxylic at 3435 cm^{-1} and carbonyl groups of AA at 1730 cm^{-1}, the hydroxyl groups of starch at 1242 cm^{-1} and the C-NH groups of B-AAm at 1409 cm^{-1}. SEVA-C thermoplastic hydrogels and bone cements formulations were also characterized by ^{1}H-NMR spectroscopy and the assignment of signals of the corresponding chemical groups: α-CH_3 of MMA at 0.9 ppm, -CH_2- and CH- of AA and –CH_2- of MMA at 1.9 ppm, -CH_2- and CH of starch and –OCH_3 of MMA and cellulose acetate at 3.4 ppm, and the –OH of starch in the range of 4.1-5.6

ppm[22,24]. The level of the residual monomer as determined by ^1H NMR Spectroscopy, was lower than the detection threshold of the technique, i.e. less than 1 mol% respect to the whole formulation[22].

3. CURING PARAMETERS OF THE BONE CEMENT FORMULATIONS

Following the ASTM F451-86[25] specifications about the curing parameters of acrylic bone cements, the peak temperature (T_{max} < 90 °C) that the mass can reach, the dough time (t_d, time between mixing phases and paste state < 5 min) and the setting time (t_s, the time at which the temperature equals (T_{max} + T_{amb})/2, should be between 5-15 min), were determined. These parameters are important due to the previously reported possible thermal necrosis of bone caused by the exothermal polymerization of the formulation. As the real *in vivo* maximum polymerization temperature attained by the system is much lower than the values measured by following the ASTM procedure[26], it was established that during curing the cement could not reach temperatures higher than 90 °C.

As it can be observed in Table 2, most of the 55/45 bone cement formulations meet the requirements of the ASTM specifications[22]. These formulations presented higher setting times than the others, being the 50/50 formulations in between those setting time values. Formulations with higher amounts of liquid phase (45/55) will employ less time to polymerize than 55/45, and consequently, t_s will be smaller. As it was not clear what produced this tendency (according to Table 1 it could be due to either higher solid amount or higher MMA/AA ratio, since 55/45 formulation contained only 7% of AA), further work was performed in order to clarify this point and separate the influence of S/L and MMA/AA ratios[27]. A full factorial design of experiments was performed. The obtained results shown that the increase in S/L in fact decreased t_s, but this effect was overcome by the positive effect of the MMA/AA ratio. So the lower t_s of 55/45 formulation is due to the higher MMA/AA ratio (84/16) when compared to the other two (76/24 and 69/31). Additionally, the increase in any ratio decreased T_{max}, explaining the lower polymerization temperature of the 55/45 formulation.

Table 2 also shows a tendency of t_s to increase when increasing the percentage of hydroxyapatite (HA), the bioactive filler added to the formulations in order to impart bioactive character to them. Non-sintered HA particles presented average size of 25-30 μm diameter. Sintered HA was prepared by heating HA at 1200 °C for 12 hours and subsequently crushing, milling and particle size classification (particles with an average diameter of

4.7 μm). The t_s was slightly larger for formulations containing non-sintered HA, for a given formulation and percentage[22].

Table 2. Curing parameters of 55/45 bone cement formulations. S: sintered HA; N: non-sintered HA. t_d: dough time. t_s: setting time. T_{max}: peak temperature

HA amount	t_d (min)	t_s (min)	T_{max} (°C)
0	2.00	5.39	87.7
5%S	1.90	5.60	92.2
5%N	1.87	7.04	90.4
10%S	1.58	6.50	93.2
10%N	1.27	6.92	87.9
20%S	1.75	7.20	85.4
20%N	1.47	7.53	85.5
30%S	2.47	8.48	91.8
30%N	1.80	8.17	88.9

4. DEGREE OF HYDRATION AND DEGRADATION BEHAVIOUR

The swelling behaviour of polymeric systems has a great importance when they are applied in the biomedical field as their hydration degree influences the surface properties and mobility, the mechanical properties and the type of solute transport mechanism through the hydrogels[28].

The hydration degree of the prepared hydrogels and bone cements can be modulated in the different formulations by the incorporation of different monomers (AA, AAm, MMA). In the case of the prepared SCA hydrogels it was found that the equilibrium hydration degree ranged from 43 to 1400% at 37 °C and pH 7, being the higher in those formulations more rich in AA due to its hydrophilic character. The most important characteristic of these systems was their sensitivity to both pH and temperature changes, which makes the hydration degree dependent on these type of stimuli. As it can be observed in Figure 1, the hydration degree varies in all formulations from pH 5 to 8, the highest hydration degree being observed at pH = 7. When the swelling is measured at different temperatures it was observed that the highest values are obtained at 50 °C as can be observed in Figure 1.

The kinetics of the sorption process was also studied, by means of equation (1).

$$M_t/M_{eq} = kt^n \qquad (1)$$

M_t and M_{eq} are the absorbed mass at time t and after equilibrium has been reached, respectively, k is a constant and n is the exponent, which indicates the type of diffusion transport in the hydration process. The kinetics was also dependent on the compositions of the prepared formulations as n was close to 0.5 (indicating Fickian type behaviour) for hydrogels containing MMA. When analysing formulations without MMA it was found that n was close to unity indicating non-Fickian behaviour and case II water transport mechanism, which is the most desirable kinetic behaviour for a swelling-controlled release material[23].

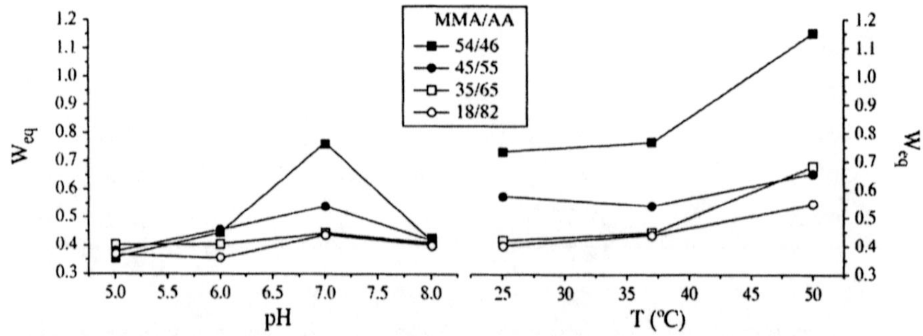

Figure 1. Dependence of the equilibrium hydration degree (W_{eq}) of bone cement formulations on the pH at 37 °C (left) and on the temperature at pH 7 (right).

SEVA-C hydrogels with AA, AAm and some crosslinked with B-AAm showed also an intimate dependence on the formulation composition as the higher hydration degrees were found for those formulations containing larger amounts of hydrophilic acrylic copolymers (40/60 and 50/50 S/L) with the equilibrium hydration degree ranging from 1150% for 50/50 (AA/AAm = 65/35) formulation to 350% for 50/50 (AA/B-AAm = 97/03) and 40/60 (97/03) at 37 °C and at pH 7.4. It was found that the hydration degree is higher in the case of the thermoplastic hydrogels with respect to the chemically crosslinked ones, as the diffusion of water molecules is more restricted in the latter. It was also observed that the hydration degree is sensitive to pH changes, being higher at physiological pH, similar to what has been reported for bone cements (Figure 1). In terms of the swelling kinetics, the exponent n becomes again near unity (transport II case) when AA increases in the formulation (50/50 and 40/60 S/L) whereas it is in between 0.5 and 1 for 60/40 S/L formulations showing a non-Fickian behaviour. In terms of degradation of these SEVA-C based hydrogels, the incorporation of polar acrylic copolymers of AA and AAm gives a resorbable character to the developed hydrogels, as they are partially soluble in water. Figure 2 represents the weight loss diagrams versus immersion time in simulated physiological solution of 50/50 (S/L) formulations. The prepared formulations exhibit a weight loss clearly dependent on the composition. In that sense, thermoplastic hydrogels show a quick initial weight loss step at 3 and 7 days of immersion, which after analysis of the residue by ^{1}H-NMR spectra was assigned to an initial solubilization of the acrylic copolymers AA-AAm and plasticizers of the SEVA-C blend. It can be concluded that as the acrylic phase is higher (40/60) in the formulation, the thermoplastic hydrogels lost more weight as a consequence of the water solubility of the AA-AAm copolymer. Glycerol was also released, whereas in the case of hydrogels richer in SEVA-C the weight loss follows a similar behaviour but starch also starts to be degraded after 60 days of immersion. In the case of all chemically crosslinked hydrogels, the weight loss is stabilized after 15 days, ranging from 8 % in 40/60 to about 6% in 60/40 samples. In all cases the weight loss is attributed to the leaching of glycerol and to the solubilization of the non-crosslinked poly-AA, as determined ^{1}H NMR analysis[24].

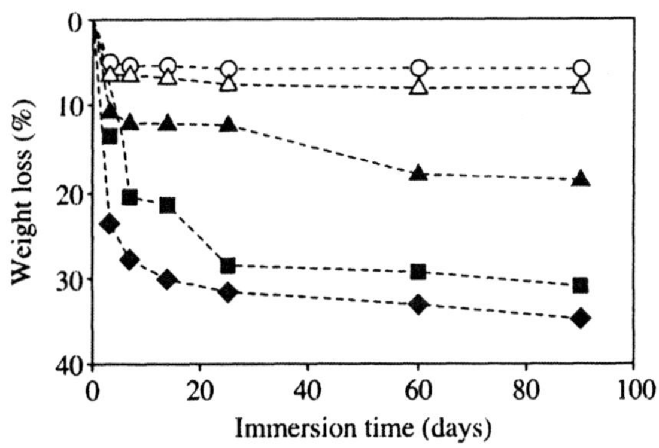

Figure 2. Plots of weight loss (WL) versus immersion time of SEVA-C 50/50 (S/L) hydrogels. AA/AAm: (◆) 65/35, (■) 80/20, (▲) 90/10. AA/B-AAm: (△) 97/03, (○) 99/01.

In the prepared bone cement formulations the hydrophilic/hydrophobic balance is important, as it will be controlled by the bone cement composition in which MMA gives the hydrophobic character, and SCA and AA the hydrophilic one. The polymerization reaction in acrylic bone cements causes a volume reduction[29], which can be compensated for with the uptake of physiological fluids. After three days of immersion in saline solution, all formulations seemed to reach equilibrium water uptake (for the specimen geometry used) that ranged from 27-33% in 45/55 (S/L), to 15-20% in 55/45 formulations[22]. For a given solid/liquid ratio, the equilibrium hydration degree was o dependent on the percentage of HA, whereas the type of HA, sintered or not, did not affect the equilibrium hydration degree. The degradation process should have a large importance in these formulations because it will be ideally associated with the generation of new bone tissue. This should allow the cement to maintain the support structure of the cemented system, while in growth of new tissue occurs. The degradation behaviour of the bone cements give a weight loss of about 12% in all prepared formulations over a period of 90 days and the composition does not affect the weight loss[22]. The weight loss was assigned to the release of glycerol (used in processing SCA blends) and low molecular weight chains of starch and cellulose acetate, observed by ^{1}H NMR spectroscopy. These weight loss profiles indicate that the developed bone cements will keep their support structures after a long period of time although enzymatic hydrolysis (due to the presence of α-amylase in human blood) will contribute to accelerate the *in vivo* degradation process of cements containing the degradable starch component.

5. MECHANICAL PROPERTIES

5.1 Quasi-static Behaviour

For the cements, the modulus at 1% strain varied between 3.2 and 5.6 GPa (tensile) and between 1.6 and 2.3 GPa (compression)[22]. This parameter tends to increase as the HA content is incremented. The ultimate tensile and compressive strengths (UTS and UCS) tend to increase as the HA content becomes higher. This is true up to 20% of HA (up to 16-20 MPa for UTS and 70-98 MPa for UCS, depending on the solid/liquid ratio). Above that values, both UTS and UCS tend to decrease, as the ceramic particles start to act as stress concentration points, giving rise to more brittle materials with lower UTS and UCS[22]. When compared with commercial acrylic bone cements, it can be concluded that the developed cements show similar (or slightly lower) UTS and similar or superior UCS and modulus.

The hydrogel materials in the dry state (xerogels) show UTS and UCS ranging from 13 to 34 MPa and from 39 to 98 MPa, respectively, and tensile and compression moduli ranging from 0.9 to 3 GPa and from 0.15 to 1.9 GPa, respectively[24]. There is a tendency for an increase of all mechanical properties with the increase of AAm content in the thermoplastic formulations, but no clear correlations are found with the solid/liquid ratio. The crosslinked materials show improved mechanical properties when compared with the thermoplastics, especially for higher crosslinking densities and higher SEVA-C contents[24].

As it is usual for hydrogels, the immersion of specimens in water (or other fluids) brings about a decrease of mechanical strength and modulus, due to the plasticizing effect of water. On the other hand, the ductility and toughness are greatly improved, making the materials able to deform to higher extents without fracturing. This could be observed in both compression and tensile tests, as shown in Figure 3. It is clearly seen that samples with brittle fracture became very ductile after seven days immersion in isotonic saline solution (ISS) and did not fracture during the tests (carried out until 60% of decrease in height). A similar behaviour was seen in tensile tests; for the shown formulation, the ductility increased four times and the total energy at break of the specimens was doubled. The fractured surfaces, analysed by SEM, proved this behaviour: before immersion, the surface was smooth, denoting a rapid failure after crack nucleation around a pore or an heterogeneity; after immersion, pores continued to be crack generators, but the surface was a rough one, typical of ductile fractures[27].

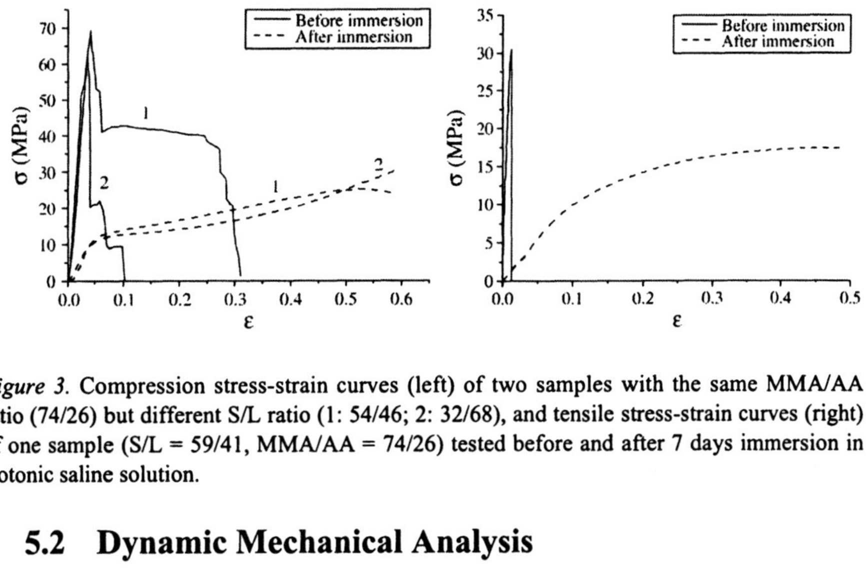

Figure 3. Compression stress-strain curves (left) of two samples with the same MMA/AA ratio (74/26) but different S/L ratio (1: 54/46; 2: 32/68), and tensile stress-strain curves (right) of one sample (S/L = 59/41, MMA/AA = 74/26) tested before and after 7 days immersion in isotonic saline solution.

5.2 Dynamic Mechanical Analysis

Dynamic mechanical analysis (DMA) is a suitable technique that allows the characterisation of the solid-state rheological behaviour of materials, including biomaterials[30], in a broad temperature and frequency ranges. Specifically, this technique has been used in the characterisation of bone cements[31-35] or hydrogels[36-38]. Such materials display usually an anelastic behaviour and DMA is able to monitor the complex mechanical modulus ($E^* = E' + iE''$, where E' is the storage modulus and E'' is the loss modulus, and the complex compliance ($D^* = D' - iD''$). The loss factor, $\tan\delta = E''/E' = D''/D'$, measure the damping capability of the material.

The storage modulus was measured at 37 °C[22], showing values between 1 and 2 GPa. This parameter can be related to the stiffness of the material.

However, it was found that the obtained values are lower than the modulus obtained from quasi-static measurements, indicating that a direct comparison between the results from both techniques is not straightforward. A weak tendency of increasing E' with increasing HA content is seen up to 20% ceramic content. As observed before, this formulation also optimised the ultimate strength, being in principle the material with better mechanical performance.

DMA experiments on the bone cements were carried out in isochronal conditions (temperature scan at 1 Hz). The glass transition was clearly observed by the peak in $E''(T)$, where the maximum temperature can quantify the glass transition temperature, T_g. An example is shown in Figure 4 where E' and E'' of two samples are shown as a function of temperature. The results for a commercial formulation (Palacos®-R, Merck, Germany) were also included for comparison. The peak at higher temperatures corresponds to the dynamic glass transition (α-relaxation), where cooperative motions arise within the polymer backbone. The peak at 20-30 °C is the thermally activated β-relaxation, assigned to local motions within the side-groups. This broad process may have influence on the viscoelastic properties of bone cements as it provides mechanisms for energy dissipation.

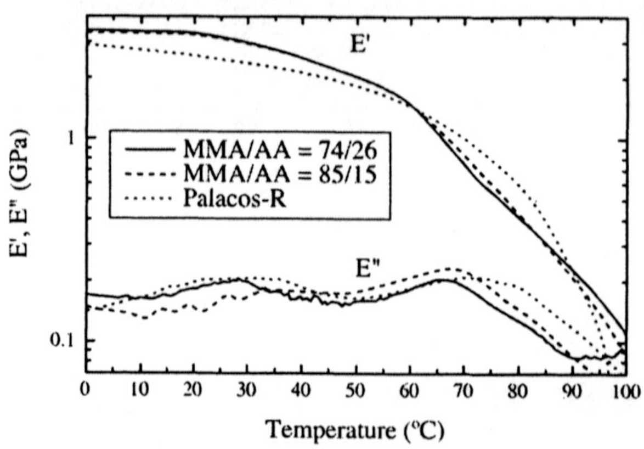

Figure 4. DMA results at 1 Hz obtained for two dry samples (S/L = 59/41) having different MMA/AA ratio and for commercial bone cement Palacos®-R.

It was found for the investigated bone cements that T_g could vary between 67 and 77 °C, with no correlation with the solid/liquid ratio or the HA content[22]. For the hydrogel systems, a stronger dependence of T_g on the composition was found in xerogels[24]. At constant solid/liquid ratio, T_g strongly decreases with increasing AA content, due to the higher flexibility of such chains relative to AAm sequences. The glass transition varied between 40 and 70 °C (maximum of tan δ). These values are much lower than the T_g of the pure AA and AAm homopolymers (~106 and ~165 °C, respectively); this result can be attributed to the effect of water, especially in depressing the formation of hydrogen bonds within the polymer chains. DMA is also suitable for performing experiments on wet materials. Those tests provide more realistic information concerning the true viscoelastic behaviour of the materials under physiological conditions. Figure 5 shows the flexural storage modulus of the hydrogels in the hydrated state at 37 °C (f = 1 Hz). The results point out a decrease of E' with increasing the AA content in the thermoplastics. As expected, the crosslinked materials show higher storage modulus. Within the experimental error, one may detect a tendency of increasing E' with increasing solid/liquid ratio, that may be explained by the stiffer nature of the solid phase.

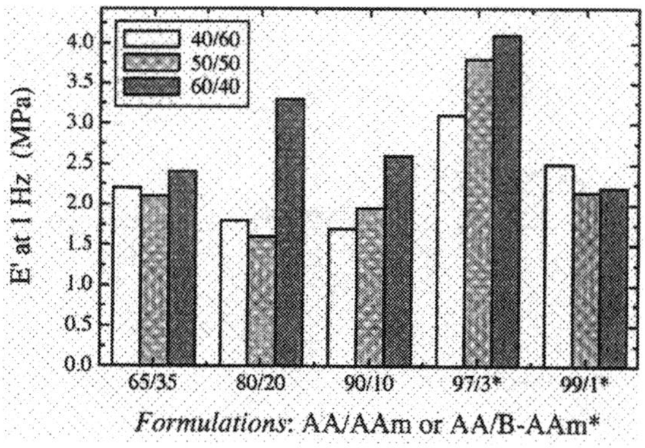

Figure 5. Flexural storage modulus of hydrated hydrogel formulations (kept in water for 24 hours) at 37 °C and 1 Hz.

5.2.1 Creep

Creep experiments may provide complementary information on the viscoelastic behaviour of biomaterials, especially if the tests are performed in simulated physiological conditions. Creep tests also allow for a better prediction of the long-term *in vivo* performance of biomaterials. It has been suggested that creep of acrylic cement allows the expansion of the cement mantle and subsequent prosthetic subsidence without causing cement fracture, besides relaxing cement stresses and creating a more favourable stress distribution at the interfaces[39]. So, this is an important property to be characterized.

In Figure 6 the creep/recovery experiments are performed while the samples are immersed in ISS at 37 °C. It can be concluded that creep occurs mainly in the formulation A. Comparatively, the commercial sample shows almost no creep, which may be related to its very low water-uptake capability. The recovery results for samples A and B shows that a fraction of the creep strain does not completely recover for long times, indicating that an irreversible viscous component takes place in the complete mechanical behaviour of the materials. This behaviour could be assigned to the presence of water (viscous material) in the systems.

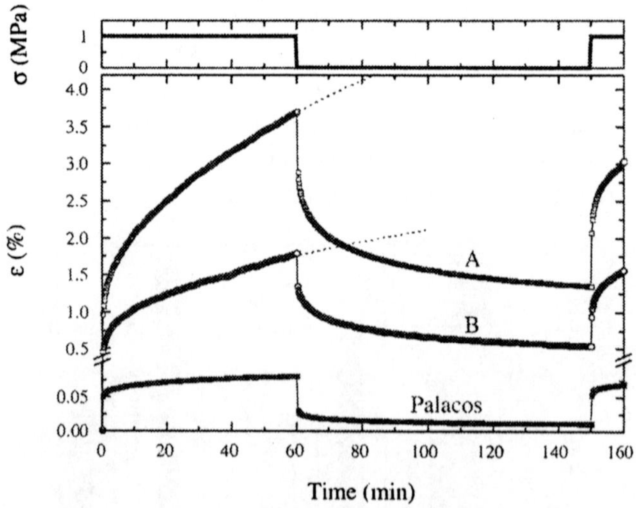

Figure 6. Creep/Recovery results obtain ed for hydrated samples at 37 °C. The stress program is shown in the top graphics. The dotted lines are the hyperbolic sine fitting[27] of A and B data. Formulations A (S/L=59/41, MMA/AA=74/26), B (S/L=59/41, MMA/AA=85/15) and Palacos®-R are the same as in Figure 4.

6. BIOACTIVITY TESTS

The eventual formation of a Ca-P layer on the surface of bone cement formulations after incubation in simulated body fluids (SBF) will indicate that the material may present a bone-bonding behaviour *in vivo.* All compositions with HA amounts below 20% exhibited a bioinert behaviour with no formation of Ca-P layer similar to PE (used as negative control). SEM micrographs of 55/45 30N surfaces (Figures 7a and 7b) show the formation of an apatite-like calcium phosphate layer (confirmed by X-Ray diffraction) after immersion in SBF for 7 days. The Ca/P ratios, as determined by energy dispersive spectroscopy (EDS), were always between 1.5 and 1.7 that is between tricalcium phosphate and hydroxyapatite. For HA amounts of at least 20% the developed cement formulations are clearly bioactive, showing a so-called "cauliflower" morphology[40], and are expected to present an *in vivo* bone-bonding nature[22].

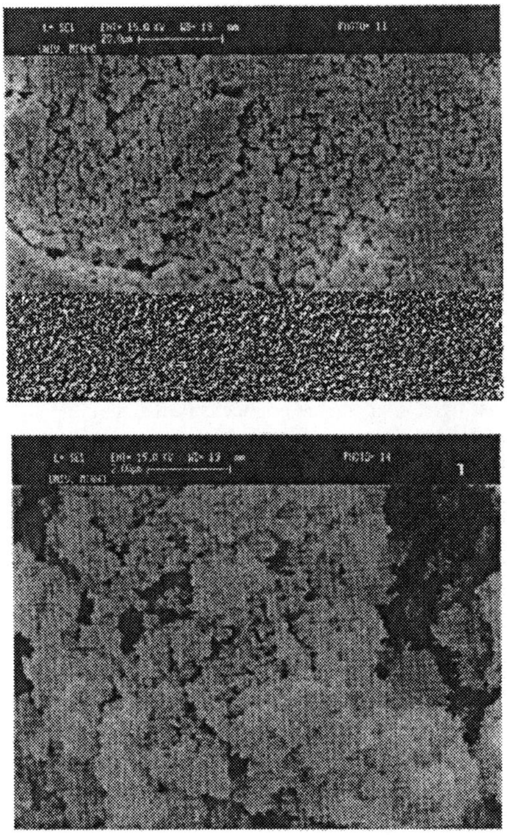

Figure 7. Scanning Electron micrographs of 55/45 (S/L) 30 N (30% of non-sintered HA), magnifications 1000X (top), and 10000X (bottom).

7. CONCLUSIONS

The systems presented in this paper have shown their suitability to be used on load-bearing applications (bone cements) or drug delivery systems (hydrogels). Changes in their composition allow the tailoring of mechanical, degradation and bioactive properties according to the desired application. Moreover, the proposed formulations were designed to be applied also as drug delivery systems; in this case, the release profile would be controlled by two mechanisms (swelling and degradation), allowing for a higher flexibility in the design and applications of the system. Concerning bone cements, their hydrophilicity, (partial) degradability and bioactivity (imparted by HA or bioactive glasses) would combine to form systems able to induce bone growth inside the material, greatly improving the adhesion to bone and the mechanical properties of the whole construct.

REFERENCES

1. Lewis, G., 1997, Properties of acrylic bone cement: state of the art review. *J. Biomed. Mater. Res.* **38**: 155-182.
2. Nafei, A., Kristensen, O., Knudsen, A. M., Hvid, I., and Jensen, J., 1996, Survivorship analysis of cemented total condylar knee arthroplasty: a long-term follow-up report on 348 cases. *J. Arthroplasty* **11**: 7-10.
3. Deb, S., 1999, A Review of improvements in acrylic bone cements. *J. Biomater. Appl.* **14**: 16-48.
4. Katz, J. L., 1996, Application of Materials in Medicine and Dentistry: Orthopedic Applications. In: *Biomaterials Science* (B.D. Ratner, A.S. Hoffman, F.J. Schoen and J.E. Lemons, eds.), Academic Press, San Diego, pp. 335-346.
5. Shinzato, S., Nakamura, T., Kokubo, T., and Kitamura, Y., 2001, A new bioactive bone cement: effect of glass bead filler content on mechanical and biological properties. *J. Biomed. Mater. Res.* **54**: 491-500.
6. Shinzato, S., Nakamura, T., Kokubo, T., and Kitamura, Y., 2002, PMMA-based bioactive cement: effect of glass bead filler content and histological change with time. *J. Biomed. Mater. Res.* **59**: 225-232.
7. Harper, E.J., Braden, M., and Bonfield, W., 2000, Mechanical properties of hydroxyapatite reinforced poly(ethylmethacrylate) bone cement after immersion in a physiological solution: influence of a silane coupling agent. *J. Mater. Sci. Mater. Med.* **11**: 491-497.
8. Peppas, N. A., 1996, Classes of Materials Used in Medicine: Hydrogels. In *Biomaterials Science* (B.D. Ratner, A.S. Hoffman, F.J. Schoen and J.E. Lemons, eds.), Academic Press, San Diego, pp. 60-64.
9. Netti, P. A., Shelton, J.C., Revell, P. A., Pirie, C., Smith, S., Ambrosio, L., Nicolais, L., and Bonfield, W., 1993, Hydrogels as an interface between bone and an implant. *Biomater.* **14**: 1098-1104.
10. Taguchi, T., Kishida, A., and Akashi, M., 1999, Apatite formation on/in hydrogel matrices using an alternate soaking process: II: effect of swelling ratios of poly(vinyl alcohol) hydrogel matrices on apatite formation. *J. Biomater. Sci. Polym. Ed.* **10**: 331-339.
11. Taguchi, T., Kishida, A., and Akashi, M., 1999, Apatite formation on/in hydrogel matrices using an alternate soaking process: III: effect of physico-chemical factors on apatite formation on/in poly(vinyl alcohol) hydrogel matrices. *J. Biomater. Sci. Polym. Ed.* **10**: 795-804.
12. Qiu, Y., and Park, K., 2001, Environment-sensitive hydrogels for drug delivery. *Adv. Drug Del. Rev.* **53**: 321-339.
13. Athawale, V.D., and Lele, V., 2000, Syntheses and characterisation of graft copolymers of maize starch and methacrylonitrile. *Carbohyd. Polym.* **41**: 407-416.
14. Athawale, V.D., and Lele, V., 1998, Graft copolymerization onto starch. II. Grafting of acrylic acid and preparation of it's hydrogels. *Carbohyd. Polym.* **35**: 21-27.
15. Clausen, A. E., and Schnürch, A. B., 2001, Direct compressible polymethacrylic acid-starch compositions for site-specific drug delivery. *J. Control. Release*, **75**: 93-102.
16. Gomes, M. E., Ribeiro, A. S., Malafaya, P. B., Reis, R. L., and Cunha, A. M., 2001, A new approach based on injection moulding to produce biodegradable starch based polymeric scaffolds. *Biomater.* **22**: 883-889.
17. Gomes, M. E., Reis, R. L., Cunha, A. M., Blitterswijk, C. A., and de Bruijn, J. D., 2001, Cytocompatibility and response of osteoblastic-like cells to starch based polymers: effects of several additives and processing conditions. *Biomater.* **22**: 1911-1917.

18. Gomes, M. E., Godinho, J. S., Tchalamov, D., Cunha, A. M., and Reis, R. L., 2002, Alternative tissue engineering scaffolds based on starch: processing methodologies, morphology, degradation, mechanical properties and biological response. *Mater. Sci. Eng. C.*, **20**: 19-26.
19. Malafaya, P. B., Elvira, C., Gallardo, A., San Román, J., and Reis, R. L., 2001, Porous starch-based drug delivery systems processed by a microwabe treatment. *J. Biomater. Sci. Polym. Ed.* **12**: 1227-1241.
20. Sousa, R. A., Kalay, G., Reis, R. L., Cunha, A. M., and Bevis, M. J., 2000, Injection molding of a starch/EVOH blend aimed as an alternative biomaterial for temporary applications. *J. Appl. Polym. Sci.*, **77**: 1303-1315.
21. Reis, R. L., Cunha, A. M., and Bevis, M. J., 1997, Using nonconventional processing to develop anisotropic and biodegradable composites of starch-based thermoplastics reinforced with bone-like ceramics. *Med. Plast. Biomater.* **4**: 46-50.
22. Espigares, I., Elvira, C., Mano, J. F., Vázquez, B., San Roman, J., and Reis, R. L., 2002, New partially degradable and bioactive acrylic bone cements based on starch blends and ceramic fillers. *Biomater.* **23**: 1883-1895.
23. Pereira, C. S., Cunha, A. M., Reis, R. L., Vázquez, B., and San Román, J., 1998, New starch–based thermoplastic hydrogels for use as bone cements or drug-delivery carriers. *J. Mater. Sci. Mater. Med.* **9**: 825-833.
24. Elvira, C., Mano, J. F., San Román, J., and Reis, R. L., 2002, Starch-based biodegradable hydrogels with potential medical applications as drug delivery systems. *Biomater.* **23**: 1955-1966.
25. ASTM specification F451-86, 1986, Standard specification for acrylic bone cement. In *Annual Book of ASTM Standards: Medical Devices; Emergency Medical Services.* American Society for Testing and Materials, Philadelphia.
26. Kühn, K.D., 2000, *Bone cements.* Springer, Heidelberg.
27. Boesel, L. F., Mano, J. F., and Reis, R.L., 2002, Optimization of the formulation and mechanical properties of starch based partially degradable bone cements. *J. Mater Sci.: Mater. Med.*, submitted.
28. Mack E.J., Okano T, Kim S.W., 1987, Biomedical applications of poly(HEMA) and its copolymers. In: *Hydrogels in Medicine and Pharmacy-Polymers*, vol. II, (N.A. Peppas, ed), CRC Press: Boca Raton, p.65-93
29. Haas S.S., Brauer G.M., and Dickson M.A., 1975, A characterization of poly(methylmethacrylate) bone cement. *J. Bone Joint Surg.* **57**: 380-391.
30. Mano, J. F., Reis, R. L., and Cunha, A. M., 2002, Dynamic mechanical analysis in polymers for medical applications. In: *Polymer Based Systems on Tissue Engineering, Replacement and Regeneration* (R. L. Reis and D. Cohn, eds.), Nato Science Series, Kluwer Academic, Dordrecht, pp.139-164.
31. Papadogiannis, Y., Lakes, R. S., Petrou-Americanos, A., and Theothoridou-Pahini, S., 1993, Temperature dependence of the dynamic viscoelastic behavior of chemical-and light-cured composites. *Dent. Mater.* **9**: 118-122.
32. Vaidyanathan, J., and Vaidyanathan, T. K., 1995, Dynamic mechanical analysis of heat, microwave and visible ligh cure denture base resins. *J. Mater. Sci. Mat. Med.* **6**: 670-674.
33. Vallo C. I., Cuadrado T. R., and Frontini, P. M., 1997, Mechanical and fracture behaviour evaluation of commercial acrylic bone cements. *Polym. Int.* **43**: 260-268.
34. Yang, J.-M., Huang, P.-Y., Yang, M.-C., and Lo, S. K., 1997, Effect of MMA-*g*-UHMWPE grafted fiber on mechanical properties of acrylic bone cements. *J. Biomed. Mater. Res. (Appl. Biomater.)* **38**: 361-369.

35. Yang, J.-M., Li, H.-M., Yang, M.-C., and Shih, C.-H., 1999, Characterization of acrylic bone cements using dynamic mechanical analysis. *J. Biomed. Mater. Res. (Appl. Biomater.)* **48**: 52-60.
36. Kalachandra, S., Minton, R. J., Takamata, T., and Taylor, D. F., 1995, Characterization of commercial soft liners by dynamic mechanical analysis. *J. Mater. Sci. Mat. Med.* **6**: 218-222.
37. Saber-Sheikh, K., Clarke, R. L., and Braden, M., 2000, Viscoelastic properties of some soft lining materials. II- Ageing characteristics. *Biomater.* **20**: 2055-2062.
38. Sheu, M.-T., Huang, J.-C., Yeh, G.-C., and Ho, H.-O., 2001, Characterization of collagen gel solutions and collagen matrices for cell cultures. *Biomater.* **22:** 1713-1719.
39. Verdonschot, N., and Huiskes, R., 1997, Acrylic cement creeps but does not allow much subsidence of femoral stems. *J. Bone Joint Surg.* **79-B**: 665-669.
40. Reis R.L., Cunha A. M., Fernandes M.H., and Correia R.N., 1997, Treatments to induce the nucleation and growth of apatite-like layers on polymeric surfaces and foams. *J. Mater. Sci. Mater. Med.* **8**: 897-905.

Segmented Polyetheresters Containing Hydrogen Bonding Units

FRANCESCA SIGNORI[1], ROBERTO SOLARO[1], EMO CHIELLINI[1], PRISCILLA A. M. LIPS[2], PIETER J. DIJKSTRA[2], and JAN FEIJEN[2]
[1]*Department of Chemistry and Industrial Chemistry, University of Pisa, via Risorgimento 35, 56126 Pisa, Italy.*
[2]*University of Twente, Dept. of Chemical Technology, P.O. Box 217, 7500 AE Enschede, The Netherlands.*

1. INTRODUCTION

In recent years, the growing interest in the delivery of protein and peptide drugs stimulated the development of new materials with tailored properties. Requisites to be fulfilled are the non-denaturation of the incorporated protein as well as the loaded drug and the possibility of controlling both the mechanism and kinetics of polymer degradation and drug release[1].

In this respect, natural and synthetic hydrogels have been extensively investigated because of their proven compatibility with protein and peptide drugs. In addition, biodegradable amphiphilic di-, tri-, and multi-block copolymers have been prepared and investigated during the past years, in order to achieve a better control over the degradation behaviour and to obtain devices specifically designed for high performance release applications (temperature dependent gel formation[2], self assembly in nanospheres[3], 3D scaffolds able to deliver growth factors[4]). In general, these polymers are made of hydrophilic flexible segments (soft blocks), and of hydrophobic blocks (hard blocks), which act as physical crosslinks in the soft hydrophilic matrix. In aqueous environments, these systems are expected to self-organize in an outer hydrophilic surface and an inner hydrophobic core. The final polymer properties, especially in terms of degradation and release behaviour are strongly dependent upon the ratio between hard and soft blocks, while the mechanical properties are expected to be influenced to a larger extent by the chemical nature of the hard blocks. Due to its unique biocompatibility, poly(ethylene glycol) [PEG] is generally elected as the hydrophilic component, whereas various biodegradable hydrophobic polymers, such as polylactide[5], polyglycolide[6], and their copolymers, poly(ε-caprolactone)[7], and poly(butylene terephthalate)[8] have been investigated during recent years as hydrophobic block.

Following these concepts, aliphatic poly(ether ester amide)s containing specifically designed short ester amide blocks have been recently synthesized[9]. These materials do combine the high compatibility properties of polyethers, the degradability of polyesters, and the good mechanical properties of crystalline H-bonded polyamides. It was envisaged that the physical crosslinking of the polymer matrix comprises both hydrophobic interactions and a strongly organized H-bonded network. Accordingly, multi-block copolymers based on PEG 1,4-butanediol, and short ester-amide segments with defined structures have been prepared. It was found that the material properties depend on and can be modulated by the PEG content. Protein release from films and microspheres made of these materials is controlled by the degree of swelling and molecular weight.

In recent years the concept of introducing H-bonding units into polyesters was used for the synthesis of poly(ester-amide)s and poly(ester-urethane)s, in order to improve the mechanical properties of already known and applied biodegradable polyesters like poly(butylene adipate)[10]. Polycondensation and interfacial polycondensation reactions were applied to obtain materials from various diacids and diamines, amino alcohols or amino acids. In general these materials are biocompatible, biodegradable, and very promising for applications in the biomedical and pharmaceutical field, such as controlled drug delivery devices[11-15].

In the present research we have been focussing on the preparation of biodegradable amphiphilic block copolymers comprising different hydrogen bonding hard segments, and soft PEG sequences. Besides the diester diamide monomers, diester dicarbamate and diester diurea derivatives were used as monomers, in order to evaluate the influence of the different hydrogen-bonding moieties on the final properties of the materials. Moreover, the effects of the type of H-bonding groups, the length of the polymethylene spacer between functional amide groups, and the PEG content on polymer physical-chemical properties were extensively investigated, in order to correlate the material structural features with both their bulk and surface properties.

2. RESULTS AND DISCUSSION

2.1 Monomer Synthesis

Symmetric diester monomers containing H-bonding moieties, such as amide, carbamate, and urea groups spaced apart by polymethylene segments of different length (Figure 1) were synthesized as reported elsewhere[16]. These amide containing monomers were prepared by $Ti(OBu)_4$ catalyzed aminolysis of methyl adipate and methyl suberate, whereas carbamate and urea containing monomers were prepared by alcoholysis and aminolysis of hexamethylenediisocyanate, respectively. Following these synthetic strategies, which make use of cheap and commercially available starting materials, monomers which are structurally similar but differ in the nature of the H-bonding units (amide in M4A4 and M6A6, carbamate in E5C6, urea in E3U6) and in the length of the spacer in between the H-bonding groups (six methylene chain in M6A6, E5C6, and E3U6, four methylene chain in M4A4) were obtained in good yields.

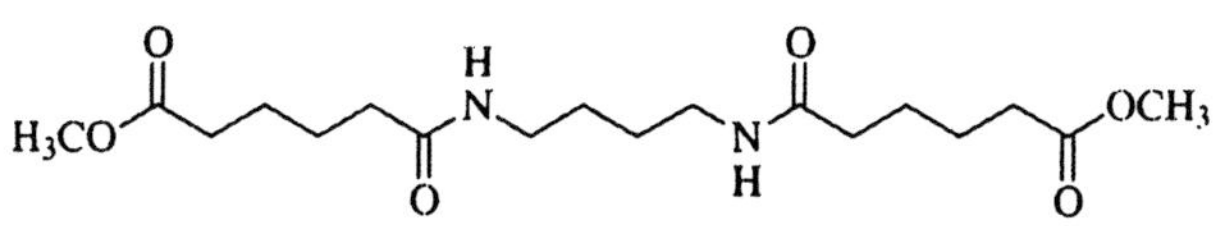

Dimethyl 7,12-diaza-6,13-dione-1,18-octadecanodioate

M4A4

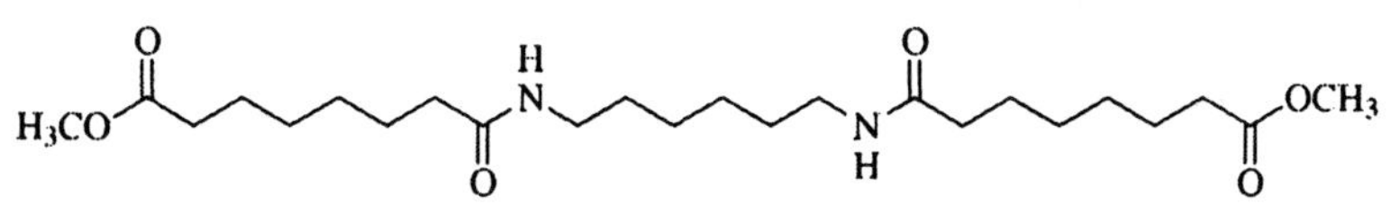

Dimethyl 9,16-diaza-8,17-dione-1,24-tetracosanedioate

M6A6

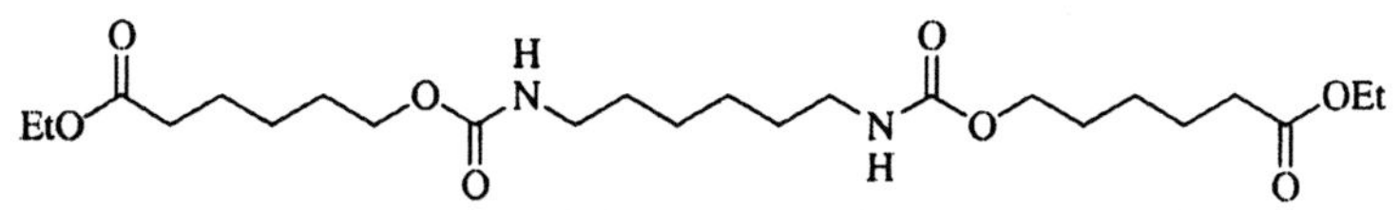

Diethyl 9,16-diaza-7,18-dioxa-8,17-dione-1,24-tetracosanedioate

E5C6

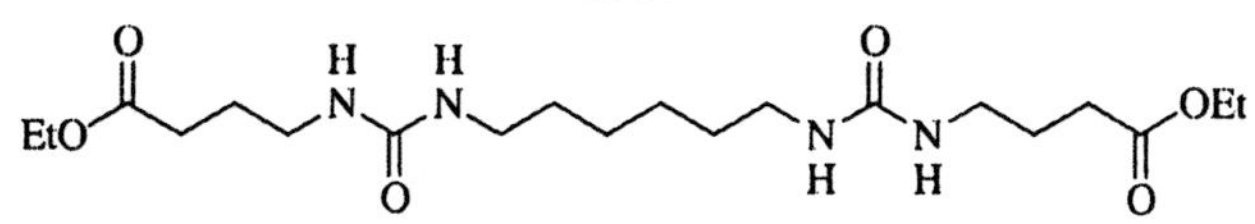

Diethyl 5, 7, 14, 16-tetraaza-6,15-dione-1,20-eicosanedioate

E3U6

Figure 1. Structures of the synthesized monomers.

Spectroscopic (^{1}H-NMR, ^{13}C-NMR, FT-IR) and thermal (DSC, TGA) properties, and the solubility behaviour of the prepared monomers were thoroughly investigated[16]. The thermal analysis of the monomers appeared important in perspective of using these materials as building block in the polycondensation reactions. The thermal stability of the synthesized monomers was determined with respect to the processing window of polycondensation reactions. Amide containing monomers resulted stable up to 300 °C, whereas the degradation of urea and especially carbamate based monomers started at temperatures lower than 200 °C.

Moreover, particular attention was paid to the melting behaviour, since the synthesized monomers are expected to determine the crystallization behaviour and hence the mechanical properties of the corresponding polymeric materials. A single sharp endothermic transition was observed for the carbamate monomer **E5C6**. On the other hand, the diurea monomer **E3U6** showed two partially overlapping peaks whose position and intensity was independent of the annealing at different temperatures. Since the samples were pure based on GC analysis, this behaviour is very likely due to the presence of less perfect crystals and/or different crystal structures.

2.2 Polymers

Polycondensation of the prepared monomers containing H-bonding units (**M4A4, M6A6, E5C6** and **E3U6**) and PEG1000/1,4-butanediol mixtures afforded the poly(ether ester amide)s, poly(ether ester carbamate)s, poly(ether ester urea)s, esteramide/urea and estercarbamate/urea copolymers, respectively (Scheme 1).

Polymerization experiments were performed in bulk by a two-step procedure. After initial transesterification of the monomers with PEG and 1,4-butanediol at 100 °C, the temperature was increased stepwise while allowing the excess of 1,4-butanediol to distil off and affording high molecular weight products. The final temperatures were optimized to avoid monomer thermal degradation and interchange reactions. Therefore, taking into account the thermal degradation behaviour of the starting monomers, the final temperature was set at 220, 160, and 180 °C for amide, carbamate, and urea based monomers, **M4A4, M6A6, E5C6** and **E3U6** respectively. Polymerization time was varied from 12 to 30 hours, while a final pressure of 0.1-0.05 mbar was always achieved. The feed compositions were set to give polymers containing 30, 40, 50, and 70% by weight of hydrophilic PEG-based soft blocks. Polymers containing 30%-wt of soft segments and 5 or 10%-wt of **E3U6** comonomer were also prepared. Data relevant to the polycondensation reactions are summarized in Table 1.

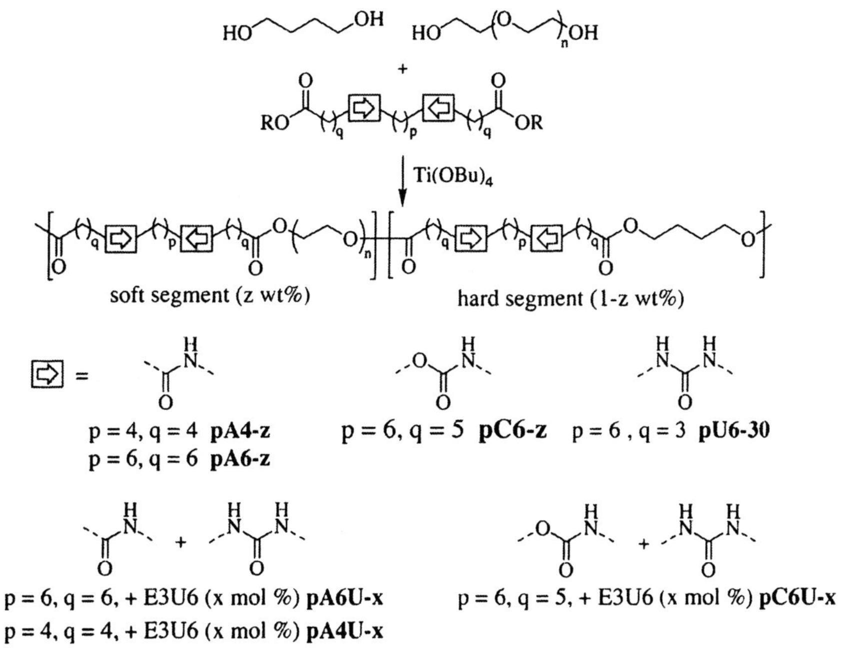

Scheme 1. Polycondensation of H-bonding monomers with PEG1000/1,4-butanediol mixtures.

Generally, the prepared polymers were obtained in good yields. The prepared materials were characterized by spectroscopic analysis (^{1}H-NMR, ^{13}C-NMR, FT-IR). In particular, evaluation of diagnostic peaks belonging to the different repeating units allowed for the determination of the polymer chemical composition. In all cases, the PEG content in the copolymers was found to be slightly higher than that of the corresponding polymerization feed, suggesting that small amounts of 1,4-butanediol distilled off the reaction mixture during the polymerization process. Moreover, in no case the presence of signals attributable to end-groups was detected, suggesting that molecular weights higher that 15000 were obtained in all cases. Amide containing polymers showed a regular sequence of monomeric units along the polymer chain, whereas irregular chain segments were found in carbamate containing materials, independent of the feed composition. This behaviour was attributed to the occurrence of carbamate alcoholysis, which took place simultaneously to the main transesterification process.[16] SEC analysis was not possible due to by the polymer solubility in organic solvents. An indication of the molecular weights could only be obtained from viscosity measurements. The polymer intrinsic viscosities were in between 0.56 and 1.47 dl/g. These values are close to those found for poly(ether ester amide)s of comparable structure that were previously prepared by similar synthetic routes[17]. It is worth noting that no significant crosslinking process occurred during the polycondensation, as indicated by the polymer solubility and viscometric measurements performed using different polymer concentrations. Indeed, a linear relationship between polymer concentration and relative viscosity was observed in all cases.

Table 1. Polycondensation reactions of H-bonding monomers (M_I) with mixtures of PEG1000 and 1,4-butanediol

Run	Polymerization conditions					Yield	Polymer	
	M_I	E3U6 (mol %) [c]	PEG1000 [a] (% mol)	Final T (°C)	Duration (h)	(%)	PEG1000[a,b] (% mol)	[η] (dl/g)
pA4-30a	M4A4	0	8	220	10	67	13	0.53
pA4-30b	M4A4	0	10	220	24	68	12	0.86
pA4-40	M4A4	0	11	165	35	84	22	0.36
pA4-50	M4A4	0	17	165	35	65	21	0.32
pA4-70	M4A4	0	35	165	35	79	39	0.33
pA4U-5	M4A4	5	14	165	35	60	14	0.29
pA4U-10	M4A4	10	14	165	35	55	14	0.21
pA6-30	M6A6	0	10	220	24	30	15	1.47
pA6-40	M6A6	0	13	165	35	89	19	0.40
pA6-50	M6A6	0	20	165	35	81	25	0.32
pA6-70	M6A6	0	36	165	35	53	50	0.42
pA6U-5	M6A6	5	17	165	35	56	18	0.41
pA6U-10	M6A6	10	14	165	35	55	13	0.42
pC6-30	E5C6	0	10	165	35	66	13	0.55
pC6-40	E5C6	0	14	165	35	35	19	0.49
pC6-50	E5C6	0	21	165	35	63	24	0.56
pC6-70	E5C6	0	38	165	35	79	45	0.61
pC6U-5	E5C6	5	9	165	35	53	11	0.39
pC6U-10	E5C6	10	9	165	35	45	11	0.53
pU6-30	E3U6	100	12	165	35	63	n.d.	n.d.

[a] Evaluated as 100 mol PEG/(mol PEG + mol 1,4-butanediol). [b] Evaluated by ^{1}H–NMR analysis. [c] Referred to M_I.

Material bulk and surface characterization was carried out in the perspective of taking into account the influence of the nature of the H-bonding unit, the PEG content, and the length of the polymethylene chain in between two H-bonding units[16,18]. The polymer solubility apparently mainly depends on the nature of the hydrogen-bonding unit, which determines the polarity and the flexibility of the polymer chain. In general, the observed behaviour is similar to that of poly(amide)s that are soluble in strong hydrogen bond breaking solvents but not in apolar solvents. It is worth noting that the polymer containing only the urea monomer (pU6-30) is insoluble in all tested solvents, including TFA; even a small percentage of urea monomers strongly decreases the polymer solubility. Finally, the solubility behaviour of amide-based materials appeared to depend more upon the length of the spacer between the amide bonds rather than on the molecular weight of the material.

Thermal gravimetric analysis (TGA) indicated that amide-containing materials were stable at higher temperatures in contrast to carbamate and urea based polymers. This trend closely corresponds to that already observed for the corresponding monomers. The decomposition temperature of all polymers was almost independent of the PEG content.

The thermal behaviour of the synthesized materials was carefully investigated by DSC analysis performed under different conditions[16,18]. Representative DSC curves are presented in Figure 2. All polymers showed a glass transition attributable to the soft segment and at least one melting peak, in accordance with a semicrystalline structure. The T_g's observed decreased from –45 and –56 °C and levelled off to that of pure PEG1000 (Figure 3). The length of the polymethylene spacer seems to only slightly affect the glass transition temperature.

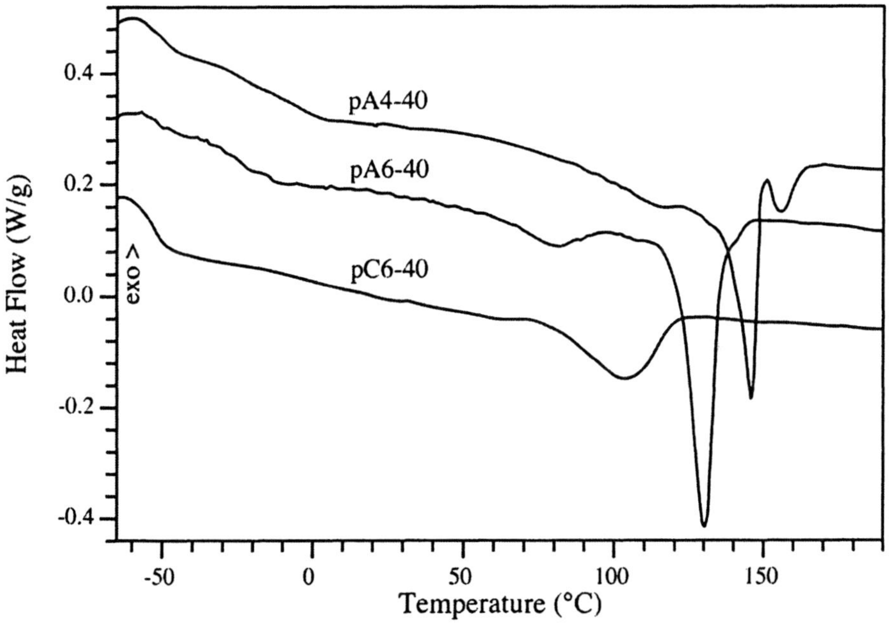

Figure 2. Representative DSC curves of the prepared polymers (second heating, 10 °C/min heating rate).

At low PEG content, the polymer melting temperature was found in all cases very close to that of the corresponding diester monomers, thus suggesting, as expected, that the H-bonding units play a major role in the polymer crystallinity. As the PEG content increases, endothermic peaks attributable to the melting of PEG crystals are observed upon heating.

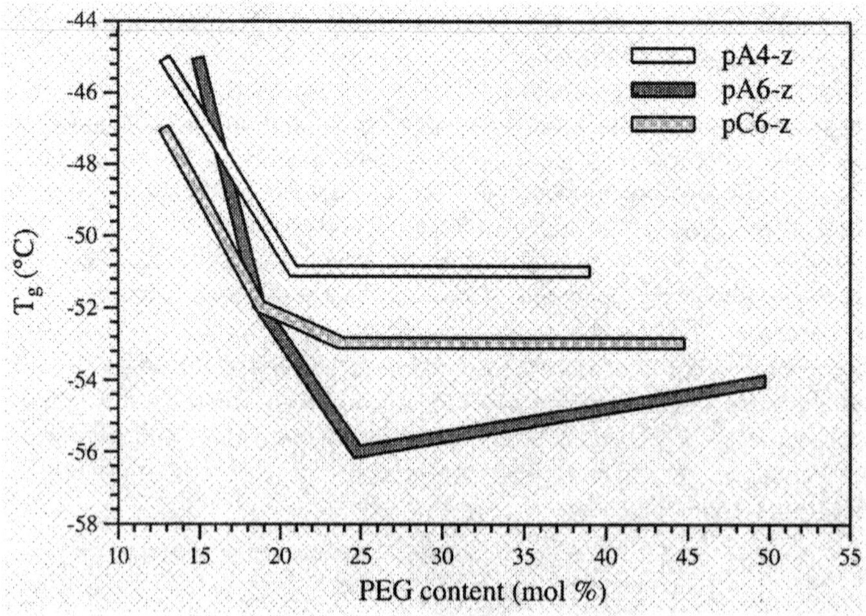

Figure 3. Dependence of the Tg on the PEG content of the prepared polymers.

The presence of small amounts of urea monomer does not strongly affect the melting behaviour of the corresponding materials. Overall, the reported thermal properties suggest that the synthesized polymers containing H-bonding units are phase-separated materials, consisting of crystalline H-bonded domains embedded in an amorphous PEG containing matrix. Additionally, PEG rich crystalline domains tend to phase separate as the PEG content increases.

The polymer surface properties of polymer cast films were evaluated by measuring the static contact angles (Figure 4) of water and CH_2I_2, according to a reported procedure[19] that allows for the evaluation of the surface free energy, both in terms of dispersive and H-bonding forces. No significant influence of the nature of the H-bonding unit was detected, whereas the hydrophilicity increased with increasing the PEG content. The reported behaviour confirms that the PEG content strongly affects both bulk and surface properties of the materials.

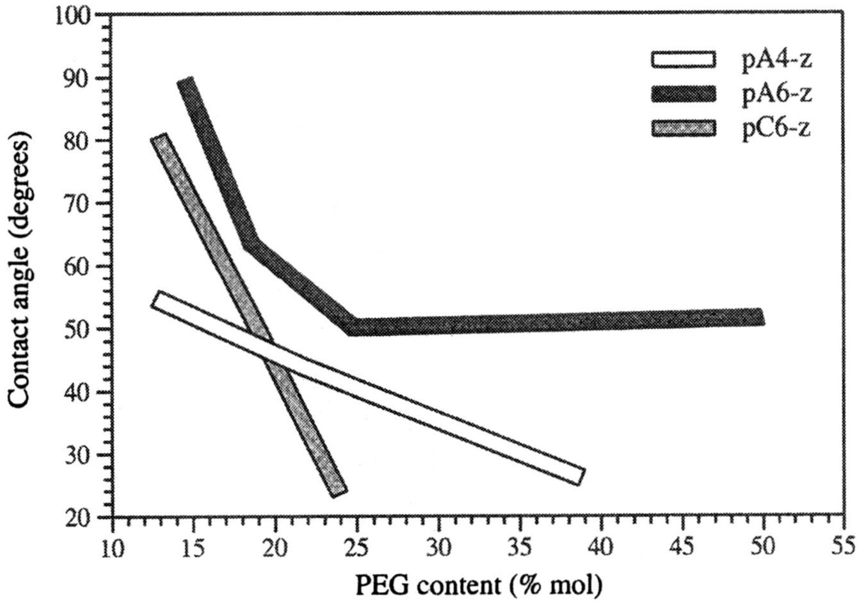

Figure 4. Dependence of water static contact angle on the PEG content of cast polymer films.

3. CONCLUSIONS

The preparation of symmetrical aliphatic diesters containing H-bonding moieties can be performed by diester aminolysis, and diisocyanate coupling with diols and diamines to yield diester diamides, diester dicarbamates, and diester diureas, respectively. The adopted procedures allow for the synthesis of end-functionalized diesters in good yields (62-95 %) and high chemical purity.

Poly(ether ester)s containing H-bonding units can be prepared in fairly good yields by bulk polycondensation of symmetric diesters containing different H-bonding moieties with mixtures of 1,4–butanediol and PEG1000. The slight enrichment in PEG content that was observed in all cases suggests that during the polymerization some 1,4-butanediol distilled off from the reaction mixture. Incorporation of different H-bonding units and suitable variations of the feed composition allowed for the tailoring of the polymer properties. Indeed, the chemical structure and the composition of the resulting polymeric materials were very effective in determining the polymer thermal and solubility properties. In particular, the presence of strongly interacting H-bonding units and of soft PEG segments is responsible of the observed strong tendency to phase separation. The polymer surface hydrophilicity, which plays a key role in material biocompatibility, can be tuned easily by changing the content of PEG segments, which are progressively exposed at the polymer surface. The reported characteristics and the presence of easily hydrolysable ester bonds along the polymer backbone make the prepared materials very promising for application in the biomedical field.

REFERENCES

1. Ulrich, K. E., Cannizzaro, S. M., Langer, R., and Shakesheff, K. M., 1999, Polymeric Systems for Controlled Drug Release. *Chem. Rev.* **99**: 3181-3198.
2. Jeong, B., Bae, Y. H., and Kim, S. W., 2000, Drug Release from Biodegradable Injectable Thermosensitive Hydrogel of PEG-PLGA-PEG Triblock Copolymers. *J. Control. Rel.* **63**: 155-163.
3. Peracchia, M. T., Gref, R., Minamitake, Y., Domb, A., Lotan, N., and Langer, R., 1997, PEG-Coated Nanospheres from Amphiphilic Diblock and Multiblock Copolymers: Investigation of their Drug Encapsulation and Release Characteristics. *J. Control. Rel.* **46**: 223-231.
4. Saltzman, W. M., and Olbricht, W. L., 2002, Building Drug Delivery into Tissue Engineering. *Nature Rev.* **1**: 177-186.
5. Cohn, D., and Younes, H., 1988, Biodegradable PEO/PLA Block Copolymers. *J. Biomed. Mater. Res.* **22**: 993-1009.
6. Casey, D. J., Jarret, P. K., and Rosati, l., 1987, Diblock and Triblock Copolymers. *U. S. Pat.* 4,716,203; *Chem. Abstr.* **108**: 132531.
7. Cerrai, P., Tricoli, M., Andruzzi, F., and Paci, M., 2000, Polyether Ester Block Copolymers by non-Catalysed Polymerization of ε-Caprolactone with Poly(ethylene glycol). *Polymer* **30**: 338-343.
8. Bakker, D., van Blitterswijk, C. A., Hesseling, S. C., Koerten, H. K., Kuijpers, W., and Grote, J. J., 1990, Biocompatibility of a Polyether Urethane, Polypropilene Oxide, and a Polyether Polyester Copolymer. A Qualitative and Quantitative Study of three Alloplastic Tympanic Membrane Materials in the Rat Middle Ear. *J. Biomed. Mater. Res.* **24**: 489-515.
9. Bezemer, J. M., Oude Weme P., Grijpma, D. W., Dijkstra, P. J., van Blitterswijk, C. A., and Feijen, J., 2000, Amphiphilic Poly(ether ester amide) Multiblock Copolymers as Biodegradable Matrices for the Controlled Release of Proteins. *J. Biomed. Mater. Res.* **52**: 8-17.
10. Stapert, H R., Bouwens, A. M., Dijkstra, P. J., and Feijen, J., 1999, Enviromentally Degradable Aliphatic Poly(ester amide)s based on Short, Symmetrical and Uniform Bisamide-Diol Bloks 1. Synthesis and Interchange Reactions, *Macromol. Chem. Phys.* **200**: 1921-1929.
11. Alla, L., Rodriguez-Galan, A., and Munoz-Guerra, S., 2000, Hydrolytic and Enzymatic Degradation of Copoly(ester amide)s based on L-Tartaric and Succinic acids. *Polymer* **41**: 6995-7002.
12. Paredes, N., Rodriguez-Galan, A., Puiggali, J., and Peraire, C., 1998, Studies on the Biodegradation and Biocompatibility of a New Poly(ester amide) Derived from L-Alanine. *J. Appl. Polym. Sci.* **69**: 1537-1549.

13. Paredes, N., Rodriguez-Galan, A., and Puiggali, J., 1998, Synthesis and Characterization of a Family of Biodegradable Poly(ester amide)s Derived from Glycine. *J. Polym. Sci. Part A: Polym. Chem.* **36**: 1271-1282.
14. Tuominen, J., and Seppala, J. V., 2000, Synthesis and Characterization of Lactic Acid Based Poly(ester amide). *Macromolecules* **33**: 3530-3535.
15. Villuendas, I., Bou, J. J., Rodriguez-Galan, A., and Munoz-Guerra, S., 2001, Alternanting Copoly(ester amide)s Derived from Amino Alcohols and L-tartaric and Succinic Acids. *Macromol. Chem. Phys.* **202**: 236-244.
16. Signori, F., Solaro, R., Chiellini, E., Lips, P.A.M., Dijkstra, P. J., and Feijen, J., 2003, Synthesis and Characterization of Segmented Poly(ether ester)s Containing H-bonding Units, *Macromol. Chem. Phys.* submitted.
17. Deschamps, A. A, Grijpma, D W, Feijen, J., 2002, Phase Separation and Physical Properties of PEO-Containing Poly(ether ester amide)s, *J. Biomat. Sci.***13**:1337-1352.
18. Signori, F., Solaro, R., Chiellini, E., Lips, P.A.M., Dijkstra, P. J., and Feijen, J., 2003, Synthesis and Characterization of Segmented Poly(ether ester)s Containing H-bonding Units and PEG segments, *Macromol. Chem. Phys.* submitted.
19. Owens, D. K., and Wendt, R. C., 1969, Estimation of the Surface Free Energy of Polymers. *J. Appl. Polym. Sci.* **13**: 1741-1747.

The Foaming Process of Biodegradable Polyesters

SALVATORE IANNACE, ERNESTO DI MAIO, YINGWEY W. DI, GIUSEPPE MENSITIERI, and LUIGI NICOLAIS
Institute of Composite Materials Technology (ITMC-CNR) & Department of Materials and Production Engineering, University of Naples "Federico II", Piazzale Tecchio 80, 80125 Naples-ITALY

1. INTRODUCTION

Polymeric foams represent a valuable class of materials with important technological applications. Due to their peculiar properties, these materials find wide applications when good mechanical properties and low weight need to be coupled for acoustic insulation and damping, thermal insulation and impact resistance.

They can be prepared from virtually any polymer by introducing or generating a gas in a polymeric matrix. However, suitable materials for industrial foaming applications must possess adequate properties, and the manufacture process must be easy and economic.

The manufacture of foamed products requires a careful selection of the proper combination of polymer/foaming agent system and the proper coordination of the individual steps in the process. The main characteristic of the suitable materials is its foamability, which is related to the rheological characteristics of the melt. Polymers whose viscosity decreases slowly with the increase of the temperature are favoured and therefore amorphous polymers are generally easier to foam than semicrystalline polymers. In the latter case, materials partially crosslinked and/or highly branched should be used in order to get suitable elongational melt viscosity to withstand the stresses on the cell walls during the growth of gas bubbles.

Biodegradable Polymers and Plastics, Edited by Chiellini and Solaro
Kluwer Academic/Plenum Publishers, New York, 2003

An important factor that controls the final morphology of foams is the crystallization kinetic of the polymer. In the extrusion process, once the melt exits the die, bubbles nucleate first and then they grow in not isothermal conditions. The viscosity increase occurring during the cooling/crystallization of the polymer is necessary to stabilize the cellular structure and to avoid the collapse and/or the coalescence of the bubbles. For this reason, the investigation of the crystallization behaviour of the polymer should be known in order to optimize the temperature profile in the extruder and at the die.

In this chapter, the methods utilized to prepare cellular polymers from biodegradable polyesters are described. Foams from poly(ε-caprolactone) (PCL) and PCL/clay nanocomposites were prepared by using two physical foaming agents: CO_2 and N_2. The solubility of the expanding gases at different pressures and temperatures in molten polymer were evaluated. The thermodynamic behaviour of the polymer melts was evaluated by performing Pressure-Volume-Temperature (PVT) analysis. This data were utilized to predict the sorption isotherms of the expanding gas in PCL and PCL/nanoclay systems, by using the "lattice" model proposed by Sanchez and Lacombe. Finally, the effect of the processing conditions on the foam morphology was evaluated.

2. THEORETICAL BACKGROUND

The optimisation of manufacturing processes of cellular polymers involves the control of fluidodynamic behaviour of macromolecular viscoelastic materials containing a dissolved gas at high concentration and at thermodynamic conditions able to promote the formation of gas bubbles in the melt. Nucleation and growth rates, which determine the final morphology of the foam are related to the physical and the rheological properties of the polymeric melt: surface tension, elongational viscosity of the polymer/gas system, solubility and diffusivity of the gas into the melt.

Most of these characteristics cannot be simply evaluated in experimental tests because of the difficulties of measuring physical and mechanical properties of polymer/gas solutions in some cases and of the long time required for the experiments (i.e. solubility and diffusivity) in other cases. Therefore, a theoretical prediction of the effect of the operative conditions on the foaming process becomes an important tool for an off-line optimisation of the entire process, and requires the knowledge of the equation of state (EOS) able to correlate pressure, volume and temperature of the material. Several empirical and theoretical equations of state for polymer melts have been proposed by different authors and have been reviewed by Zoller[1] and Bhateja and Pae[2]. Among the several EOS proposed in the literature, the theory of Sanchez and Lacombe (SLEOS)[3-4] is one of the most popular and efficient. It was used in this work to predict the solubility of the expanding gas.

2.1 Bubble Nucleation and Growing

In order to form polymeric foam, bubbles must first nucleate and grow within the molten or plasticized viscoelastic material. Subsequently, setting of the structure must occur due to the increase of viscosity during cooling and/or reduction of plasticization and finally solidification of the continuous phase. The initial nucleation is induced by a change in thermodynamic conditions, generally a change of temperature and/or pressure. During this stage a second phase is generated from the metastable polymer/gas homogeneous mixture.

Based on the classical theory on foam nucleation, macroscopic properties such as solubility S, diffusivity D, gas concentration C, surface tension σ, temperature and degree of supersaturation are the parameters controlling the nucleation rate J. There are several equation proposed for J, but they can be summarized in the following general expression[5].

$$J = MB\exp\left[-\frac{16\pi\sigma^3}{3k_BT(P_V - P_L)^2}\right] \tag{1}$$

M and B are function of the gas concentration C and of the gas diffusivity D, respectively; k_B is the Boltzmann constant; P_V e P_L are the equilibrium vapour pressure and the pressure in the liquid phase, whose difference describes the supersaturation of the expanding gas in the solution.

Once the bubble is nucleated, it will grow due to the diffusion of gas molecules from the solution to the gas phase. The phenomena governing the expansion of the bubble will be therefore related to the transport properties of the gas in the polymeric melt and to the rheological properties of the matter around the bubble. Bubble growth is then governed by the following differential equations[6].

$$\frac{d}{dt}\left(\rho_G R^3\right) = 3\rho D R^2\left(\frac{\partial c}{\partial r}\right)_{r=R} \tag{2}$$

$$\frac{\partial c}{\partial t} + v_r\frac{\partial c}{\partial r} = \frac{D}{r^2}\frac{\partial}{\partial r}\left(r^2\frac{\partial c}{\partial r}\right) \tag{3}$$

$$\rho\left(\frac{\partial v_r}{\partial t} + v_r\frac{\partial v_r}{\partial r}\right) = -\frac{\partial P}{\partial r} + \frac{1}{r^2}\frac{\partial}{\partial r}\left(r\tau_{rr}\right) - \frac{\tau_{\theta\theta} + \tau_{\phi\phi}}{r} \tag{4}$$

Equations 2-4 represent the mass balance in the bubble, the mass balance of the gas in the solution, and the force balance around the bubble, respectively. For the solution of these three equations, it is necessary to know the dependence of diffusivity, elongational viscosity and surface energy on gas concentration.

In summary, nucleation and growth rates are related to the surface tension, to the viscosity of the polymer-gas solution and to the diffusivity of dissolved gas molecules into the melt. In general, the formation of a gas bubble is determined by the degree of the gas supersaturation when pressure is suddenly reduced. Its growth is kinetically governed by the gas diffusion from the polymer/gas solution to the gas phase and by the viscoelastic forces around the expanding bubble. By controlling nucleation rate and growth rate of the bubble it is possible to optimise the morphology of the foam (number and size of cells) and, as consequence, the performances of the cellular structure.

2.2 Sanchez-Lacombe Lattice Theory for Polymer Mixtures

The Sanchez and Lacombe theory for polymer mixtures is based on an appropriate mixing of rules, which allows for the prediction of the thermodynamic behaviour of a binary mixture, once the pure fluids parameters are known and an interaction parameter is determined.

The model appears to describe accurately sorption isotherms when the equation of state parameters of both polymer and penetrant are determined. Like the Flory-Huggins model[7], the Sanchez-Lacombe model assumes that the different components mix randomly in a lattice. Unlike the Flory-Huggins model, the Sanchez-Lacombe model permits some lattice sites to be empty, which allows holes or free volume in the fluid. The addition of free volume to the lattice permits volume changes upon mixing components. The amount of absorbed penetrant in the polymer is determined by equating the chemical potential of the penetrant and the chemical potential of the penetrant in the mixture and by satisfying the equation of state of the pure penetrant phase and of the polymer-penetrant mixture. At fixed temperature and pressure, these conditions are met by equations 5-7.

Equation of state for pure penetrant:

$$\tilde{\rho}_1 = 1 - \exp\left[-\frac{\tilde{\rho}_1^{\,2}}{\tilde{T}_1} - \frac{\tilde{P}_1}{\tilde{T}_1} - \left(1 - \frac{1}{r_1}\right)\tilde{\rho}_1\right] \qquad (5)$$

Equation of state for the penetrant-polymer mixture:

$$\tilde{\rho}=1-\exp\left[-\frac{\tilde{\rho}^2}{\tilde{T}}-\frac{\tilde{P}}{\tilde{T}}-\left(1-\frac{\Phi_1}{r_1}\right)\tilde{\rho}\right] \tag{6}$$

By equating the chemical potential of the pure penetrant and of the penetrant in the mixture:

$$\left[-\frac{\tilde{\rho}_1}{\tilde{T}_1}+\frac{\tilde{P}_1}{\tilde{T}_1\tilde{\rho}_1}+\frac{(1-\tilde{\rho}_1)\ln(1-\tilde{\rho}_1)}{\tilde{\rho}_1}+\frac{\ln\tilde{\rho}_1}{r_1^0}\right]=$$
$$\ln\phi_1+1-\phi_1+\tilde{\rho}\, r_1^0 X_1(1-\phi_1)^2+\left[-\frac{\tilde{\rho}}{\tilde{T}_1}+\frac{\tilde{P}_1}{\tilde{T}_1\tilde{\rho}}+\frac{(1-\tilde{\rho})\ln(1-\tilde{\rho})}{\tilde{\rho}}+\frac{\ln\tilde{\rho}}{r_1^0}\right]r_1^0 \tag{7}$$

$\tilde{T}$, $\tilde{P}$, and $\tilde{\rho}$ are the reduced temperature, pressure, and density defined as:

$$\tilde{P}=P/P^*$$
$$\tilde{T}=T/T^*$$
$$\tilde{\rho}=\rho/\rho^*$$

X_1 is a measure of the deviation of the mixture from the ideal geometric mean combination rule and r^o is the number of lattice sites occupied by a fluid molecule.

The scale factors P^*, T^* and ρ^* can be calculated by PVT data and they were experimentally evaluated for PCL and PCL/nanoclay systems. PVT data of CO_2 were taken from the literature[8].

3. EXPERIMENTAL METHODS

3.1 Materials

PCL (commercial name CAPA 680) was kindly supplied from Solvay Interox Ltd, UK. The organoclay (commercial name Cloisite 30B) used in the preparation of nanocomposites was purchased from Southern Clay Products Inc, USA and was used as received. This organoclay consists of 2:1 montmorillonite (MMT) treated with methyl tallow bis-2-hydroxyethyl (MT2EtOT) as organic modifier.

3.2 Preparation and Characterization of PCL/Clay Nanocomposites

A Haake Rheomix®600 internal mixer was used for the preparation of the PCL/organoclay nanocomposites. The processing temperature was set at 80 °C. The rotating speed and mixing time were fixed at 100 rpm and 12 minutes. After the mixing, the samples were compressed into *ca*. 2 mm thick plates for further characterization.

X-ray diffraction was performed at room temperature to evaluate the dispersion of the silicate layers in the PCL matrix. The thermal properties of the samples were then analysed by using a TA Instrument DSC 2910. Dynamic and isothermal crystallization analysis were performed on samples kept at 80 °C for 5 minutes to eliminate the previous thermal history and to allow for complete melting of the crystalline phase.

Rheological tests were performed in shear at 80 °C in order to analyse the effect of composition on the complex viscosity and their elastic and dissipative components.

3.3 Gas Solubility and PVT Measurements

Gas transport and equilibrium absorption properties were determined by using two different balances, a Cahn D110 microbalance and a quartz spring balance.

The Pressure-Volume-Temperature (PVT) behaviour of the polymeric matrix and of the nanocomposite materials was measured at pressures up to 200 MPa and temperatures from 25 to 120 °C. The analysis was performed by using the classical bellows technique, in which pressure is applied to the samples through a confining fluid, and the volume is measured by an LVDT mounted beneath the pressure vessel.

3.4 Foaming Process

The effect of gas composition and of the operative conditions on the structure of the foams was evaluated by using a batch process. The experiments were conducted using the following procedure. PCL cylinders (10 mm thick, 10 mm diameter) were saturated at T_{sat} with the foaming agent. The saturation pressure (p_{sat}) varied in the range 20-140 bar, depending on the gas. The reactor was then cooled to the foaming temperature (T_{foam}) and finally the pressure was released with a controlled pressure gradient in order to promote gas nucleation and growth.

A continuous extrusion process was also used to produce foamed sheets. The extrusion line is composed by a lab-scale twin-screw extruder (Haake CTW100), a hot heat exchanger, and a capillary system acting as nucleation nozzle. The latter device allowed for a good control of the pressure profile in the extrusion line. Detailed monitoring of the process parameters was obtained by means of four pressure transducers and five melt thermocouples.

4. RESULTS AND DISCUSSION

4.1 PCL/Clay Nanocomposites

The X-ray diffraction (XRD) patterns of the organoclay and of PCL/organoclay composites are shown in Figure 1.

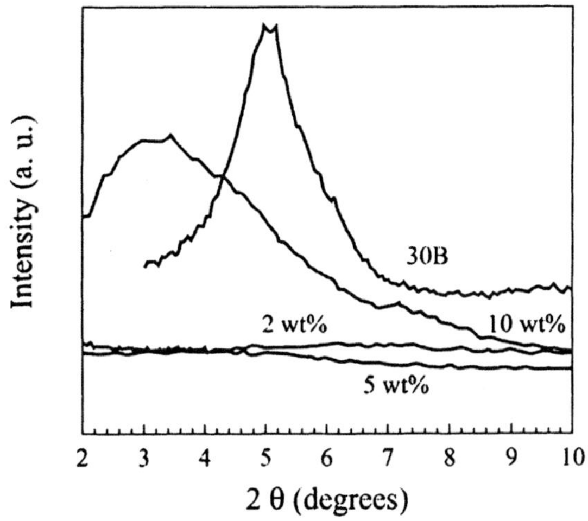

Figure 1. XRD patterns of Cloisite 30B and Cloisite/PCL composites with various contents of Cloisite 30B.

The primary silicate reflection at $2\theta = 4.92°$ in the organoclay corresponds to a layer spacing of 18.3 Å. For 2 wt% and 5 wt% Clay-PCL composites, we did not see any peak of 30B at low angle range, suggesting the possibility of having exfoliated the silicate layers of organoclay dispersed in PCL matrix. Recently an exfoliation mechanism for organoclay has been proposed[9]. In summary, during the mixing process of polymer matrix and organoclay, the fracturing process of the clay particles would take place first, i.e. external platelets are subjected to dynamic high shear forces which ultimately cause their delamination from the stack of layers building the organoclay particles, and then an onion-like delamination process will continue to disperse the silicate platelets into the PCL matrix. In the PCL/organoclay composites, these two steps should take place as indicated by the X-ray pattern because of the interaction between the hydroxyl groups of the organic modifier of 30B and the carbonyl groups of PCL. During intercalation, the polymer chains, which are initially in an unconstrained environment, must enter the constrained environment of the narrow silicate interlayer, whereas the organic chains gain configurational freedom as the interlayer distance increases. Accordingly, processing of highly interacting systems in an efficient shearing field is an important parameter in the fracturing and delamination steps of organoclays, leading to the formation of nanocomposites structures. However, at constant processing condition, the exfoliation showed a composition dependent behaviour. At high clay content (10%), a reflection at lower angle than in the organoclay was observed (Figure 1), indicating that both intercalation and exfoliation of organoclay exist in the PCL matrix.

The crystallization kinetics observed under dynamic cooling (Figure 2) and during isothermal experiments was affected by the presence of the mineral. The crystallization rate v_c increased at lower clay concentration (2% and 5%) and it was delayed at higher concentration (10%). The increase of v_c can be attributed to the action of the nanoparticles as nucleation agents while the lower v_c can be related to the reduced mobility of polymer chains in the presence of strongly interacting nanoparticles. In other words, the competition between nucleation rate and growth rate determines the increase or the decrease of the overall crystallization rate.

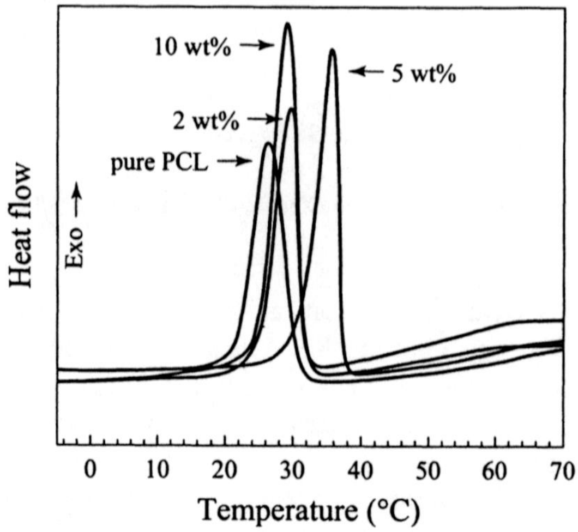

Figure 2. Crystallization peaks (cooling at 10 °C/min) of PCL and PCL/Cloisite systems with various content of Cloisite 30B (2, 5 and 10%).

The reduced molecular mobility induced by the interactions between macromolecular chains and nanoparticles had a strong effect on the melt viscosity of PCL, which increased up to two orders of magnitude at low shear rate (Figure 3). The dependence of complex viscosity, and its components, on the shear rate, showed a very different behaviour from pure PCL. At low shear rate, the pure polymer had a typical pseudo Newtonian behaviour while the nanocomposites showed a yielding behaviour, typical of multiphase polymeric systems[10].

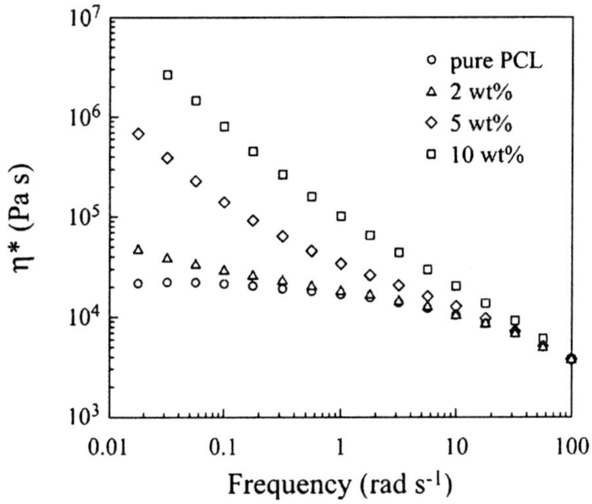

Figure 3. Complex viscosity of PCL and various PCL/Cloisite systems at 80 °C.

4.2 Thermodynamic Behaviour of the Polymer/Gas Systems

Equilibrium sorption concentrations of CO_2, N_2 and a 50%wt mixture of CO_2 and N_2 in pure PCL at 80 °C are reported in Figure 4. The solubility of CO_2 is higher than that of N_2 and there is no interaction between the two gases since the solubility of the mixture is very close to the weighed average of each component solubility.

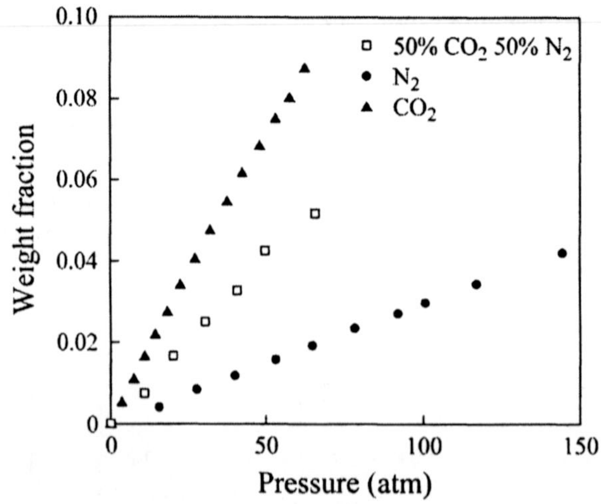

Figure 4. Solubility of CO_2, N_2 and CO_2/N_2 in PCL at 80 °C.

In order to predict solubility data of the polymer/gas systems investigated, the SLEOS parameters (P*, T* and ρ*) were calculated from the experimental PVT data. The experimental PVT curves relative to pure PCL are reported as an example in Figure 5.

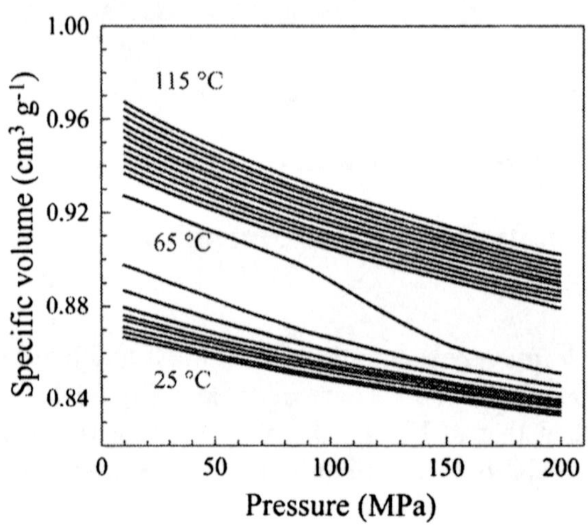

Figure 5. Pressure-Volume-Temperature (PVT) curves of PCL.

PVT data of PCL and PCL/nanoclays were analysed with the Sanchez and Lacombe model (the analogue of equation 5 for pure PCL) in the melt region, at temperatures higher than 65 °C; the results are summarised in Table 1.

Table 1. SLEOS parameters for PCL and PCL/nanoclay systems

Sample	P^* (MPa)	V^* (cm^3/g)	T^* (K)
PCL	548.6	494.2	520.3
5% nanoclay	0.8635	0.8686	0.8474
10% nanoclay	637.7	687.2	653.9

Solubility data for the systems PCL/CO_2 and PCL/nanoclays/CO_2 were then predicted by means of the equations 5-7 and they are reported in Figure 6. The amount of gas absorbed in the polymeric matrix was reduced by the presence of nanoclay particles and this effect was more pronounced in nanocomposites that showed complete clay exfoliation. As a matter of fact, the highest reduction of solubility was observed in samples containing 5% of nanoclay. In this case, the reduction of the solubility of CO_2 in PCL was about 12%. The calculated sorption isotherms in this work were sensitive to the value of the interaction parameter ψ, contained in X_l[3-4] The best fitting was obtained by using a value of 0.98 for the interaction parameter ψ ($\psi = 1$ for non polar solvents).

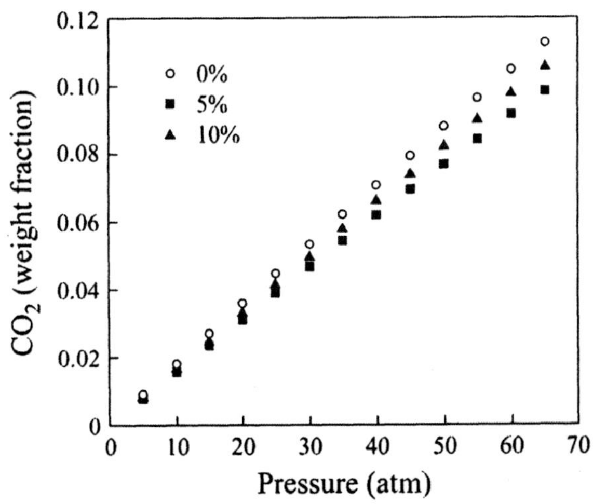

Figure 6. Prediction (SLEOS) of CO_2 solubility in PCL and PCL/nanoclay systems at 80 °C.

4.3 Processing and Characterization of Foams

4.3.1 Batch Process

Foamed products are typically manufactured by employing extrusion lines with high L/D. Long residence time is necessary to obtain the complete solubilisation of the gas in the polymer melt and to allow for subsequent cooling of the melt at the optimum temperature for bubble nucleation and growth. In the batch process, these steps can be controlled to evaluate the effect of the process parameters on the foamability of the polymer and on the cellular morphology.

With the latter type of process it is possible to prepare foams under free expansion conditions, which is not possible to obtain in continuous extrusion processes. During the free expansion, the material does not experience extensional and/or shear stresses, and the analysis of cell nucleation and growth is more rigorous. The system is constituted by a cylinder, where the temperature is kept constant by means of an oil bath and/or an electrical resistance. Gas, under different pressure conditions is inserted in the chamber and, after complete solubilisation in the polymeric melt, the pressure is released at controlled drop rate. There are several parameters that influence the entire process, the most important ones are: a) the saturation pressure, that determines the amount of gas dissolved in the polymer and hence the final density; b) the foaming temperature, that affects the final density and the structure of the foam; c) the pressure drop rate, that affects the nucleation rate and determines the cell density.

Figure 7 shows a comparison of foams produced at the same saturation pressure and temperature but with different pressure gradient. As evidenced by the SEM micrographs, finer structures with more cells and smaller size were obtained by using higher pressure gradients. At 30 bar/s (Figure 7b) the cellular structure was characterized by a density of about 0.01 g/cm^3 and average cell diameter of 100 μm.

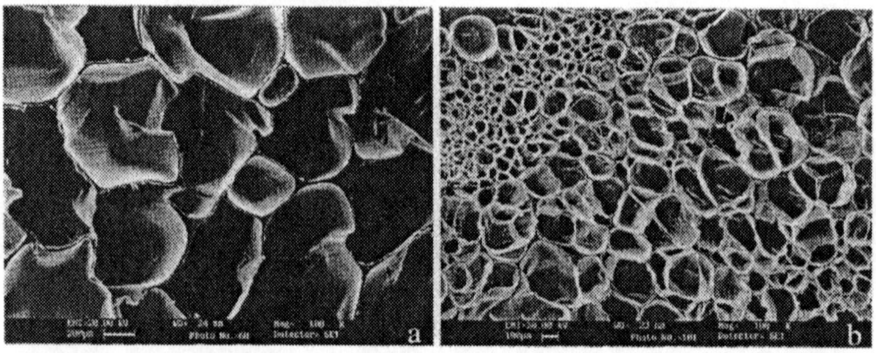

Figure 7 PCL foamed with CO_2 at different pressure gradients. T_{foam} = 30 °C, p_{sat} = 60 bar. a) 7 bar/s, b) 30 bar/s.

The main problem associated to the use of CO_2 as foaming agent is related to the loss of gas from the cellular structure. At room temperature, the diffusivity of CO_2 is still high and in a few hours the gas escapes from the cells, leading to shrinkage of the structure.

Compared to CO_2, foams prepared by using N_2 as foaming agent showed higher density (0.3 g/cm^3), due to the lower solubility of the gas in the polymeric melt. However, the high saturation pressure (150 bar) and the high pressure gradient (50 bar/s), employed in these cases, leaded to structures characterized by very fine cellular morphology, with cells having a mean diameter of about 10 μm.

Low-density microcellular foams were achieved by using a mixture of CO_2 and N_2. Foams of PCL and PCL containing 2% of nanoclay, prepared with CO_2/N_2 mixture are shown in Figure 8. Even though the final density of these foams is almost the same, the cellular structure of the nanocomposite foam is characterized by higher number of cells of smaller size.

4.3.2 Extrusion-Foaming

The continuous extrusion process is important from industrial and productivity points of view. So far, little interest was given to biodegradable polymers, and especially polyesters, mainly because they are, in general, considered poorly foamable. However, as shown above, it is possible to achieve foamed structures if the choice of the expanding matter and the processing conditions are properly optimised. The extrusion-foaming process deals with the same problems of shrinkage, density and morphology and the methods to be used to improve the extruded foams have the same basic idea than those utilized in the batch foaming.

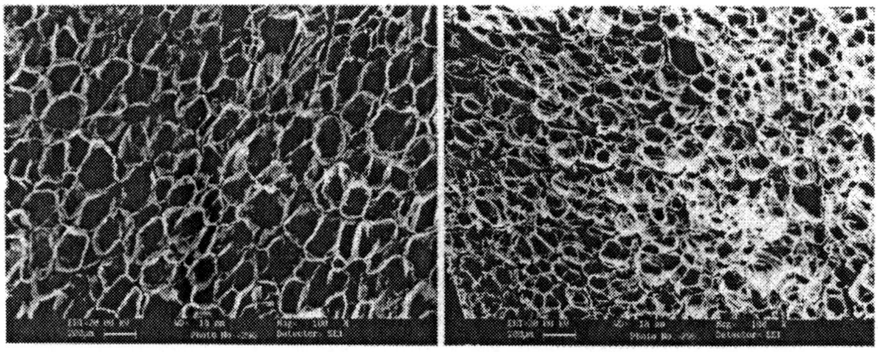

Figure 8. PCL (left) and PCL/nanoclay (2%) (right) foamed with a mixture of N_2 (80%) and CO_2 (20%). T_{foam} = 45 °C, p_{sat} = 150 bar, 50 bar/s. Magnification is 100x.

Even though the batch process has provided useful information on the effects of the foaming processing conditions, the extrusion process is so much different that some points for discussion hold. What in the batch process has to be controlled and enhanced on the time scale, in the extrusion is on the screw axis and distance scale. In the first three screw zones, we have melting, solubilisation, and pressurization. In the static hot mixer further solubilisation is allowed by increasing the residence time and the contact between the two phases. When the solubilisation step is complete, the polymer/gas solution is forced to pass through a narrow capillary (nucleation nozzle), where a fast pressure drop provokes nucleation.

Experimental tests, performed by changing the configuration of the capillary (length and diameter), showed that the pressure drop and the pressure gradient was varied and this resulted in different nucleation rate. Therefore, foams prepared with the longer capillary (l = 30 mm) had a very fine cellular structure with 80 μm average diameter of cells as compared to those (300 μm diameter) obtained by using a shorter capillary (l = 10 mm).

5. CONCLUSIONS

PCL and PCL/nanoclay foams were obtained by using CO_2 and/or N_2 as foaming agents. The different solubility and diffusivity of the two gases leaded to different morphology, in terms of foam density and cell size. The most interesting cellular morphology, characterized by a low-density microcellular structure, was obtained when a mixture of the two gases was employed.

The presence of nanoclays led to materials with different thermal and rheological properties that affected nucleation and growth phenomena of gas bubbles. Compared to PCL foams, the cellular structure of nanocomposites was characterized by a higher density of lower size cells.

ACKNOWLEDGEMENTS

The work was supported by grants from the Italian Ministry of University and Research (MIUR-Cluster 26).

REFERENCES

1. Zoeller, P., 1989, PVT Relationships and Equations of State of Polymers. In *Polymer Handbook*, Wiley-Interscience, Ney York, Chap. VI.
2. Bhateja, S. K., and Pae, K. D., 1975, Effects of hydrostatic pressure on the compressibility, crystallization, and melting of polymers. *J. Macromol. Sci., Rev. Macromol. Chem.* **C13**(1): 77-133.
3. Sanchez, I. C., and Lacombe, R. H., 1976, Statistical Thermodynamics of Fluid Mixtures. *J. Phys. Chem,* **80**: 2568-2580.

4. Sanchez, I. C., and Lacombe, R. H., 1978, Statistical Thermodynamics of Polymer Solutions. *Macromolecules* **2**: 1145-1156.
5. Han, J. H., and Han, C. D., 1990, Bubble nucleation in polymeric liquid. II. Theoretical consideration. *J. Polym. Sci Part B* **30**: 743-761.
6. Arefmanesh, A., Advani, S. G., and Michaelides, E. E., 1992, An accurate numerical solution for mass diffusion-induced bubble growth in viscous liquids containing limited dissolved gas. *Int. J. Heat Mass Transfer* **35**(7): 1711-1722.
7. Flory, P. J., 1953, *Principles of Polymer Chemistry*. Cornell University Press, Ithaca, NY.
8. Hariharan, R., Freeman, B. D., Carbonell, R. G., and Sarti, G. C., 1993, Equation of State Predictions of Sorption Isotherms in Polymeric Materials. *J. Appl., Polym. Sci.* **50**: 1781-1795.
9. Dennis, H. R., Hunter, D. L., Chang, D., Kim, S., White, J. L., Cho, J. W., and Paul, D. R., 2001, Effect of melt processing conditions on the extent of exfoliation in organoclay-based nanocomposites. *Polymer* **42**: 9513-9522.
10. Han, C.D., 1981, *Multiphase Flow in Polymer Processing*. Academic Press, New York.

PART 4

NEW TRENDS AND VISIONS

Significance and Implications of Green Polymer Chemistry

GRAHAM SWIFT
GS Polymer Consultants, 1078 Eastchurch, Chapel Hill, North Carolina 27517, USA

1. INTRODUCTION

The goal of green chemistry described recently in the June 2002, issue of the Chemical and Engineering News[1] quite obviously pertains to polymer chemistry and the future of the Polymer Industry as well. This multifaceted goal was recorded as:

> ***The design, development, and implementation of chemical products and processes to reduce or eliminate the use and generation of substances hazardous to human health and the environment.***

The goal builds on the twelve principles of green chemistry first promulgated by Paul Anastas and John Warner[2] in the late 1990s and paraphrased below:

- Prevention of waste is better than treatment or clean up.
- Atom economy should be maximized such that all materials are used in the final product.
- Less hazardous chemicals for human health and environmental effects should be used and generated in chemical synthesis.
- Design chemicals for effect and minimize their toxicity.
- Solvents and synthesis auxiliaries such as catalysts should be avoided or innocuous.

- Design new chemical processes for minimal economic and environmental impacts.
- Use renewable resources wherever technically and economically practicable.
- Reduction or elimination of temporary blocking derivatives, as they are waste generators.
- Catalysis should be a as specific as possible
- Design chemicals so that they are degradable to innocuous and non-persistent products in the environment.
- Real time analytical methodologies needed to monitor processing to minimize hazardous by-products.
- Inherently safer chemicals should be used to minimize potential chemicals accidents such as release, explosions, and fires.

These principles were proposed as guidelines for the chemical industry as it moves towards a sustainable technology base. Interestingly, almost all of the principles are included in several major initiatives such as Responsible Care, emissions control, renewable resources, waste management, and Life Cycle Assessment that have been reshaping the polymer industry over the last half of the last century and continue to evolve in this century. Hence, the principles are expected to be a cogent and major driver in the future development of polymer technology. It is noteworthy that they stop short of a call for immediate sustainability, though their intent is clear for the long range. This is an astute recognition that cost/performance, the guiding principle of the polymer business and any successful business, should not be compromised to implement unacceptable technical changes for the sake of change; for example to legislate the use of renewable resources or to enforce biodegradability in waste management. The principles put forward by Anastas and Warner do not call for such radical changes recognizing that science evolves slowly, especially from an optimized and successful industrial base already in existence. Radical changes for the sake of change would be of little benefit to the consumer or for the definition of the best path forward for the industry.

This paper represents an attempt to briefly overview where the polymer industry is now and where it is likely to be headed, and where some answers may lie, as it struggles with the inevitable progression towards sustainability. A major message that emerges is that we currently have a vastly improved polymer industry over yesteryear, which is already very green and we have time for science to move it forward without fear of disaster to humans or the environment. The advances anticipated will take time and patience is needed to save many wasted efforts due to asking the wrong questions, developing wrong answers, and jumping to conclusion too quickly.

2. STATUS OF THE POLYMER INDUSTRY

The schematic shown in Figure 1 is a convenient way of depicting the current status of the polymer industry, it encompasses raw materials, processing, formulating or compounding, use and waste management on the completion of useful life. All these aspects are under careful assessment by detailed life cycle assessment (LCA) to ensure that green principles are a driving force for now and in the future as changes are implemented. These aspects have been broadly covered in Responsible Care initiatives, emissions control, effluent controls, opportunities for renewable resources, and waste management in general. Life Cycle Assessment, though, is emerging as the new tool to assess the merits of new chemistry and to tighten current chemistries, processes, and disposal with respect to the environment. It should ultimately evolve to the state of predicting cradle to grave impacts on the environment of all chemicals and plastics, allowing selection and implementation of best technology.

2.1 Raw Materials and Processing

Raw materials in the polymer and plastics industry continue to be predominantly based on non-renewable or so-called fossil feedstocks with only minor utilization of renewable feedstocks. This is not surprising as a whole infrastructure is in place from inexpensive raw materials to processes and products that have been highly optimized to meet a wide array of markets with very acceptable and very favourable cost/performance characteristics. The production of monomers from fossil feedstocks and their polymerization is under excellent quality control with many characteristics called for in green chemistry principles such as good atom efficient, few hazardous raw materials, limited solvents, and tight emission controls.

2.2 Waste Management

The waste management of plastics and polymers was one of the first areas to receive attention as the industry moved towards an environmental awareness.

This was brought on by the growing pressure to avoid or reduce environmental pollution from waste polymers and plastics, which were accumulating widely on land and in oceans due to careless disposal, and because of the decreasing availability of landfill space for controlled disposal. Options available then, and now, were landfill, incineration, recycle, and controlled environmental degradation. Environmental degradation at that time was almost always referred to as biodegradation even though the latter is the end stage in environmental assimilation of all products designed for environmental degradation by several mechanisms including oxidation, radiation, etc.

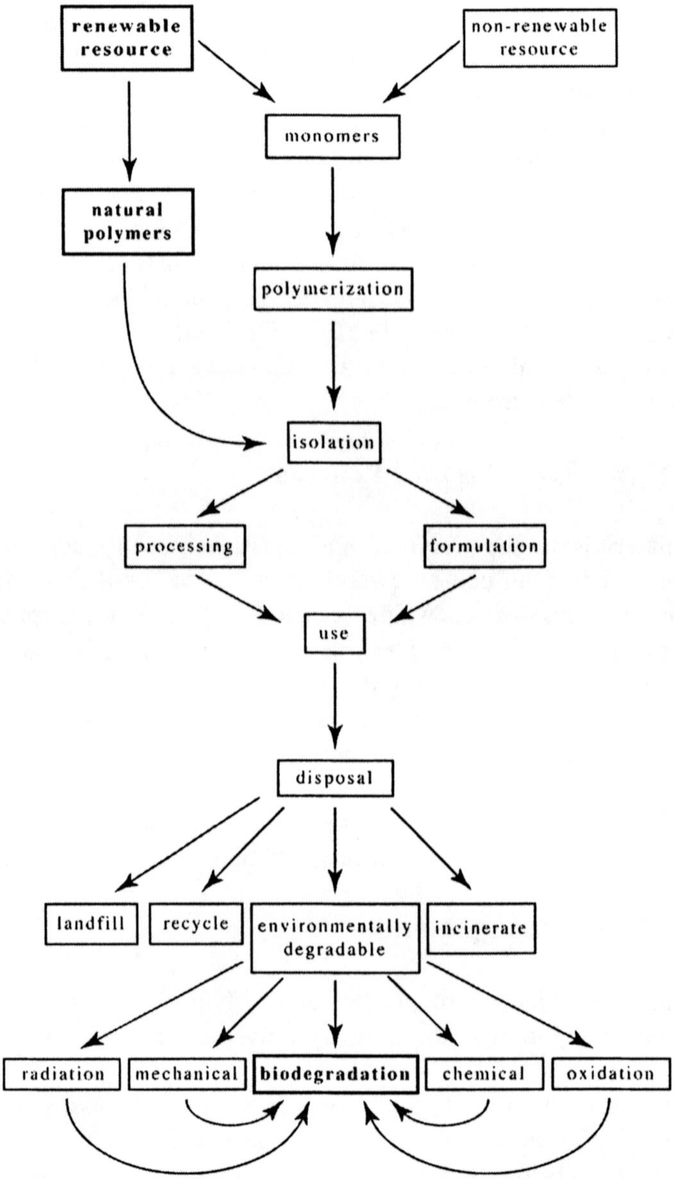

Figure 1. Schematic of polymerization process and polymer disposal.

Recycle is still a high priority choice but there are many plastic and polymer applications where recovery after use for recycle is not easy to accomplish such as in water-soluble polymers and plastics and in one use plastics in the food packaging and agricultural industries. In the former case, the plastic and polymers are lost as dilute solutions in water and in the latter case recovery and cleaning from mixed food waste or from soil are labour intensive and expensive.

Incineration of waste plastics and polymers, though still widely practiced in many countries, is presently not favoured due to the potential for toxic off-gases such as dioxins from the use of mixed polymer feedstocks, which include halogenated polymers such as poly(vinyl chloride). If this issue can be resolved, incineration and fuel value recovery from waste plastics becomes very attractive and probably the best disposal option.

Landfill continues to be widely used, also, particularly in North America where land space is plentiful. However, it must be considered a waste of resources and, therefore should be minimized at the expense of the other options for waste management wherever possible.

Environmentally degradable polymers and plastics are a very viable option for the waste management of plastics and polymers that are difficult to recover and include one-use plastics in the food, packaging, and agricultural industries, and water-soluble polymers. The latter are often a forgotten group because they generally pass unseen into the water environment after their use. Not in sight, not a problem!

Plastics used in the food and packaging industry are being designed for disposal in composting facilities and water-soluble polymers are being designed for disposal in wastewater treatment plants. Though, at this time, no great success has been achieved in either case with a commercially viable environmentally degradable product, niche markets are, however, growing in the designated market areas and this is a sure sign of future acceptance and of a need for further development of other opportunities.

Industry's growing attention to green chemistry is an indicator that its benefits to people, the environment and business are being recognized. It is possible to balance green chemistry and successful new developments, and, perhaps more important, to recognize that a sustainable polymer industry is eventually inevitable with the limited supply of fossil feedstocks. Some of the newer developments in the degradable plastics area are already based on renewable resources.

3. IMPLICATIONS OF GREEN POLYMER CHEMISTRY

As already mentioned, the implications of green chemistry in the polymer industry are ultimately far reaching requiring an increase in renewable resources, development of new highly specific natural and synthetic polymerization and processing chemistry, and continued advances in the balancing of waste management options. Much of the implementation of these innovations will be managed through improved life cycle assessments.

3.1 Renewable Resources

Renewable resources are resources that renew on a regular basis, generally annually, and for the most part are plant based, though animal waste should not be forgotten. Renewable resources have always formed part of the polymer industry feedstocks; from the beginning of civilization they served as vital raw materials and, before the middle of the last century, they were predominant feedstocks prior to the petroleum-based industry emerging after World War 2. In a sense, we are going to witness the re-emergence of an old industry with much refinement, of course, and more clearly defined goals and many challenges. Figure 2 indicates a relationship between polymer industries built on fossil and renewable resources and how they are very similar and complementary in their down line chemical production of monomers and polymers. This, of course, will be beneficial to the current polymer industry in the period of change over of feedstocks that we are beginning to go through. A smooth transition is possible while we continue to produce the same or similar polymers and plastics.

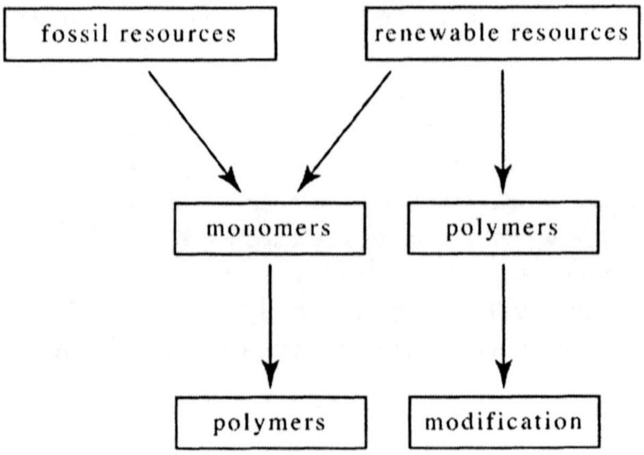

Figure 2. Polymers from renewable and fossil resources.

The fact that similar monomers are available from renewable and fossil resources permits the polymers currently in vogue to be continued. The monomers are identified as those predominantly used in condensation polymerization. Monomers from natural resources currently available by fermentation include for example lactic acid, aspartic acid, glutamic acid, lysine, serine, 1,3-propanediol, and citric acid. Itaconic acid, a free radically polymerizable vinyl monomer is available from citric acid indicating further developments that are possible from first derived monomers. Saturated and unsaturated fatty acids, alcohols, and amines are available from fats, oils, and protein sources. The possibilities are limitless, the opportunities large, but progress should be with caution.

Polymers from nature may be used as isolated or modified. Modification was generally used in the past and continues to produce protein, starch and cellulose derivatives for many applications. New opportunities are becoming apparent for surplus by-products from the food industry such as the proteins from wheat, corn, soy and milk. Lignin continues to be a cheap resource from the paper industry. All of these polymers may be chemically modified or blended with plastics from fossil or renewable origin to produce promising new materials.

A unique family of polymers based on β-hydroxyacids produced by fermentation has been an evolving industry for several years since the early attempted commercialization of poly(β-hydroxyalkanoates) by Imperial Chemical Industries in the UK. Remarkable advances have been made in optimizing polymer production in bacteria in the fermentation process and in the beginnings of transgenic production in plants. Fermentation optimization to produce a range of copolymers with a variety of controllable properties to fit specific high value applications and markets appears to be evolving. As production gains momentum and costs come down, more market penetration is likely especially with blends to upgrade polymers with less desirable mechanical properties.

3.2 Natural Processes

Nature has developed many chemicals and polymers over billions of years using mild and very controlled processes that are not available to us in the laboratory at this time. Efforts are underway to understand, copy, and modify these processes to develop new polymers and materials suited to our needs.

Enzyme catalysis is specific, controlled, gives few by-products and is generally conducted in water under mild conditions of temperature and pressure. An ideal protocol for the polymer industry. Now, we have pioneers in the laboratory utilizing enzymes to produce addition polymers from vinyl monomers, condensation polymers from alcohols, amines and acids. One addition polymer is in commercial production in the UK utilizing specific enzyme condensation polymerization of primary alcohol groups in the presence of secondary alcohols to synthesize functional polyesters. Recently, a polyurethane has been produced in the laboratory.

Genetic modifications are being used to produce polymers in plants. Similar polyesters polyhydroxyalkanoates, are produced in bacteria, as are poly(amino acid)s.

Finally, self-assembly from ionic interaction, hydrophobic interactions, and hydrogen bonding used widely by nature to build composites is receiving very wide attention in academia and industry to build new composites. Simple analogs already exist as in micelles, vesicles, rheology modifiers, and thickeners.

3.3 Chemical Processes

In recognition of the specificity of natural processes, much synthetic catalysis research is focused on developing new catalysts to control functional group placement, monomer sequence control, molecular architecture, and molecular weight in polymer synthesis. These catalysts may be considered natural mimics.

3.4 Waste Management

Recycle ha a major opportunity to maintain and increase its use in green chemistry and as part of a sustainable industry through the preservation of resources. Problems to be addressed include separation of plastics for sole recovery, recycling of blends, and how to recover feedstocks or monomers.

Incineration is an obvious opportunity for many situations to recover fuel. Issues are safety with respect to incomplete combustion and removal of all toxic by-products.

Environmental degradation is still at a difficult stage. Polymers defined for this disposal method, with a few exceptions, have not been commercially attractive due to cost/performance relative to non-degradable polymers that they are designed to replace. There is no doubt that many polymers degrade in the environment by one or more of several pathways, oxidation, hydrolysis, mechanical, biodegradation, and radiation. The final step in all cases for environmental acceptance is for assimilation and complete biodegradation in the disposal site of choice such as compost facility, soil, landfill, or wastewater treatment plant. Much has been accomplished, simple test methods are established which are perfectly adequate for identifying and classifying fast biodegrading polymers such as starch or cellulose and synthetic hydrophilic condensation polymers. A major issue and point of concern to be addressed is how to accommodate slower biodegrading polymers, especially as these are usually based on current inexpensive commodity hydrophobic polyolefinic polymers. It is a fact that nature degrades many seemingly recalcitrant polymers that are hydrophobic, lignin and rubber are prime examples, by sequential processes of oxidation and biodegradation. Activated polyolefins degrade by a similar process, but too slowly to pass current standard test methodology designed to control market access and the consumer from non-biodegradable polymers. These glaring deficiencies in definition and testing methodology deserves rapid resolution so that we do not exclude old and new synthetic polymers based on their longer time to degrade in the environment. The green principles listed at the beginning of this paper do not give a time for degradation, but insist on no toxicity to the environment. If activated polyolefins can meet this requirement, time to degrade should not be an issue. The consumer should not be held to ransom by standards that are artificially limiting, we have the scientific ability to develop more meaningful standards for slower degrading polymers and we must do this.

4. CONCLUSIONS

The significance of green polymer chemistry is clearly a more environmentally acceptable and desirable technology; and the major implication is ultimately a sustainable polymer industry based on green chemistry, renewable resources, and natural processes. The transition will be implemented deliberately and slowly over many years as new technology evolves based on renewable resources to replace fossil resources with a minimum of economic penalty to the consumer.

The pathforward from fossil feedstocks and the implementation of renewable resources offers multiple choices and a very difficult selection process requiring life cycle assessments for direction. These techniques are not highly developed at this time, tend to be somewhat subjective, and we must move with caution. We must recognize that fossil fuels will not run out overnight and renewable resources need vast infrastructure development. It should be recognized, also, that environmental issues are still present and need to be addressed even for renewable resources and their utilization. Cost/performance will control market success, and rightly so. Specialty applications of renewables will be easier for market entry. From there, good science will develop new opportunities and markets as the technology base grows.

REFERENCES

1. *Chem. Eng. News*, June, 2001
2. Anastas, P., and John C. Warner, J.C., 1998, *Green Chemistry Theory and Practice.* Oxford University Press, New York.

Artificial and Natural Functionalized Biopolyesters: From Macromolecular Skeleton Selection to Property Design by Ester Pendant Groups

ESTELLE RENARD, CHRISTEL BARBAUD, VALÉRIE LANGLOIS, and PHILIPPE GUÉRIN
LRP, UMR 7581, CNRS, Université Paris 12-Val de Marne, 2/8 rue Henry Dunant 94320-Thiais, France

1. INTRODUCTION

In polymer science, chemical modification has been early a large area of research and industrial development. The possibility of chemically fitting properties to specific applications has been often used in the case of natural macromolecules, such as cellulose or dextrans. In the biomedical and therapeutic domains, the necessity for smart temporary devices leads to an increase in the macromolecular structures complexity. The building of such polyvalent polymers can be achieved by copolymerization of monomers, by the tailor making of multiblock macromolecules or by chemical modification starting from a functional parent compound. The lactic and glycolic acids polymers and copolymers family, which have been the pioneer class of biocompatible and hydrolysable polymeric medical materials must be affixed by other polymers types for adjusting hydrophilic/hydrophobic balance, interactions, targeting, bioactive molecules attachment, degradation rate, suitable properties. Therefore, polymers with hydrolysable backbone and side functional groups have been selected: poly(amino acids), polyesteramides and polyesters[1]. Our choice has been concerned with synthetic, natural and artificial functionalized poly(hydroxyalkanoate)s. They have a cleavable ester bond in the macromolecular chain, one or more

Biodegradable Polymers and Plastics, Edited by Chiellini and Solaro
Kluwer Academic/Plenum Publishers, New York, 2003

chiral centres in the monomer unit and a reactive group in the side chain. This group can hold an unsaturated bond, which will be turned to epoxy or carboxyl function by chemical modification, or an ester function, which will be turned to carboxyl group. Further reactions are conceivable.

Three routes are possible to prepare these functionalized polyesters (Figure 1): the first one is a chemical way and starts from racemic or optically active malic acid[2] or aspartic acid[3]. The ultimate step of the monomer synthesis is a ring closure, leading to β-substituted β-lactones, without any racemization[4]. Anionic ring opening polymerization of malolactonic acid esters leads to high molecular weight polymers, random or block copolymers according to the experimental process. Ability to strictly control the synthesis of benzyl malolactonate and other β-substituted β-lactones has allowed the preparation of a large family of poly(malic acid) derivatives, having alkyl pendant groups, lateral functional groups or more sophisticated groups such as menthyl, adamantyl, cholesteryl, and lactic acid oligomers[5-7]. The sole limitation is the possibility to introduce the ester group during the lactone preparation. The access to new monomers has been opened through the catalytic hydrogenolysis of the benzyl group of benzyl malolactone leading to malolactonic acid[8]. This molecule, stable despite the presence of a free carboxyl group, has been enhanced through the coupling of bioactive molecules and the preparation, by copolymerization, of macromolecular drugs[9].

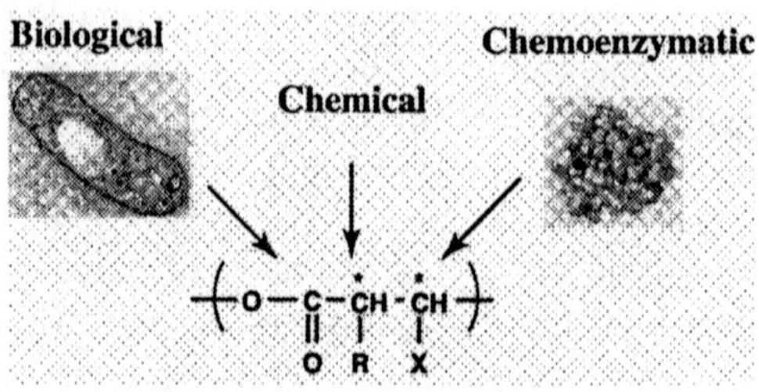

Figure 1. Preparation of functionalized polyesters.

The two driving forces in the chemoenzymatic route have been chirality and metabolism. In synthetic polymers, chirality leads to supramolecular structures and make easier the formation of highly crystalline materials. It is possible to take advantage of stereogenic centres to control the polymer configurational structure and hence their properties. The stake was to introduce a second stereogenic centre in the monomer unit in order to have a series of crystalline polyesters with a large range of thermal and degradation properties. The best way was to start from biomolecules related to aspartic acid. 3-Methylaspartic acid is present in the metabolism of bacterium *Clostridium tetanomorphum*. This molecule appears in the catabolism of (S)-

glutamic acid[10] and it is of great interest for its ability to be transformed in lactone and incorporated in a bioassimilable polyester as 3-methylmalic acid monomeric units. 3-Methylaspartase, a *C. tetanomorphum* enzyme, is known to catalyse the elimination of ammonia from the (2S,3S)-3-methylaspartic acid to give mesaconic acid. It has been shown that this enzyme can be efficiently used in the reverse reaction[11,12]. Moreover, this enzyme is useful in providing access to different 3-alkykaspartic acids via direct amination of the corresponding alkylfumaric acids[13]. After extraction and purification of 3-methylaspartase, bioconversion of methyl-, ethyl-, and *iso*propyl-fumarates have been carried out in batch, yielding large amounts of the corresponding (2S,3S)-3-alkylaspartic acid[14]. The presence of L-*erythro*-3-methylaspartase activity, under particular experimental conditions has been used in the preparation of a 60/40 mixture of (2S,3S)/(2S,3R)-3-methylaspartic acid from mesaconic acid[14]. Further chemical conversion of these natural molecules led to optically pure lactones[15]. Moreover, unnatural stereoisomers of 3-methylaspartic acid are available by enzymatic resolution[16]. Stereo(co)polymers displaying a wide range of melting points (140 to 270 °C) were prepared by polymerization and copolymerization[14,15].

The third route is biological, that is the synthesis of poly(hydroxy-alkanoate)s (PHAs) by using prokaryotic microorganisms. These polymers are now very well known and serve as intracellular and energy reserve materials in bacteria. They can be produced in large quantities by a fermentation process[17,18]. Today, one may consider biological synthesis of bacterial polyesters as an alternative or complementary method of macromolecular chemistry. Among bacteria, which have the ability to produce PHAs with C_4 to C_{12} alkyl pendant groups, *Pseudomonas sp. GPO1* has been intensively investigated[19]. This bacterium can metabolize a variety of carbon substrates, including *n*-alkanes and alkanoic acids to produce very unusual PHAs[20]. The production of poly(hydroxyoctanoate)s by 40 different strains of *Pseudomonas* has been studied in order to better understand the relationship between the strain and the production of short to medium chain length PHAs. Moreover, *Pseudomonas sp. GPO1* develops a unique performance: this strain is able to incorporate a wide spectrum of functional substrates through the cellular membrane and to produce polyesters including these molecules[21,22].

Whatever the synthetic route, we have chosen all described polymers that fulfil the requirements as biomedical and hydrolyzable materials. Moreover, the control and the powerful adaptability of their architecture have been exploited to expand their properties by chemical modification. In this paper, the access to functionalized materials, their chemical modification and its consequences on properties, and some examples of potential applications will be presented.

2. RESULTS AND DISCUSSION

2.1 Preparation of Unsaturated Polyesters

2.1.1 Poly(Malic Acid) Derivatives

Propen-3-ol and 3-methyl-3-butenol have been used to introduce an unsaturation in β-substituted β-lactones according to the aspartic route[23] (Figure 2). These new malolactonic esters have been polymerized or copolymerized with benzyl malolactonate leading to high molecular weight functional polyesters.

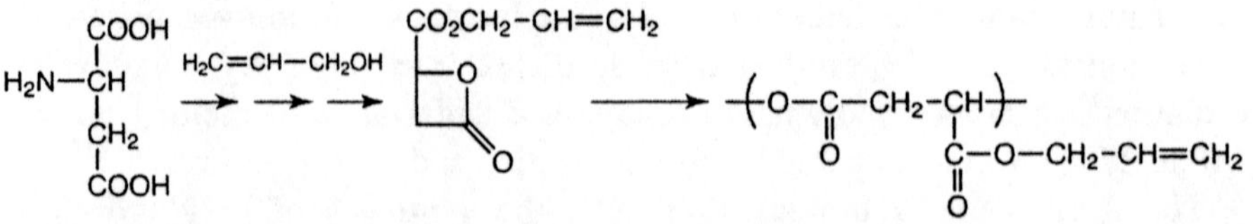

allyl PMLA polymer: M_w (SEC) = 80,000 g mol^{-1}
30/70 allyl/benzyl PMLA copolymer; M_w (SEC) = 100,000 g mol^{-1}

Figure 2. Synthesis of poly(malic acid) (PMLA) derivatives.

2.1.2 Unsaturated Bacterial Polyesters

Two types of polyesters have been prepared by using *Rhodospirillum rubrum* (ATCC25903) and *Pseudomonas sp. GPO1* (ATCC29347) microorganisms. In the case of the first bacterium (a phototropic purple, non sulphur bacterium) cells were grown on pentenoic acid under anaerobic conditions, in the light at 31 °C[24]. The extracted polymer was a terpolymer P(HB-HV-HP) with 3-hydroxybutyrate (23%), 3-hydroxyvalerate (67%) and 3-hydroxypentenoate (10%) monomer units (Figure 3). The appeal of this polymer resides in the presence of very short pendant chains and therefore, the double bond is very close to the macromolecular skeleton.

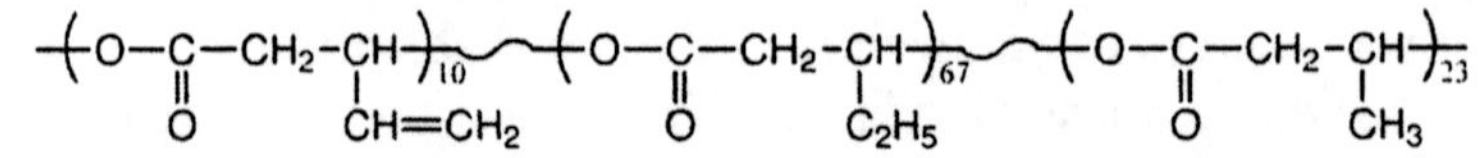

P(HB-HV-HP)

Figure 3. Structure of P(HB-HV-HP) terpolymer.

In the case of *P. sp. GPO1,* the ability of this bacterium to grow on sodium octanoate (Oct)/10-undecenoic acid (Und) mixtures and on pure sodium octanoate has been used[24]. A series of random copolyesters noted as PHOU (Figure 4) have been synthesized from different Oct/Und nutrient compositions (from 100/0 to 0/100).

It is worth noting that the fraction of unsaturated repeating units in the copolymers corresponds to the proportion of the unsaturated substrate in the feed. In the obtained PHA from cells grown with 10-undecenoic acid as sole nutrient, all repeating units contained terminal alkene groups, including 3-hydroxy-10-undecenoate, 3-hydroxy-8-nonenoate, and 3-hydroxy-6-heptenoate units.

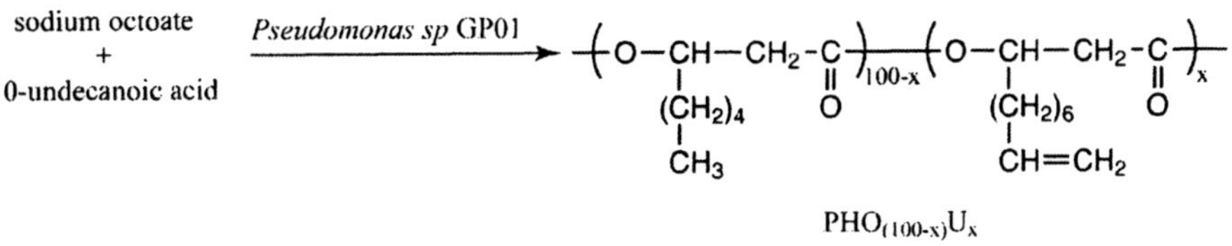

Figure 4. Bioconversion route to PHOU copolymers.

3. EPOXIDATION REACTION

Two chemical reagents have been used for this reaction: *m*-chloroperbenzoic acid (MPCBA) and dimethyldioxirane (DMD)[23,24]. MPCBA is commercial, not very aggressive, selective and easily used. In all cases, epoxidation was carried out in CH_2Cl_2 by changing temperature, reaction time, and equivalents of chemical reagent (Figures 5 and 6). All (co)polymers have been totally epoxidized (Table 1). The epoxidation reaction of different materials was investigated by ^{13}C NMR and ^{1}H NMR. ^{13}C NMR spectra presented a similar behaviour; that is disappearance of peaks corresponding to unsaturated carbons (140 and 115 ppm) and the presence of two peaks assigned to the oxirane carbon atoms (47 and 52 ppm). Moreover, no crosslinking reaction was observed. The different epoxidized polymeric materials were still soluble in common organic solvents. It is important to observe that the polymer molecular weight, as measured by SEC and with light scattering detection was not affected by the epoxidation reaction (Table 2). The small increase of molecular weight could be explained by the change of the polymer chemical structure. It can be concluded that this chemical modification does not *traumatize* unsaturated poly(hydroxyalkanoate)s, at least under the adopted experimental conditions. Epoxy groups are reactive and can lead to polymer networks, hydrophilic materials (presence of diols), and macromolecular prodrugs by therapeutic molecules attachment.

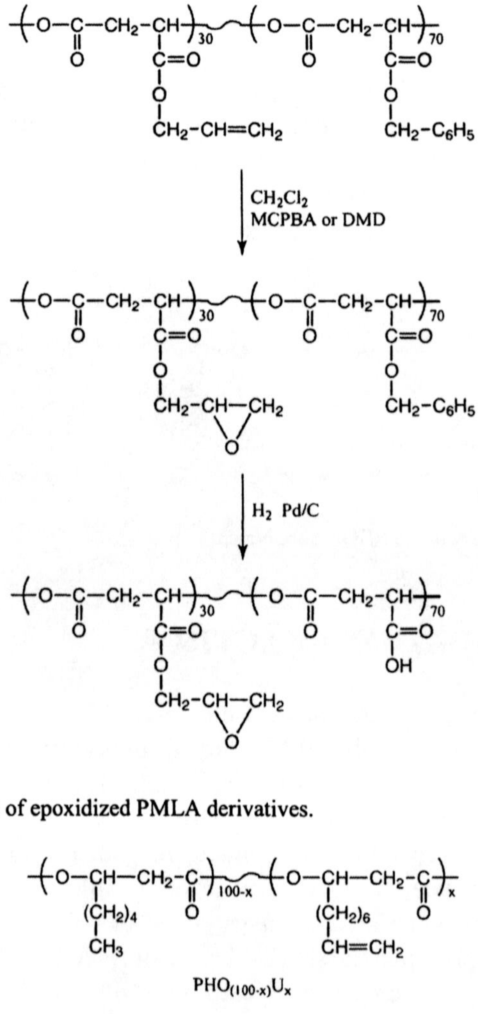

Figure 5. Preparation of epoxidized PMLA derivatives.

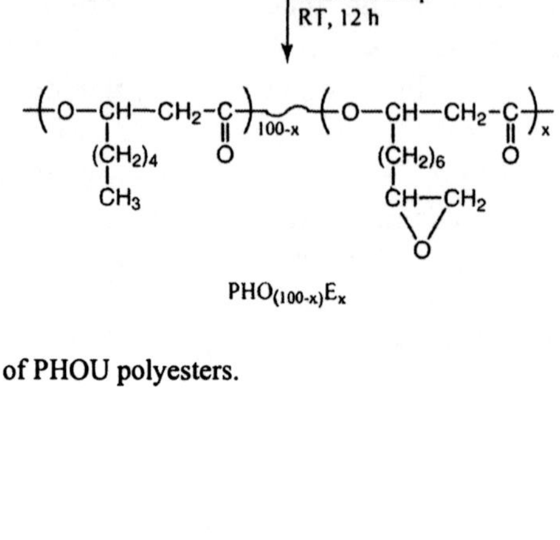

Figure 6. Epoxidation of PHOU polyesters.

Table 1. Results of the epoxidation of PMLA derivatives

Polymer	Reagents (parts)	Reaction time (days)	Yield (%)
PMLA MeBu	MCPBA (1.3)	1.5	100
PMLA (Be-*co*-Al) (70/30)	MCPBA (2)	4	100
PMLA (Be-*co*-MeBu) (70/30)	MCPBA (1.3)	1.5	100
PMLA Al	DMD (6)	1	100

PMLA MeBu : poly(3-methyl-3-butenyl β-malate); PMLA(Be-*co*-Al) (70/30) : poly(benzyl β-malate-*co*-allyl βmalate) (70/30); PMLA(Be-*co*-MeBu) (70/30): poly(benzyl β-malate-*co*-3-methyl-3-butenyl β-malate) (70/30); PMLA Al : poly(allyl β-malate)

Table 2. Molecular stability of macromolecular chains after epoxidation

Polymers	M_n	M_w	M_w/M_n
$PHO_{90}U_{10}$	83500	148500	1.8
$PHO_{90}E_{10}$	**74600**	**147500**	**2.0**
$PHO_{50}U_{50}$	58900	79200	1.3
$PHO_{50}E_{50}$	**53500**	**101000**	**1.9**
PHU_{100}	79900	123500	1.5
PHE_{100}	**77700**	**133700**	**1.7**

4. 10-EPOXYUNDECANOIC ACID AS NUTRIENT

Because of *P. sp. GPO1* versatility to produce particular PHAs containing functional groups in side chains, and of the interest in polyesters with reactive epoxy pendant groups, it was important to investigate the ability of this strain to growth on an epoxy substrate or co-substrate. This incorporation was not evident due to the metabolism of this microorganism. Bioconversion of epoxy groups to glycols by *Pseudomonas* strains has been reported in organic chemistry. For this reason, a new carbon source, 10-epoxyundecenoic acid, prepared by chemical modification of 10-undecenoic acid, has been tested either alone or in mixture with sodium octanoate in different proportions (Figure 7). The fermentation conditions have been changed for this series of experiments. The oxygen flow was very important in order to force bacteria to use oxygen gas instead of the oxirane oxygen for their metabolism and to increase the polyesters production with respect to the bacteria development[24].

Experimental results show that both cell and polymer yields depended on the percentage of 10-epoxyundecanoic acid in the feed mixture. *P. sp. GPO1* is unable to use the epoxidized nutrient as sole source of carbon for energy storage. The percentage of repeating units with epoxy groups has been determined by ^1H and ^{13}C NMR. These results demonstrate that all possibilities to produce *exotic* functionalized polyesters have not been explored.

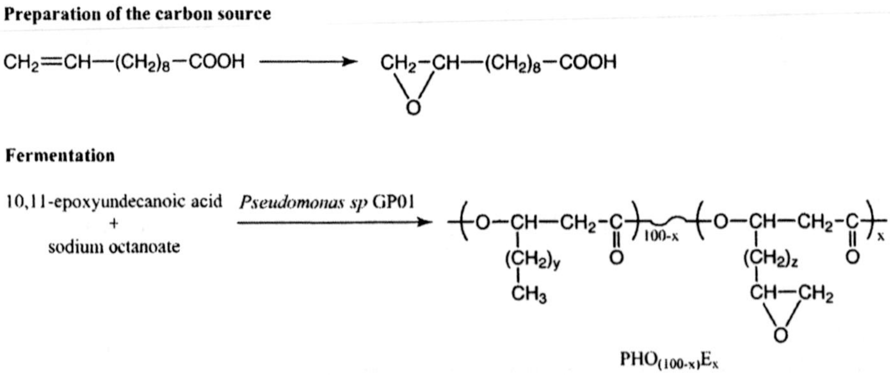

Figure 7. Synthesis of epoxidized PHAs from epoxidized substrates.

5. RADICAL ADDITION TO DOUBLE BONDS

Double bonds have the ability to react with thiols according to a radical process. Generation of radical can be obtained by using azobis*iso*butyronitrile (AIBN)[25]. Two thiols were reacted with poly(malic acid) esters, mercaptoethanol and N-acetylcysteine (Figure 8). The interest of these molecules is the possibility to have non-ionic hydrophilic side-chains in the case of mercaptoethanol. Moreover hydroxyl function can serve for the attachment of bioactive or targeting agents. N-acetylcysteine is attractive in the preparation of bioadhesives for surgery. The best results were obtained in anhydrous tetrahydrofuran with 2.7 equivalents of mercaptan and 0.1 equivalent of AIBN with respect to double bonds. Polymer solubility was changed after modification; methanol, ethanol, and acetone are possible solvents.

6. OXIDATIVE REACTION OF UNSATURATED PHAS

Carboxylic groups are of the greatest importance in regard to the attachment of bioactive compounds, hydrolyzable, hydrophilic oligomers or targeting proteins. Moreover, carboxyl groups will change the behaviour of the modified polymers in water. These groups cannot be directly introduced by biosynthesis. Therefore, a chemical route has been carried out starting from unsaturated bacterial polyesters (Figure 9). Two different PHAs have been used for this chemical modification, containing respectively 10% ($PHO_{90}U_{10}$) and 25% ($PHO_{75}U_{25}$) of unsaturated pendant groups[24].

groups was confirmed for all samples by ^{1}H NMR, and significant modification of the macromolecular chain was observed in the case of $PHO_{75}D_{25(COOH)}$; however, rather high molecular weight (M_w=54000 g mol^{-1}) were still present (Table 3)[26].

Table 3. Physical characteristics of $PHO_{100-x}D_{x(COOH)}$

Polymer	M_n (g mol^{-1})	M_w (g mol^{-1})	Ip
$PHO_{90}U_{10}$	43000	109000	2.5
$PHO_{90}D_{10(COOH)}$	48000	103000	2.1
$PHO_{75}U_{25}$	80000	179000	2.2
$PHO_{75}D_{25(COOH)}$	54000	135000	2.5

The introduction of polar groups such as carboxylic acid groups modified polymer physical properties. As reported in Table 4, $PHO_{100-x}U_x$ samples are soluble in organic solvents such as dichloromethane, chloroform, tetrahydrofuran, but are insoluble in polar solvents such as methanol, acetone/water 85/15 (v/v). After oxidation $PHO_{90}U_{10}$ and $PHO_{90}D_{10(COOH)}$ behaved differently in acetone/water solvent. More clearly, contrary to $PHO_{75}U_{25}$, $PHO_{75}D_{25(COOH)}$ was soluble in methanol, acetone, and in different acetone/water mixtures.

Table 4. Solubility of $PHOD_{(COOH)}$

Solvent	$PHO_{90}U_{10}$	$PHO_{90}D_{10(COOH)}$	$PHO_{75}U_{25}$	$PHO_{75}D_{25(COOH)}$
Chloroform	+	+	+	+
Acetone	–	–	–	+
Water	–	–	–	–
Acetone/water (v/v)				
90/10	–	±	–	+
85/15	–	±	–	+
83/17	–	–	–	±
80/20	–	–	–	–
Methanol	–	–	–	+

(+) soluble, (±) partially soluble, (-) insoluble (insoluble sticky feature)

7. CONCLUSIONS

Several types of chemical modification of polyesters have been presented in this paper. This smart chemistry leads to very well defined materials and can be used for specific applications. Polyesters with three different functional groups have been prepared[27] and tested in bone repair and muscle regeneration[28]. Micelles have been prepared and studied for the release of therapeutic agents[29]. Functionalized nanoparticles are under investigation for cancer therapy[6]. Combination of different chemistry domains with biology know-how succeed in perfecting more and more sophisticated and smart devices in the biomedical field.

REFERENCES

1. Vert, M., 1986, Biomedical Polymers from chiral and functional lactones: properties and applications. *Makromol. Symp.* **6**: 109-122.
2. Cammas-Marion, and S., Guerin, Ph., 2000, 4-Alkyloxycarbonyl-2-oxetanones and 3-alkyloxycarbonyl-2-oxetanones as versatile chiral precursors in the design of functionalized polyesters with a controlled architecture. *Des. Monomers Polym.* **3** (1): 77-93.
3. Cammas, S., Renard, I., Boutault, K., and Guerin, Ph., 1993, A novel synthesis of optically active 4-benzyloxy- and –4-alkoxycarbonyl-2-oxetanones. *Tetrahedron Asym.* **4** (8): 1925-1930.
4. Guerin, Ph., Vert, M., Braud, C., and Lenz, R.W., 1985, Optically active poly(β–malic acid). *Polym. Bull.* **124**: 187-192.
5. Cammas-Marion, S., and Guerin, Ph., 2000, Design of malolactonic acid esters with a large spectrum of specified pendant groups in the engineering of biofunctional and hydrolyzable polyesters. *Macromol. Symp.* **153**: 167.
6. Moine, L., Amiel, C., Brown, W., and Guerin, Ph., 2001, Associations between a hydrophobically modified, degradable poly(malic acid and a β-cyclodextrin polymer in solution. *Polym. Int.* **50**: 663-676.
7. Bizzari, R., Chiellini, F., Solaro, R., Chiellini, E., Cammas-Marion, S., and Guerin, Ph., 2002, Synthesis and characterization of new malolactonate polymers and copolymers for biomedical applications. *Macromolecules* **35**: 1215-1223.
8. Leboucher-Durand, M.A., Langlois, V., and Guerin, Ph., 1996, 4-Carboxy-2-oxetanone as a new chiral precursor in the preparation of functionalized racemic or optically active poly(β–malic acid) derivatives. *Polym. Bull.* **36**: 35-41.
9. Cammas, S., Bear, M.M., Moine, L., Escalup, R., Ponchel, G., Kataoka, K., and Guerin, Ph., 1999, Polymers of malic acid and 3-alkylmalic acid as synthetic PHAs in the design of biocompatible hydrolyzable devices. *Int. J. Biol. Macromol.* **25**: 273-282.
10. Archer, C.H., and Gani, D., 1993, Kinetics and mechanism of syn-elimination of ammonia from (2S, 3R)-3-methylaspartic acid by methylaspartase. *J. Chem. Soc. Chem. Commun.* 140-142.
11. Akhtar, M., Botting, N.B., Cohen, M.A., and Gani, D., 1987, Enantiospecific synthesis os 3-substituted aspartic acids via enzymic amination of substituted fumaric acids. *Tetrahedron* **24**: 5899-5908.
12. Barker, H.A., Smyth, R.D., and Weissbach, R., Purification and properties of β–methylaspartase. 1959, *J. Biol. Chem.* **234**: 320-328.
13. Bear, M.M., Cammas, S., Langlois, V., and Guerin, Ph., 1997, Chemoenzymatic synthesis of poly((2R,3S)-benzyl-β–3–methyl malate): 3- methylaspartase as a versatile enzyme in the praparatio of the chiral precursor. *C. R. Acad. Sci. Paris* **325** (IIb): 165-172.
14. Monne, C., Robic, D., Campion, G., Bourbouze, R., Rimbault, A., Masure, M., Langlois, V., Hemery, P., and Guerin, Ph., 1996, Enantiospecific enzymic preparation of (2S,3S)-3-alkylaspartic acids of current interest in the synthesis of stereoregular poly(β(2S,3S)-3-alkylmali acids) as new optically active functional polyesters. *Chirality* **8**: 300-304.

15. Bear, M.M., Langlois, V., Masure, M., and Guerin, Ph., 1998, Optically active polymers of 3-alkylmalic acids. Contribution of the bioconversion for diversifying the chiral precursors. *Macromol. Symp.* **132**: 337-348
16. Bear, M.M., Langlois, V., Robic, D., Bourbouze, R., and Guerin, Ph., 1999, Natural and artificial functionalized polyesters : I Chemoenzymatic mediated synthesis of chiral precursors. *J. Env. Polym. Degrad.* **7**: 127-134.
17. Steinbüchel, A., and Valentin, E.H., Diversity of polyhydroxyalkanoic acids. 1995, *FEMS Microbiol. Lett.* **128**: 219-228.
18. Fritzsche, K., Lenz, R.W., and Fuller, R.C., 1990, An unusual bacterial polyester with a phenyl pendant group. *Makromol. Chem.* **191**: 1957-1965.
19. Lenz, R.W., Kim, Y.B., and Fuller, R.C., 1992, Production of unusual bacterial polyesters by *Pseudomonas Oleovorans* through cometabolism. *FEMS Microbiol. Lett.* **103** (2-4): 207-214
20. Doi, Y., and Abe, C., 1990, Biosynthesis and characterization of a new bacterial copolyester of 3-hydroxyalkanoates and a 3-hydroxy-ω–chloroalkanoates, *Macromolecules* **23**: 3705-3707
21. Leboucher-Durand, M.A., Langlois, V., and Guerin, Ph., 1996, 4-carboxy-2-oxetanones as new chiral precursors in the preparation of optically active and racemic poly(malic acid) derivatives. *Polym. Bull.* **36**: 35-41.
22. Bear, M.M., Leboucher-Durand, M.A., Langlois, V., Lenz, R.W., Goodwin, S., and Guerin, Ph., 1997, Bacterial poly-3-hydroxyalkenoates with epoxy groups in the side chains. *React. Funct. Polym.* **34**: 65-77.
23. Leboucher-Durand, M.A., 1994, PhD Thesis *Façonnage de polyesters dégradables par voie chimique et biologique : apport de la modification chimique à la diversification des produits*. Université Paris VI.
24. Kurth, N., Renard, E., Brachet, F., Robic, D., Guerin, Ph., and Bourbouze, R., 2002, Poly-3-hydroxyoctanoate containing pendant carboxylic groups for the preparation of nanoparticles aimed at drug transport and release. *Polymer* **43**: 1095-1101.
25. Jeanbat-Mimaud, V., Barbaud, C., Caruelle, J.P., Barritault, D., Cammas-Marion, S., Langlois, V., and Guerin, Ph., 2000, Bioactive functionalized polymers of malic acid for bone repair presumably by regulating heparin growth factor bioavailability. *J. Biomat. Sci. Polym. Ed.* **11**: 979-991.
26. Jeanbat-Mimaud, V., Barbaud, C., Caruelle, J.P., Barritault, D., Cammas-Marion, S., Langlois, V., and Guerin, Ph., 1999, Functionalized and degradable polymers of malic acid stimulate bone repair. *C. R. Acad. Sci., Paris* **2** (IIc): 393-401.
27. Cammas, S., Bear, M.M., Harada, A., Guerin, Ph., and Kataoka, K., 2000, New macromolecular micelles based on degradable amphiphilic block copolymers. *Macromol. Chem. Phys.* **201**: 355-364.

polyurethanes from a kraft lignin -polyether triol-polymeric MDI system. *J. Appl. Polym. Sci.* **40**: 1819-1832.

16. Nakamura, K., Hatakeyama, T., and Hatakeyama, H., 1991, Formation of the glassy state and mesophase in the water-sodium alginate system. *Polym. J.* **23**: 253-258.
17. Morck, R., Reimann, A., Kringstad, K., and Hatakeyama, H., 1991, Mechanical properties of solvolysis lignin derived polyurethanes. *Polym. Adv. Technol.* **2**: 41-47.
18. Hatakeyama, T., and Hatakeyama, H., 1992, Molecular Relaxation of Cellulosic Polyelectrolytes with Water. *ACS Symp. Ser.* **489**: 329-340.
19. Hirose, S., Kobashigawa, K., Izuta, Y. and Hatakeyama, H., 1998, Thermal degradation of polyurethanes containing lignin studied by TG-FTIR. *Polym. Int.* **41**: 247-256.
20. Hatakeyama, H., and Hatakeyama, T., 1998, Interaction between water and hydrophilic polymers. *Thermochim. Acta* **308**: 3-22.
21. Tanaka, R., Hatakeyama, T., and Hatakeyama, H., 1998, Formation of locust bean g um hydrogel by freezing thawing. *Polym. Int.* **45**: 118-126.
22. Hatakeyama, H., Hirose, S., Nakamura, K., and Hatakeyama, T., 1993, New Types of Polyurethanes Derived from Lignocellulose and Saccharides. In *Cellulosics: Chemical, Biochemical and Material Aspects* (J. F. Kennedy, G. O. Phillips and P. A. Williams, eds.), Ellis Horwood, Chichester, pp. 524-536.
23. Morohoshi, N., Hirose, S., and Hatakeyama, H., Tokashiki, T. and Teruya, K., 1995, Biodegradability of polyurethane foams derived from molasses. *Sen-i Gakkaishi* **51**: 143-149.
24. Hatakeyama, H., Hirose, S., Hatakeyama, T., Nakamura, K., Kobashigawa, K., and Morohoshi, N., 1995, Biodegradable polyurethanes from plant components. *J. Macromol. Sci., Pure Appl. Chem.* **A32**: 743-750.
25. Donnely, M. J., 1995, In Vitro enzymic synthesis of polymers containing saccharides, lignins, proteins or related Components: a Review. *Polym. Int.* **47**: 257-266.
26. Nakamura, K., Nishimura, Y., Zetterlund, P., Hatakeyama, T., and Hatakeyama, H., 1996, TG-FTIR Studies on biodegradable polyurethanes containing mono- and disaccharides Components. *Thermochim. Acta* **282/283**: 433-441.
27. Zetterlund, P., Hirose, S., Hatakeyama, T., Hatakeyama, H. and Albertsson, A-C., 1997, Thermal and mechanical properties of polyurethanes derived from mono- and disaccharides. *Polym. Int.* **42:** 1-8.
28. Hatakeyama, H., Kobahigawa, K., Hirose, S., and Hatakeyama, T., 1998, Synthesis and physical properties of polyurethanes from saccharide-based polycaprolactones. *Macromol. Symp.* **130**: 127-138.
29. Hatakeyama, T., Tokashiki, T., and Hatakeyama, H., 1998, Thermal propeties of polyurethanes derived from molasses before and after biodegradation. *Macromol. Symp.* **130**: 139-150.
30. Gandini, A., and Belgacem, N. M., 1998, Recent advances in the elaboration of polymeric materials derived from biomass components. *Polym. Int.* **47**: 267-276.
31. Hatakeyama, T. and Quinn, F. X., 1994, *Thermal Analysis, Fundamentals and Applications to Polymer Science*, John Wiley and Sons, Chichester. Pp. 81-87.
32. T. Hatakeyama, T. and Liu, Z., 1998, *Handbook of Thermal Analysis*, John Wiley, Chichester, pp. 206-210.

Environmentally Degradable Plastics Based on Oxo-Biodegradation of Conventional Polyolefins

NORMAN C. BILLINGHAM[1], MICHELA BONORA[2], and DAVID DE CORTE[3]

1: Department of Chemistry, University of Sussex, Brighton, BN1 9QJ, UK and EPI (Europe) Ltd., Unit 7, Dunston Place, Dunston Road, Chesterfield, Derbyshire, S41 8NL, UK; 2: Ciba Specialty Chemicals S.p.A., Via Pila 6/3, I-40044 Sasso Marconi, Italy; 3: Ciba Specialty Chemicals, CH-4002 Basel, Switzerland

1. INTRODUCTION

To be considered as environmentally degradable, a plastic must satisfy at least two requirements. On outdoor exposure or soil burial it must become brittle rapidly enough to disappear visually, and the degraded material must be susceptible to biological attack giving complete conversion to biomass without release of toxic products.

Among the numerous applications for commodity thermoplastics, films and molded containers are very important. Polyethylene (PE) and polypropylene (PP) are commonly used in these areas because of their low cost, easy processing and good mechanical properties. Although extending the lifetime of plastics has dominated plastics technology in the last 50 years, there are many products that have a relatively short use-life (weeks, months), following which the articles are no longer needed and are discarded. At this point the durability and persistence of man-made plastics become disadvantages. There are a number of recognized real-life situations when biodegradability would be a very useful property, but polyolefins (and most other man-made polymers) are bioinert.

For microflora (fungi, bacteria etc) to convert and assimilate the carbon in any substrate, a number of criteria must be met. The substrate must be water–wettable, and the constituent molecules must be sufficiently small that a very large number of their chain ends are accessible at the surface of the material. Hydrocarbon thermoplastics are bioinert because they are hydrophobic, and because their good mechanical properties require very high molecular weights, leading to very few accessible chain ends.

One major approach to developing biodegradable packaging materials is the use of intrinsically biodegradable polymers, typically aliphatic polyesters. These may be laboratory products, synthesized from fermentation products, or produced directly by fermentation. An obvious example is the poly(hydroxyalkanoate)s, produced by bacterial fermentation. The aliphatic polyesters poly(lactic acid) (PLA), polycaprolactone (PCL), and the poly(hydroxyalkanoate)s hydrolyze under relatively mild conditions, to yield small molecular fragments with acid or alcohol end groups, which are biodegradable. Thus bio-assimilation is a synergistic interaction between hydrolysis and biodegradation. Despite many years of research and development, these plastics have still to make major impact in the marketplace. Indeed, life-cycle analyses[1] suggest that the energy costs of producing most of the so-called "green" plastics will make them uncompetitive for many years to come.

Conventional polyolefins are still much the best solution for many applications requiring tough films, because PE and PP are cheap, easy to process and both mechanically tough and bio-inert. A particularly good application is as agricultural films, which are widely used in the form of greenhouses, tunnels, mulch and silage films, and bale wrap, to improve crop cultivation and protect agricultural products after harvesting. Around 2.8 million tonnes of agricultural plastics are consumed annually worldwide and cover more than 5 million hectares of land. The majority of this surface is covered by mulch film, accounting for 4.5 million hectares.

Mulch films are extensively used to modify soil temperature, limit weed growth, prevent moisture loss, and improve crop yield and precocity. They mostly require a relatively short life span and normally have less contact with critical agrochemicals during their service life than e.g. greenhouse films. Like all plastics, ordinary PE mulch films undergo photo/thermal oxidation characterized by a steady decline in physical properties. As a result, they may fail to protect the growing crops for a sufficiently long period. Conversely, the use of pigments and stabilizers can provide a long service life but the mechanical breakdown at the end of the growing season may be too slow for convenience in cultivation and re-cropping. Collection and recycling of used film has been found to be impractical.

The increase in productivity and intensity of agriculture demanded by labour and fixed costs, leads to downgauging of mulch films and there is increasing demand for functionalization of films to enhance value. Examples of functional films vary from photo-selective coloration to enhance particular aspects of plant growth, to anti-fogging effects to control surface water distribution and flow. Whatever the film, it must also be cost effective.

Although both PE and PP will degrade naturally, the timescale is too long for them to be considered environmentally "friendly" and the increasing demand for such materials requires ways of converting them into water-wettable, mechanically weak material in short periods. The solution lies in accelerating the natural oxidative degradation of the polymers. In the case of degradable agricultural films, the target is that the properties will be quickly affected after the useful lifetime. Finally, upon total mechanical degradation, the residual plastic should be taken up into the bio-cycle without any negative influence on the environment. Under these conditions, the use of degradable agricultural films can be considered as a valid alternative to re-collection and traditional waste disposal routes.

We have developed additive packages, which can be incorporated into conventional polyolefins during normal processing to induce accelerated oxidation on exposure to UV light or, more importantly, to heat. The resulting degradation leads to products which are biodegradable. This paper describes some studies and applications of these oxo-biodegradable products with emphasis on agricultural and compostable film applications.

2. POLYOLEFIN OXIDATION AND BIODEGRADATION

The mechanisms of oxidative degradation of polymers, have been extensively studied and reviewed.[2] It is generally accepted that the key intermediates are hydroperoxides, which are always present because of oxidation during preparation or processing, and decompose under the influence of heat, light or transition metal catalysis to produce free radicals. Once radicals are produced they enter into a chain reaction with oxygen and C-H bonds in the polymer, to produce a range of oxidation products. Although the primary products are hydroperoxides, their decomposition yields alkoxy radicals, which are responsible for many secondary products. β-elimination of alkoxy radicals competes with H-abstraction, and leads to chain scission and formation of a variety of carbonyl products.

Since linear polymers derive their mechanical properties from chain entanglement, they can tolerate only limited scission before becoming brittle. It is well known that polyolefins that have undergone oxidative degradation have hydrophilic surfaces and greatly reduced molar masses. Reduction of the molecular weight of PE to around 40,000, combined with introduction of oxygen-containing functional groups, leads to biodegradable products.[3-5]

In a natural environment microorganisms on a substrate tend to form a biofilm, consisting of bacteria and fungi in a highly hydrated (85-98% water) matrix of extracellular polymers. Both hydrolysis and oxidation of the substrate can be mediated by the biofilm, by release of extracellular enzymes or free radicals. Fungi in particular can spread rapidly by secreting enzymes and free radicals. The extracellular nature of these substances enables fungi to tolerate higher concentrations of toxic chemicals than would be possible if the compounds had to be brought into the cell before degradation. In addition, insoluble compounds that cannot cross a cell membrane are also susceptible to attack. The mycelial growth habit of fungi also gives a competitive advantage over single cells, especially in the colonization of insoluble substrates. Hyphal penetration provides a mechanical complement to the chemical breakdown, and the high surface-to-cell ratio characteristic of the growing fungi maximizes both mechanical and enzymatic contact with the environment.

Cell enzymes, and particularly cytochrome P450 which is produced by many bacteria, continue peroxidation by reducing ground-state oxygen to the free radical superoxide ($O_2^{\cdot-}$). When protonated, this species is converted to the much more reactive peroxyl radical and hydrogen peroxide, which can be reduced by transition metal ions in the polymer to give the highly reactive hydroxyl radical. OH initiates further peroxidation leading to continued biodegradation and ultimate bioassimilation to biomass and CO_2 as long as environmental oxygen and cell nutrients are available. Thus, the bio-assimilation of degraded polyolefins is a synergistic oxo-biodegradation (Figure 1). In that sense it is totally analogous to the two-stage, hydro-biodegradation, by which linear polyesters are microbially assimilated.

3. DEGRADABLE PLASTIC ADDITIVE TECHNOLOGY

EPI has developed a series of Totally Degradable Plastic Additive (TDPA™) formulations that, when compounded with conventional polymers at appropriate levels, control the lifetimes of plastic films and articles. Stability is maintained during processing, storage and short-term end use. Once the material is discarded, oxidative degradation (initiated by heat, UV light or mechanical stress in the environment) is accelerated by as much as several orders of magnitude. The oxidized molecular fragments are hydrophilic, have molar mass values reduced by a factor of 10 or more, and are biodegradable. For agricultural applications EPI formulations are developed and marketed by Ciba Specialty Chemicals, under the trade name "Envirocare™".

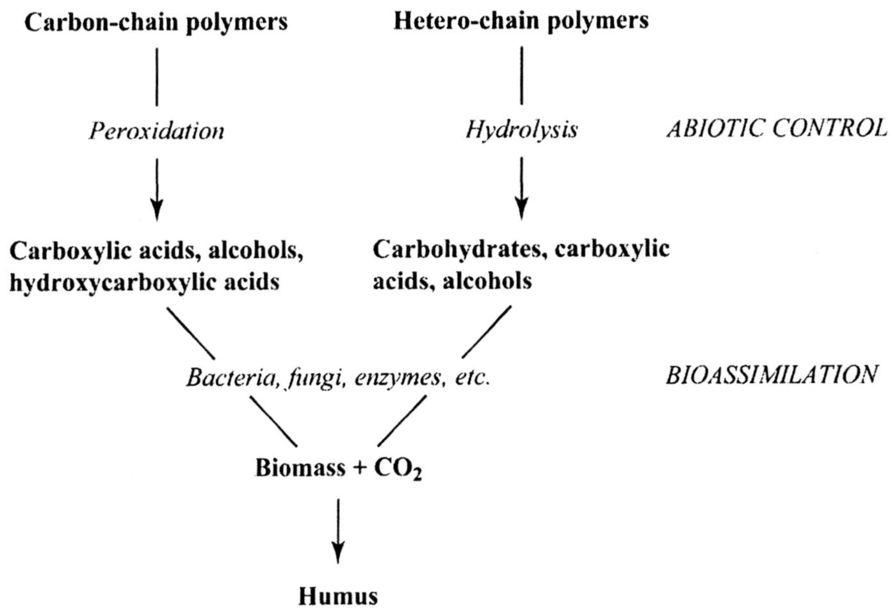

Figure 1. Biodegradation routes for oxo- and hydro-biodegradable polymers[6].

These products are typically incorporated into the final formulation as additives at levels of a few percent. They are proprietary combinations of additives, which, with appropriate compositional adjustments, allow for a wide range of storage, use, and degradation times, depending on the end use and the environment. Polyolefin pellets, which have been compounded with these additives, are processed on conventional equipment at normal speeds.

An important feature of these additives is that they are activated both by the action of sunlight and by heat. Thus degradation of mulch films can occur both on the surface that is exposed to light as well as on the parts of the film that are covered by the soil.

4. LABORATORY STUDIES OF DEGRADATION

Polyolefins compounded with Envirocare™ additives degrade rapidly in laboratory ageing. Experiments using FTIR spectroscopy, tensile testing and molecular weight measurement by size exclusion chromatography (SEC) demonstrate rapid loss of mechanical strength and chain length and formation of oxidation products.

Typical results from IR spectroscopy are presented in Figure 2, which shows the carbonyl region of the spectrum for two samples degraded in an air oven for the same time under identical conditions, with and without additive. The extensive degradation of the additive-containing sample is clear from the growth of the IR bands between 1700 and 1750 cm^{-1} associated with carbonyl groups of oxidation products.

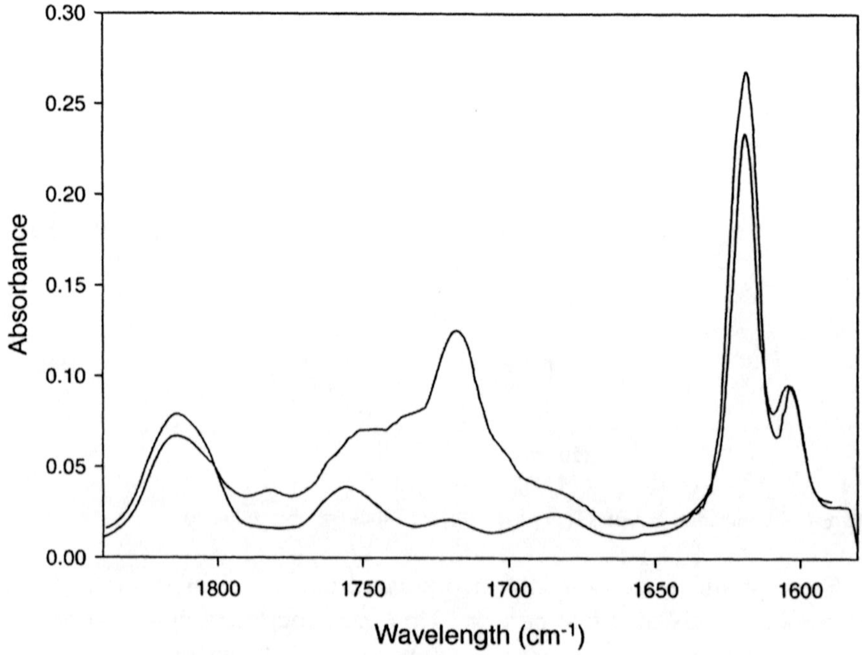

Figure 2. FTIR spectra of samples after air-oven ageing without (lower trace) and with (upper trace) Envirocare™ additive.

Figure 3 shows failure data for samples of a transparent LLDPE mulch film formulated with Envirocare™ additives and exposed in a circulating air oven at 50°C. Failure was taken as the point of embrittlement of the film; it can be varied by choosing the appropriate amount of additive and can be accelerated significantly and controllably.

Figure 4 shows data for samples of PE mulch film exposed to artificial UV ageing in a Xenon weatherometer. The lifetime to embrittlement of transparent films can be reduced significantly by using the appropriate additive. Addition of carbon black to unmodified film produces a major stabilizing effect, which can easily be overcome by the additive.

It is important to emphasize that the TDPA™ and Envirocare™ additive packages do not change the mechanism or products of the degradation of polymers. They simply accelerate the normal reactions, leading to the same final products in shorter times.

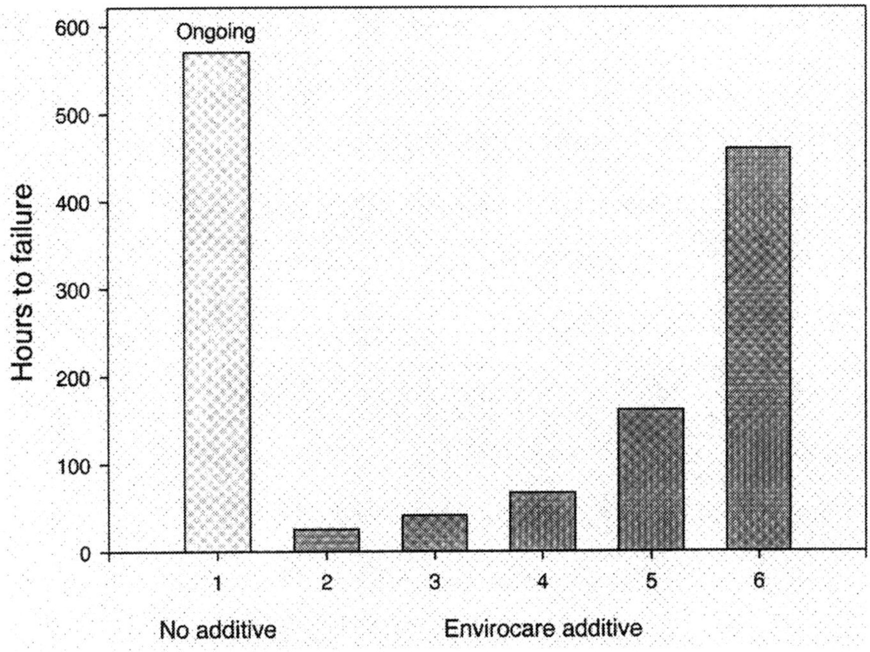

Figure 3. Performance of Envirocare™ systems in long term heat ageing. 25 μm LLDPE transparent mulch film oven exposed at 50°C.

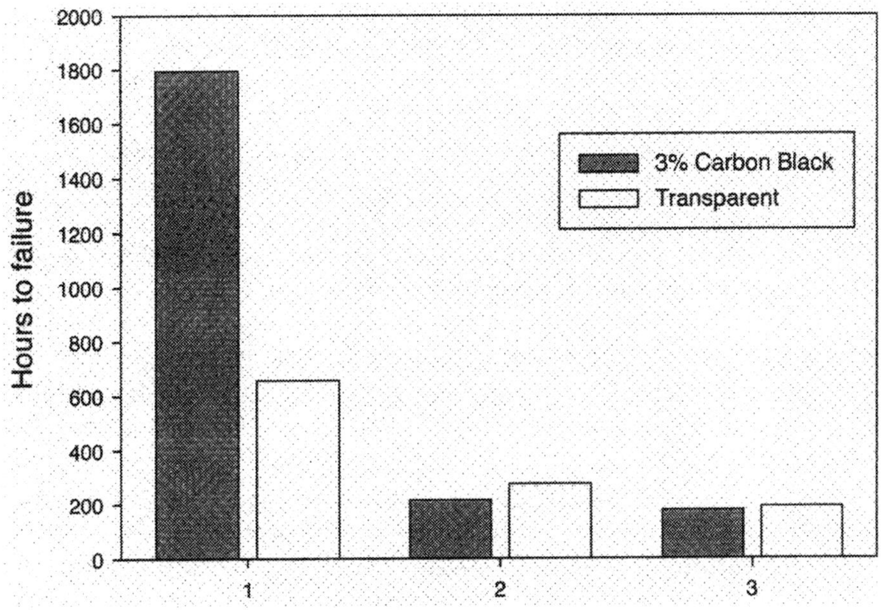

Figure 4. Performance of three different Envirocare™ systems in UV ageing. 25 μm LLDPE mulch film with and without carbon black (3 wt%). Exposure: WOM (BPT: 63 ± 3°C)

Table 1 shows molecular weight data for some degradable PE formulations measured by SEC before and after either UV or thermal ageing. It is clear that the molecular weights are reduced well into the range required for biodegradability of the degraded fragments.[3]

Table 1. Molecular weights of degradable PE films before and after accelerated degradation

Sample	Exposure	M_w
Compostable LDPE bag (17 month life)	None	190,000
"	96 h QUV	13,400
Degradable LDPE bag (6 month life)	None	302,000
"	120 h QUV	24,500
Degradable HDPE bag	None	279,000
"	59 days outdoors	18,100

5. OUTDOOR DEGRADATION

Because of the complexity of degradation and stabilization, a basic understanding is necessary to develop the additive system appropriate for a given application and location. Many different Envirocare™ films have been evaluated in the Application Centre of Ciba SC in Bologna, Italy, where the variables that affect the film durability have been simulated.

To model the complexity of real life conditions, films were laid on plastic boxes filled with soil, one part of the film being exposed to the light and another being covered by the soil. The boxes were exposed outdoors in Pontecchio Marconi (BO), Italy. This set-up simulates mulch film applications and allows the monitoring of the film degradation both on the surface and under the soil. The contribution of different parameters, like different additive systems based on Envirocare™, different pigments and different film thickness has been evaluated.

Time to embrittlement and time to complete visual disappearance on and under the soil have been measured. These criteria are significant for the estimation of the service life, of the time needed to embrittle the film enough for re-working the soil and of the time needed to get the field 'plastic free'.

Figure 5 shows that, by appropriate selection of additive, it is possible to achieve film lifetimes to embrittlement ranging from 30 days (or less than 20 kLys) upwards.

Figure 6 shows the times to embrittlement and complete visual disappearance for some mulch film samples. The latter range between less than two months up to several months, depending on the Envirocare™ formulation and film structure. The propagation of degradation to the parts of the film covered by the soil was evident in all cases. Although the film could still be detected it was heavily degraded and friable. These results confirm the effectiveness of Envirocare™ additives under the action of heat and are in agreement with the oven aging data previously shown.

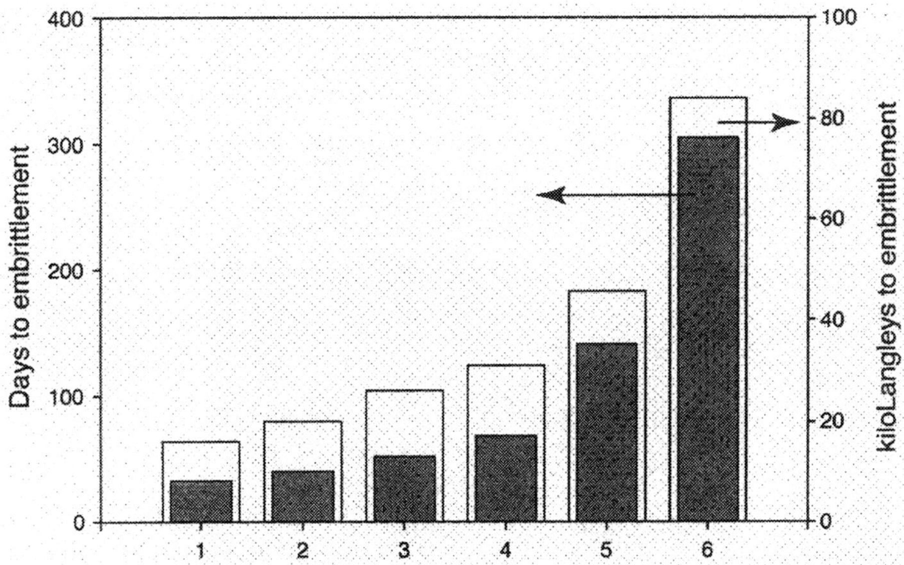

Figure 5. Embrittlement times for 25 μm LLDPE transparent mulch film with six different Envirocare™ additives exposed outdoors in soil contact in Pontecchio Marconi, Italy (110 kLy yr^{-1}). Data are times (dark bars) and absorbed energy (light bars) for embrittlement.

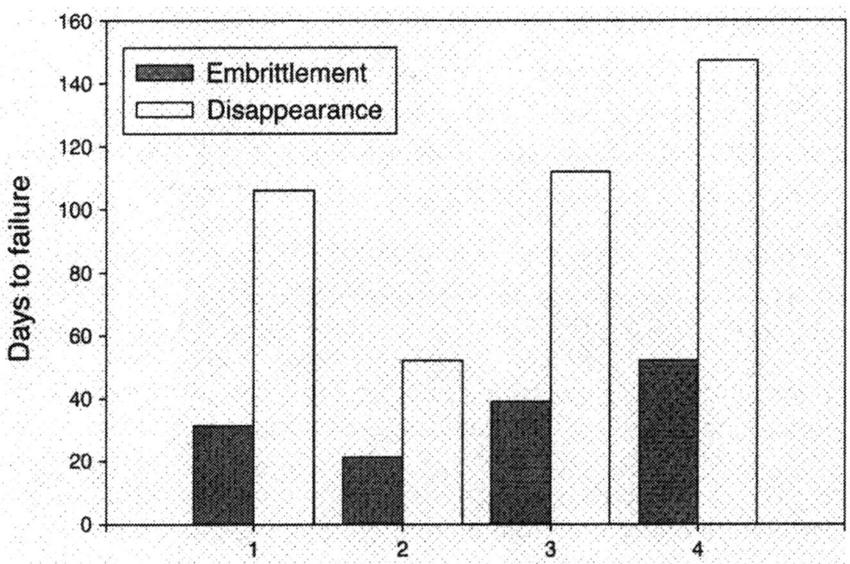

Figure 6. Embrittlement and disappearance times for 25 μm transparent LLDPE mulch films with four Envirocare™ additives exposed in soil contact in Bologna, Italy (110 kLy yr^{-1}).

6. FIELD EXPERIENCE

Laboratory and outdoor tests have been combined with large-scale field trials. Because mulch films are exposed to a wide variety of conditions and the requirements change from region to region, the assessment of film lifetime is complex. The basic laboratory data can be used to define the appropriate additive system for specific needs. A complete set of information is needed relevant to the plastic article (type, polymer, thickness, pigment), the conditions of use (crop, way of application, expected lifetime and time to degradation) and the local environmental conditions.

Mulch films containing Envirocare™ additives have been tested extensively in field trials in different countries and for different crops. Successful results have been obtained for instance, in the following applications:

- Maize - 3 months life time and 5 months degradation time
- Melons - 4 months lifetime and 5 months degradation time
- Cotton - about 1.5 month lifetime and 4 months degradation time

Field trials of films with longer required lifetime are ongoing for crops like strawberries (6-10 months), pineapple (10 months), tomato (6 months) and watermelon (6 months). In addition to mulch films, other agricultural applications where the use of Envirocare™ additive is beneficial have been identified. Hence field trials are running in solarization films, small tunnel films, seeding bags and banana bags.

7. COMPOSTING AND BIODEGRADATION

A major problem for all studies of biodegradation of polymers is the lack of clear, unambiguous and generally acceptable standards for defining or testing biodegradability. The superficially simple problem of defining biodegradability is fraught with hazards. All organic materials are biodegradable in some timescale, though complete conversion to CO_2 and water (mineralization) may take centuries. It is hard to see how, for example, the conditions in a landfill or composting operation could be simulated in small-scale laboratory conditions. Tests which require complete conversion of organic carbon to CO_2 in short periods (such as are currently being proposed by many bodies, like ASTM and CEN) can be counter-productive since a) many materials commonly recognized as biodegradable, like much plant tissue, will fail such tests and b) it is far more sensible for the organic carbon to be converted into useful biomass rather than released as greenhouse gases.

Manufacturers and users of oxo-biodegradable polyolefins view with concern the development of standards for degradable polymers which demand a high level of mineralization as the primary criterion. This protocol was originally developed for hydro-biodegradable polymers, which will primarily end up in sewage. For these polymers and in this application, such test methods are entirely acceptable but they are totally inappropriate for compost, litter and agricultural applications.

Large-scale composting operations are well established in many countries, and are an efficient way of producing useful material from what at present is largely garden and agricultural waste. Food wastes may also be used, and this would likely become much more common if inexpensive "one-way" containers were available. Such containers would need to have the low cost and the serviceability of conventional PE bags but, in addition, would need to be compostable. The ASTM definition of compostable is "capable of undergoing biological decomposition in a compost site as part of an available program, such that the material is not visually distinguishable and breaks down into carbon dioxide, water, inorganic compounds, and biomass, at a rate consistent with known compostable materials."

Biodegradability of polyethylenes modified by TDPA™ and Envirocare™ additives has been assessed by a variety of laboratory-scale and field-scale composting tests. Most recently, an extensive commercial-scale composting trial of TDPA™ additives has been carried out in the municipal composting plant of Vienna Neustadt, Austria, directed by Dr. B. Raninger (Leoben University).

This plant serves a population of about 100,000 people. It typically treats about 10,000 tons of mixed household and green garden waste annually. Composting occurs in two stages: an in-vessel, forced aeration "tunnel" process, followed by an outdoor, open-pile windrow composting stage on a paved area with weekly watering and turning. The compost produced is used mainly for landscaping and gardening.

The highly instrumented tunnels in the composting plant hold 90 m^3 of waste. In the trials of our material the input to one of the tunnels contained just over 1 wt% of LDPE bags (10,000 bags), which contained the TDPA™ additive but were not pre-aged. The compost was examined after the main maturation period (2 weeks in the tunnel), after post-maturation (12 weeks outdoors) and after six months, all according to Austrian National Standard ON S 2200. Test protocols included mass loss, analysis for heavy metals and tests of seed germination and survival of daphnia and earthworms.

The results all show that PE films modified by our additives are oxidatively biodegradable under composting conditions, yielding high-quality compost. No toxic effects could be detected on either seed germination or organism survival.

Samples of the final compost were subjected to standard ecotoxicity testing. Tests included seed germination and survival of daphnia and earthworms and were carried out according to DIN V 54900-3, ON S 2200 and ON S 2023. All tests showed absolutely no toxic or harmful by-products. As an example, Table 2 shows germination rates and plant yields for cress and summer barley on standard compost and compost containing TDPA™ formulations, demonstrating no significant differences between samples. The final conclusion of this extensive test was that products using PE and TDPA™ technology meet all requirements to be classified as degradable compostable plastics and the compost end product is fully acceptable as land fertilizer.

It has also been demonstrated that undegraded agricultural plastics based on Envirocare™ are non-ecotoxic for the environment and meet all international standards (e.g. EC OJL,219,7.8.98 for soil improvers). Ecotoxicity tests have also been carried out on soils after use of agricultural films containing Envirocare additives. Tests include *Daphnia magna* immobilization according to ISO 6341, Earthworm, acute toxicity test according to ISO 11268-1, Cress test according to ISTA, and Oat & lentils test according to ISO 11269-2. In all cases the materials were found to be non-toxic.

Table 2. Germination Rates and Plant Yields on Regular and Modified Composts

Compost	Species	Germination /%	Plant yield/g
Blank	Cress	32±5	1.4±0.2
Degradable	Cress	33±5	1.7±0.6
Blank	Barley	92±8	14.0±2
Degradable	Barley	94±3	14±2

In another series of studies, samples of PE film treated with TDPA additives have been oxidized by oven ageing, and the fragments subjected to biodegradation studies by respirometry. The results of these tests are reported elsewhere,[7] and show that the thermally oxidized material is completely biodegradable.

8. CONCLUSIONS

Macromolecules can be degraded using abiotic processes such that the molecular fragments become biodegradable. In the case of aliphatic polyesters, for example, the first stage in this two-stage process is hydrolysis. The second stage is bioassimilation of the hydrolysis products. In the case of hydrocarbon polymers, such as PE, the first stage is oxidation initiated by heat or UV light or mechanical stress. The oxidized fragments of the polymer chains are biodegraded in the second stage by the complex mixtures of microorganisms found in soil, in composting, or in landfill sites.

All of our data show that the TDPA™ and Envirocare™ additive packages can accelerate the initial degradation of many hydrocarbon polymers, to the point where they become biodegradable, in timescales which are acceptable for many practical applications, ranging from packaging to landfill cover. Control of the rates of the two stages, in the case of various commercial PE's, is achieved through a balance of appropriate additives. In this way, end-use performance can be altered to fit specific markets without altering the normal degradation pathways and products.

Independent testing has shown that full, direct food contact is permitted for both degradable and compostable film products. The additive formulations can provide sensitivity to near-UV light as well as to heat. The essential feature of the additive packages is control of the lifetime of the material.

The major benefit for the user of degradable films is convenience. After use, the plastic film or thicker part does not need to be recollected, transported to a collection center and disposed of by burial, landfill or incineration. A second important benefit is that TDPA™ and Envirocare™ additives can be used with 'commodity' plastics, with standard processing equipment and, last but not least, with standard processing conditions without affecting the mechanical or the optical properties of the plastic.

REFERENCES

1. Gerngross, T.U., and Slater, S., 2000, How green are green plastics? *Scientific American* **283** (2): 36-41.
2. See e.g Scott, G., ed., 1993, *Atmospheric Oxidation and Antioxidants*. Elsevier, London.
3. Arnaud, R., Dabin, P., Lemaire, J. Al-Malaika, S., Chohan, S., Coker, M., Scott, G., Fauve A. and Maaroufi, A., 1994, Photooxidation and biodegradation of commercial photodegradable polyethylenes. *Polym. Degr. Stab.* **46**: 211-224.
4. Weiland, M., Daro A., and David, C., 1995, Biodegradation of thermally oxidized polyethylene. *Polym. Degr. Stab.* **48**: 275-289.
5. Scott, G., 1997, Abiotic control of polymer biodegradation. *Trends Polym. Sci.* **5**: 361-368.
6. Scott, G., 1999, *Polymers and the Environment*. Royal Society of Chemistry, Ch. 5.
7. Chiellini, E., Corti, A., and Swift, G., Biodegradation of thermally oxidised fragmented low-density polyethylene. *Polym. Degr. Stab.* in press.

New Aliphatic Polyester Layered-Silicate Nanocomposites

ERIC POLLET, MARIE-AMÉLIE PAUL, and PHILIPPE DUBOIS
Laboratory of Polymeric and Composite Materials, University of Mons-Hainaut, 20 Place du Parc, 7000 MONS, Belgium

1 INTRODUCTION

Nowadays, the plastic industry occupies a predominant and growing place in our everyday life. However, many of the plastics only have application for a determined time. That is the reason why their after-use valorisation draws more and more attention from scientists and industrials. One solution to at least partially resolving the problem of plastics accumulation in the environment is to develop the sector of the biodegradable polymers. In this domain, aliphatic polyesters play a key-role[1]. These polymers exhibit many advantages in comparison to commodity plastics such as for instance polyolefins. In addition to their biodegradability, it is possible to modulate their lifetime by changing the nature of the ester repeating units or to enhance their thermo-mechanical properties by copolymerization, blending, and filling techniques.

Although this family of polyesters, like poly(ε-caprolactone) or poly(lactic acid) have been commonly used for a long time in the biomedical domain, e.g., as biomaterials, medical devices, drug delivery systems, they present some drawbacks (for example poor thermal and mechanical resistance or limited gas barrier properties) to have access to other industrial sectors such as packaging, where their use is completely justified by their biodegradability. Interestingly, it has been recently reported that significant improvements of the polymeric materials performances can be reached by dispersing at the nanometre scale very tiny amounts of layered silicate fillers

Biodegradable Polymers and Plastics, Edited by Chiellini and Solaro
Kluwer Academic/Plenum Publishers, New York, 2003

within the polymeric matrix. These new heterogeneous materials attract much attention and are often called polymer layered silicate (clay) nanocomposites[2].

Filling a polymer matrix with inorganic microparticles is a well-known process, but it may lead to fragile and opaque materials, as the filler behaves like defects at the microscopic scale. The polymer layered silicate nanocomposites, because of the nanometre size of silicate sheets, even at low filler content (1-5 wt%) exhibit markedly improved properties, as mechanical, thermal, barrier or flame retardant properties, in comparison to the unfilled matrix and more conventional microcomposites.

The field of nanocomposites was discovered in the 1960's by the Toyota research group working on nylon-6 matrix, but only about ten years ago it was extended to other polymers including, e.g., polyolefins, polystyrene, poly(meth)acrylates, or polyesters, like poly(ε-caprolactone)[2]. In the case of poly(ε-caprolactone), Messersmith and Giannelis[3] firstly evidenced the improvement of mechanical and barrier properties brought by the fine dispersion of the silicates within the polyester matrix.

The synthetic pathways to this type of nanocomposite materials can be summarized in two main processes: intercalation of the preformed polymer matrix within the silicate layer either by mixing the molten polymer with the clay, or by *in-situ* polymerization, consisting in the intercalation of the monomer into the silicate interlayer spacing followed by polymerization initiation by thermal or catalytic activation.

Combining the biodegradability of the polymer matrix with the possibility to improve its physical-chemical and thermo-mechanical performances represents a real opportunity. In addition to poly(ε-caprolactone), which is derived from the petrochemical industry, this nanocomposite technology has been extended to other aliphatic polyesters such as poly(α-hydroxyacid)s, the most representative being poly(lactic acid) issued from sugar and (poly)saccharides fermentation. Poly(lactic acid) clay nanocomposites produced from renewable (non fossil) feedstock should allow for an interesting valorisation of surplus agricultural products.

This chapter aims at reviewing the production of such aliphatic polyester layered silicate nanocomposites and the related improvements in terms of physical, mechanical, and thermal properties. Special attention will be focused on a selected polyester, i.e., poly(ε-caprolactone) (PCL), considered as a model of the family of aliphatic polyesters. These results will be then briefly extrapolated to another well-known polyester, poly(lactide) (PLA). Materials performances will be discussed and analyzed in terms of the investigated production process and related nanostructural morphology.

2 LAYERED SILICATE AS NANOFILLER

The clays most often used in the preparation of polymeric layered silicate nanocomposites belong to the structural family of 2:1 phyllosilicates (smectites) with montmorillonite as the main representative. Their crystallographic structure consists of a central octahedral layer of either alumina or magnesia sandwiched in between two tetrahedral silica layers forming sandwiched sheets characterized by a thickness of about 1 nm and lateral dimensions from 100 nm to 1 micron. These aluminosilicate layers organize themselves to form stacks separated by a regular Van der Waals gap called the interlayer or the gallery. Isomorphic substitution within the layers (for example, Al^{3+} replaced by Mg^{2+}, or Mg^{2+} replaced by Li^{+}) generates negative charges that are counterbalanced by, e.g., hydrated alkali cations such as sodium cations located in the interlayer space.

In order to render the silicate layer surface more organophilic and thus more compatible with the polymer matrix, these intercalated alkaline cations can be exchanged with cationic surfactants such as alkylammonium or alkylphosphonium ions[4-6].

Depending on the nature of the used montmorillonite (nature of the intercalated cations), the polymer matrix, and the method of preparation, three main types of composites can be obtained[2]. When the polymer is unable to intercalate between the silicate sheets, a phase separated composite is obtained, whose properties stay in the same range as traditional microcomposites. Beyond this classical family of composites, two types of nanocomposites can be recovered. *Intercalated* nanocomposites in which a (mono)layer of polymer chains is intercalated between the silicate platelets resulting in a well ordered multilayer morphology built up with alternating polymeric and inorganic layers. When the silicate layers are completely and uniformly dispersed in a continuous polymer matrix, an *exfoliated* or *delaminated* structure is formed. Two complementary techniques are used for characterizing those structures. X-ray diffraction (XRD) is used to identify intercalated structures. In such nanocomposites, the repetitive multilayer structure is well preserved, allowing the interlayer spacing to be determined. The intercalation of the polymer chains usually increases the interlayer spacing, in comparison with that of the used organoclay, leading to a shift of the diffraction peak at lower angles (angle and layer spacing being related by the Bragg's relation). As far as exfoliated structures are concerned, no more diffraction peaks are visible in the XRD diffractograms because the nanocomposite does not present ordering anymore. In the latter case, transmission electronic spectroscopy (TEM) is most often used to characterize the nanocomposite morphology. Besides these two well-defined structures, other intermediate organizations can exist presenting both intercalation and exfoliation. In this case, a broadening of the diffraction peak is often observed and one must rely on TEM observation to define the overall structure.

Essentially two strategies have been considered to prepare polymer-layered silicate nanocomposites[7]. In the so-called *intercalative polymerization*, the layered silicate is swollen within the liquid monomer (or a monomer solution) so as the polymer formation can occur in between the intercalated sheets. Polymerization is usually promoted either by heat or an appropriate catalyst. In another technique, the layered silicate is mixed with the polymer matrix in the molten state. Under these conditions and if the layer surfaces are enough compatible with the chosen polymer, the polymer can crawl into the interlayer space and form either an intercalated or an exfoliated nanocomposite. In this *melt intercalation* technique, no solvent is required.

In this study, four different clays (from Southern Clay Products) have been investigated in order to prepare PCL- and PLA-based nanocomposites: unmodified montmorillonite (Cloisite®Na^+) and three organo-modified montmorillonites, surface-modified by either bis(2-hydroxyethyl)methyl tallow ammonium ions (Cloisite®30B), or dimethyl hydrogenated tallow ammonium ions (Cloisite®20A), or dimethyl(2-ethylhexyl) tallow ammonium ions (Cloisite®25A). Details about organic modifier content (as determined by thermogravimetry) and interlayer distance (measured by XRD) of these aluminosilicate fillers are reported in Table 1. The aforementioned procedures, melt intercalation and intercalative polymerization have been approached for producing aliphatic polyester-based nanocomposites. PCL will be discussed in details first, then attention will be paid to PLA as polyester matrix. Interestingly enough, it will be reported that depending on the surface-modification of montmorillonite, nano- or microcomposites can be obtained with specific mechanical and thermal properties. Furthermore, it will be shown that the formation of PCL (or PLA)-based nanocomposites depends not only on the ammonium cation and related functionality but also, for a given cation, on the adopted synthetic route, either by melt intercalation or by *in situ* intercalative polymerization.

Table 1. Interlayer spacing and organic content of the studied (organo-modified)clays

Filler	Interlayer cation	(organic fraction wt%)	Interlayer spacing (nm)
Cloisite®Na^+	Na^+	(0)	1.21
Cloisite®30B	$(C_{18}H_{37})N^+(C_2H_4OH)_2CH_3$	(20.1)	1.84
Cloisite®25A	$(C_{18}H_{37})N^+(CH_3)_2(CH_2CH(C_2H_5)C_4H_9)$	(26.9)	2.04
Cloisite®20A	$(CH_3)_2N^+(C_{18}H_{37})_2$	(29.1)	2.36

3 PCL - BASED NANOCOMPOSITES

3.1 Melt Intercalation

The preparation of nanocomposites by melt intercalation is a very attractive environmentally friendly process since no solvent is required. In this method, the polymer matrix is blended in the molten state with a known amount of layered silicate. Under these conditions, if the layered surfaces are sufficiently compatible with the chosen matrix, the polymer can crawl into the interlayer space and forms an intercalated nanocomposite. In the case of poor compatibility between the silicate host and the polymer matrix, polymer intercalation is not allowed and micro-size clay particles are randomly dispersed in the matrix, forming a microcomposite.

Recently, the preparation of biodegradable polyester-based nanocomposites by melt intercalation has been achieved[8,9]. For that purpose, commercial PCL (CAPA®650 from Solvay Interox, M_n = 50,000 g mol^{-1}) was melt blended at 130 °C with several montmorillonites either non-modified or organo-modified by ammonium cations. Depending on the surface modification of the clays, micro- or nanocomposites with very specific properties were recovered.

3.1.1 Morphology

The morphology of the PCL-based composites was determined by X-ray diffraction (XRD) and by transmission electronic microscopy (TEM).

As far as organo-modified clays are concerned, the XRD patterns show a significant increase in the interlayer distance (Table 2), attesting for the effective polymer chain intercalation in between clay platelets.

Table 2. Influence of clay nature on the interlayer spacing before and after melt blending with PCL (3 wt% of inorganics)

Filler	Interlayer spacing in clay (nm)	Interlayer spacing in PCL composite (nm)
Cloisite®Na+	1.21	1.23
Cloisite®30B	1.84	3.10
Cloisite®25A	2.04	2.70
Cloisite®20A	2.36	3.60

TEM microphotographs of the so-obtained samples show the typical morphology of intercalated nanocomposites. Furthermore with organo-modified fillers, in addition to small stacks of intercalated clay particles, a limited amount of completely exfoliated/delaminated silicate sheets coexists with the intercalated aluminosilicates layers.

The XRD analysis of sodium montmorillonite (Cloisite®Na^+) shows that the interlayer spacing remains unchanged, that is to say that a microcomposite is formed rather than a nanocomposite. This structure has been confirmed by TEM where a clear microphase separation is observed. The properties of such a composite (filled with 3 wt% inorganics) remain in the same range as traditional microcomposites and very similar to the properties of the PCL matrix alone, as it will be shown further on.

3.1.2 Mechanical Properties

PCL is a ductile polymer able to sustain large deformations. Unfortunately, the elastic modulus is rather low making it useless for any application that requires higher stiffness that however can be in proved by the addition of filler.

The effects of the content and nature of the clay on the mechanical properties of the PCL (nano)composites have been studied by tensile testing and the major results are summarized in Table 3.

Table 3. Tensile properties of PCL before and after filling with 3 wt% of layered silicate by melt blending at 130 °C

Sample	Young's modulus (MPa)	Elongation at break ε_b (%)	Stress at break σ_b (MPa)
PCL	216 ± 5	745 ± 43	37 ± 2
PCL + Cloisite®Na^+	200 ± 9	715 ± 44	35 ± 3
PCL + Cloisite®20A	282 ± 9	625 ± 51	26 ± 3
PCL + Cloisite®25A	282 ± 9	530 ± 58	26 ± 3
PCL + Cloisite®30B	272 ± 16	560 ± 62	25 ± 3

The presence of a tiny amount of filler, i.e., as low as 3 wt% inorganics, allows to increase the elastic modulus from 216 MPa for unfilled PCL to around 280 MPa for intercalated nanocomposites obtained with organo-modified clays. It is worth pointing out that increasing the clay content enhances the material stiffness (Figure 1).

For instance, an elastic modulus as high as 399 MPa has been measured for a PCL nanocomposite filled with 10 wt% of Cloisite®30B, attesting for an almost twofold increase of the material rigidity. By contrast, the microcomposites obtained from non-modified clays (Cloisite®Na^+) show an almost constant value of Young's modulus independent of the clay content. The mechanical properties of these microcomposites remain in the same range of unfilled PCL.

For filler contents below 5 wt%, PCL retains a good ductility with an elongation at break only slightly reduced and higher than 550%. The stress at break, which represents the ultimate strength that the material can sustain before breaking, decreases with clay concentration but is maintained at an acceptable level. All these observations show that PCL nanocomposites present a high stiffness while retaining a good ductility at least up to 5 wt% filling level. Above this threshold, a drastic loss of ultimate tensile properties is observed.

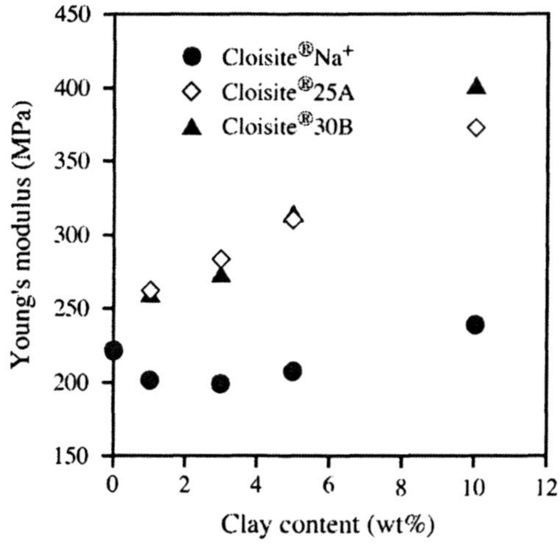

Figure 1. Evolution of Young's modulus with clay content for composite samples prepared with Cloisite®Na$^+$, Cloisite®25A, and Cloisite®30B.

3.1.3 Thermal Properties

The thermal stability of PCL-based composites can be examined by thermogravimetric analysis (TGA) under airflow. The nanocomposites show higher degradation temperature than unfilled PCL. The thermal degradation (recorded at 50% weight loss) shifts by about 50 °C towards higher temperature in the presence of a clay content as little as 1 wt% (Figure 2).

In case of microcomposites, the gain in stability is less important. Thus, the observed increase of the thermal stability of nanocomposites is to be related to the nanodispersion of the clay. The silicate layers are thought to oppose an effective barrier to the permeation of oxygen and combustion gases. However, increasing the amount of clay leads to a continuous decrease in the temperature shift which is only 30 °C for 10 wt% organo-modified clay.

Furthermore, the burning behaviour of the nanocomposites has been visually observed. It was found out that PCL nanocomposites exhibit remarkable flame retardant properties. Although unfilled PCL and PCL-based microcomposites (PCL filled with Cloisite®Na$^+$) continuously release burning droplets (flamed dripping) able to propagate the fire to surrounding materials when they are exposed to flame, PCL nanocomposites (obtained by organo-modified clays) show a totally different behaviour. No burning drop is formed anymore, rather an intensive charring of the nanocomposites is observed and this effect is more pronounced when increasing the clay content.

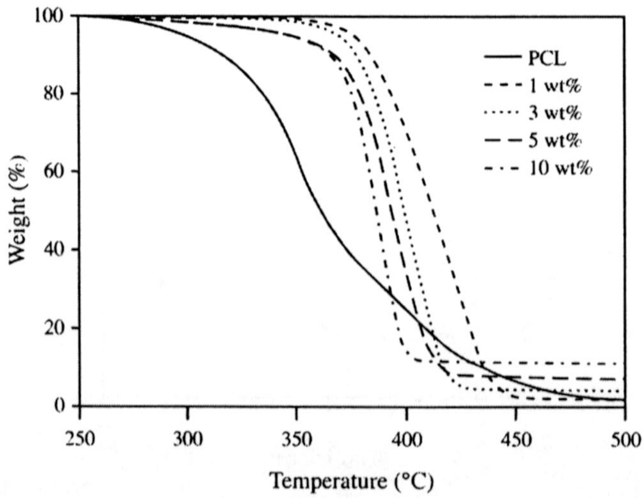

Figure 2. TGA traces of unfilled PCL and PCL-based nanocomposites prepared with 1, 3, 5 and 10 wt% of Cloisite®25A (at 20 K/min).

3.1.4 Barrier Properties

The barrier properties of the PCL-based composites were investigated. The transport properties, sorption and diffusion, were measured by a microgravimetric method[10]. The studied model permeants were methylene chloride and water vapour for which the zero concentration diffusion coefficient D_0 was determined. The presence of clay (hydrophilic platelets) in the composite gives rise to specific sites on which water molecules can be entrapped and immobilized, thus the water sorption increases on increasing the clay content, particularly for microcomposites containing Cloisite®Na^+. It was found out that the microcomposites as well as the intercalated nanocomposites have diffusion parameters for water vapour very near to those of pure PCL.

In case of CH_2Cl_2 organic vapour, the sorption at low vapour pressure is mainly dominated by the amorphous fraction present in PCL. At high vapour content, the sorption curves show an exponential increase of sorption due to the plasticization of the matrix by methylene chloride. The diffusion parameters of the microcomposites are very close to unfilled PCL while the intercalated samples show lower diffusion parameters confirming that it is not the content of clay but the size dispersion of the filler in the polymer phase that is important for improving the barrier properties of the composites.

3.1.5 Conclusions

The preparation of PCL nanocomposites by melt blending leads to microcomposites when Cloisite®Na^+ is used whereas intercalated structures are obtained with organo-modified clays such as Cloisite®25A and Cloisite®30B. As expected, the mechanical and barrier properties of the conventional microcomposites are in the same range of unfilled PCL. In contrast, all main properties of the material are improved by intercalating polymer chains between silicate sheets.

The direct melt blending process display the great advantage of processing ease but the major drawback is the sometimes-observed difficulty to intercalate preformed long polymer chains in clays displaying a poor compatibility with the matrix. Thus, the easier intercalation of the monomer could be used as the first step of an alternative process followed by the *in-situ* polymerization of the monomer in between the silicate platelets.

3.2 *In-situ* Intercalative Polymerization

In this section, PCL-based nanocomposites have been synthesized by *in-situ* intercalative polymerization of ε-CL in the presence of various organo-modified (Cloisite®25A, Cloisite®30B) or non-modified (Cloisite®Na^+) layered silicates[11,12]. As previously mentioned, this synthetic approach involves dispersion of the aluminosilicate platelets in the liquid monomer followed by polymerization (in bulk or in solution) by either thermal or catalytic activation using organometallic compounds (aluminium or tin alkoxides for instance). Depending on the nature of the filler and/or the activation mode, different composite morphologies can be obtained.

By comparison with the thermal activation procedure (by simple heating of the reaction medium at 180 °C), catalyzed Ring Opening Polymerization (ROP) allows to reach much higher monomer conversion in acceptable reaction time. Several catalysts such as tin(II) carboxylates and aluminium or tin (IV) alkoxides can be successfully used for promoting the ROP of ε-caprolactone (and lactides) according to the so-called "coordination-insertion" mechanism[13]. The polymerization proceeds through the insertion of the monomer by selective acyl-oxygen cleavage of the lactone ring. Undoubtedly, tin-based derivatives represent the most widely used catalyst in lactone polymerization. In the present study, $Bu_2Sn(OMe)_2$ has been considered as catalyst/initiator of the intercalative polymerization in order to prepare nanocomposites with various inorganic filler contents. This derivative is not only a catalyst but also the true initiator of the polymerization by lactone insertion in the tin-methoxide bonds.

3.2.1 Non-Functional Layered Silicates

For the nanocomposites prepared with non-modified clay (Cloisite®Na^+) and with Cloisite®25A, it is clear that the clay content does not affect the polymer molar masses and the polydispersity index. For all the composites, the number average molar weights (M_n) of extracted PCL chains are around 20,000 g mol^{-1} with polydispersity indexes of about 2.0. This range of molecular weight fits well the expected values computed from the initial monomer-to-tin molar ratio by assuming selective and quantitative initiation by the tin alkoxide groups.

The morphological characterization of the composites has been performed by both XRD and TEM. The XRD patterns of Cloisite®Na^+ based materials show an increase of the interlayer spacing from about 1.2 nm (in the native clay) to 1.6 nm for the *in-situ* polymerized nanocomposites, attesting for the formation of an intercalated structure. Thus, it comes out that *in-situ* polymerization of ε-CL allows to prepare intercalated nanocomposites from non-modified clay. This is a major advantage of this process since it has been observed that the direct melt blending of preformed PCL chains with the same Cloisite®Na^+ only leads to conventional microcomposites (without any intercalation).

For the nanocomposites prepared with non hydroxy-functionalized Cloisite®25A, the interlayer distance shifts from about 1.86 nm to 2.68 nm for the nanocomposites. Such a significant increase of the interlayer distance is a clear indication for the formation of an intercalated structure. This is further confirmed by TEM, which in addition shows some extent of exfoliation with the presence of individual silicate platelets spread over the PCL matrix.

3.2.2 Hydroxy-Functionalized Layered-Silicates

PCL nanocomposites prepared from Cloisite®30B, a clay modified with a functional ammonium salt bearing two primary hydroxyl groups, exhibit a continuous decrease of molar masses from 16,100 to 4,600 g mol^{-1} with increasing the clay content from 1 to 10 wt% while keeping constant the $Bu_2Sn(OMe)_2$ concentration. This behaviour can be explained by the presence of hydroxyl functions at the clay surface, which can participate in the polymerization as co-initiator/chain transfer agent.

Indeed, a fast exchange reaction between tin alkoxides species and alcohol molecules must be taken into account. In such a system, the propagating species (metal alkoxides) are temporarily converted into dormant hydroxyls and vice-versa. This rapid and reversible chain transfer reaction contributes to the polymerization control by all available hydroxyl and alkoxide functions. A decrease in the polydispersity is observed when the content of Cloisite®30B is increased and the time dependence of the PCL molar masses displays a linear increase of M_n with monomer conversion, in perfect agreement with a controlled polymerisation. These observations indicate that all hydroxyl groups available at the clay surface are potential active initiators. In this way, two types of polyester chains coexist, i.e., chains initiated by the methoxy groups of $Bu_2Sn(OMe)_2$ and chains initiated by the hydroxyl groups, thus grafted onto the clay surface.

As far as Cloisite®30B is concerned, when triethylaluminium ($AlEt_3$) is used as the initiator precursor, it reacts with the hydroxyl groups present all over the organo-modified montmorillonite to form aluminium alkoxides anchored at the surface of clay platelets. By this way, all the *in-situ* polymerized PCL chains are grafted to the clay surface by ammonium cations. An increase of the filler content leads to an increase of the number of hydroxyl groups and thus of initiating species. Therefore, increasing the clay level induces a decrease of the average molecular weight of the surface-grafted PCL chains.

For the nanocomposites prepared with Cloisite®30B (from 1 to 5 wt%), XRD analysis gives evidence for a fully exfoliated structure as demonstrated by the absence of diffraction peaks (Figure 3). The evolution of XRD patterns with the relative clay content shows that at higher filler level (10 wt%) a broad diffraction peak appears in the small angle region. This more likely indicates the formation of a partially exfoliated/partially intercalated structure. At higher clay content, the extent of exfoliation appears to be limited by steric hindrance.

These different observations are also supported by TEM analyses. A typical TEM micrograph of nanocomposites containing 3 wt% Cloisite®30B shows that individual clay platelets are randomly distributed within the PCL matrix (Figure 4). The absence of aggregates clearly confirms the completely exfoliated nature of the nanocomposites.

Initiation of ε-CL polymerization from the clay surface accounts for the tethering of the *in-situ* grown chains onto the clay surface and for clay delamination. As reported elsewhere[14], grafted and/or tethered polymer chains have a key role in preventing the exfoliated silicate layers from aggregating during post-thermal and processing treatments of the composites.

As far as Cloisite®Na$^+$ and Cloisite®25A are concerned, the obtained PCL-based nanocomposites present an intercalated structure. For these intercalated nanocomposites prepared by *in-situ* polymerization, the thermal and barrier properties are similar to those of intercalated nanocomposites obtained by melt blending. In contrast, the exfoliated nanocomposites show a dramatic improvement of the aforementioned properties, as reported by following.

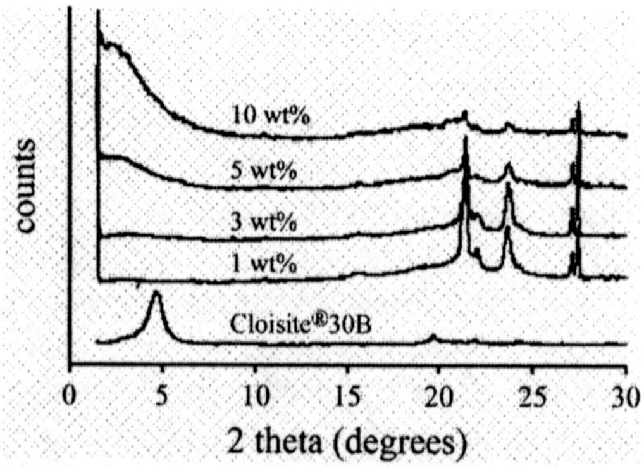

Figure 3. XRD pattern of Cloisite®30B and of PCL-based nanocomposites prepared by *in-situ* ROP with 1, 3, 5 and 10 wt% of this organo-modified clay.

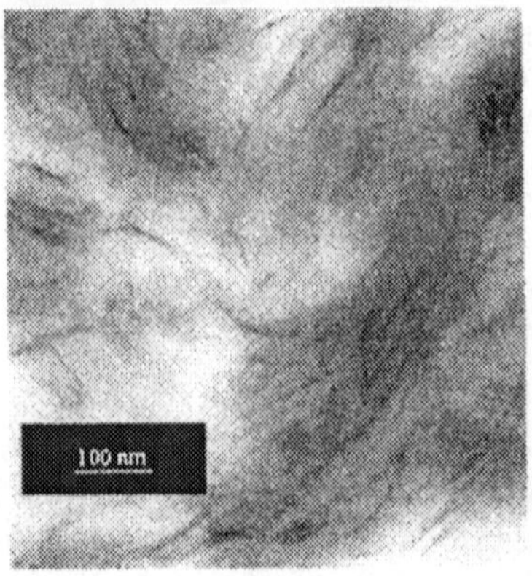

Figure 4. TEM image of PCL-based nanocomposite prepared by *in-situ* polymerization in the presence of 3 wt% Cloisite®30B.

3.2.3 Thermal Properties

Both intercalated and exfoliated nanocomposites are degraded at higher temperatures as compared to PCL and PCL/clay microcomposites. Furthermore, the resistance to thermal degradation is improved when the clay content is increased up to 5 wt%. Beyond that content, the effect levels off independent of the clay type.

The thermal degradation of PCL nanocomposites containing 3 wt% clay has been investigated by TGA. The lower thermal stability is observed for the natural sodium layered silicate Cloisite®Na$^+$ (Figure 5). When nanocomposites containing the organo-modified clays are compared, the exfoliated nanocomposites prepared in the presence of Cloisite®30B are degraded at much higher temperature.

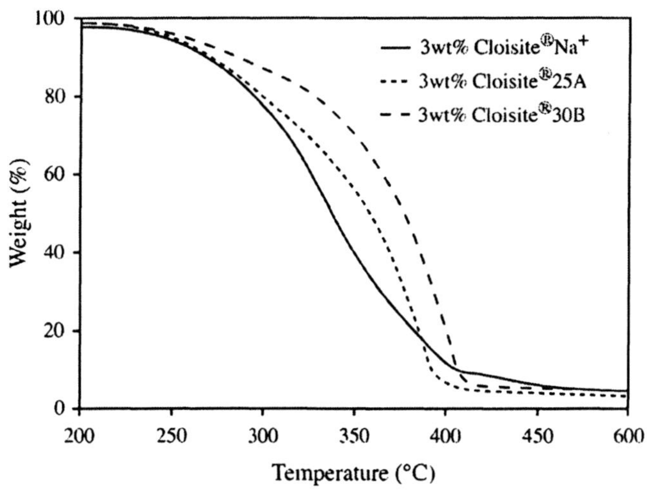

Figure 5. Temperature dependence of the weight loss of nanocomposites prepared by *in-situ* polymerization with 3 wt% of Cloisite®Na$^+$, Cloisite®25A and Cloisite®30B (at 20 K/min.).

This difference in thermal stability cannot be assigned to the PCL molecular weight, which is comparable for the three studied samples (M_n ~ 15,000 g mol^{-1}). Morphology differences as previously observed by X-ray diffraction and TEM, would be a reasonable explanation. The complete delamination of clay platelets whose surface is grafted by PCL chains is at least partly responsible for degradation at higher temperature of the nanocomposites containing Cloisite®30B[2].

3.2.4 Barrier Properties

As expected, all intercalated nanocomposites show very similar "gas barrier" behaviour, whether they have been prepared by melt blending or by *in-situ* intercalative polymerization. In contrast, the exfoliated samples obtained by *in-situ* polymerization of ε-CL in presence of Cloisite®30B show enhanced barrier properties, as evidenced by determining the zero concentration diffusion coefficient (D_0) by microgravimetry[10] (Figure 6). In fact, microcomposites and intercalated nanocomposites have diffusion parameters close to the value calculated for unfilled PCL, whereas the values of exfoliated nanocomposites are much lower even at low filler content, e.g. 1 wt% Cloisite®30B.

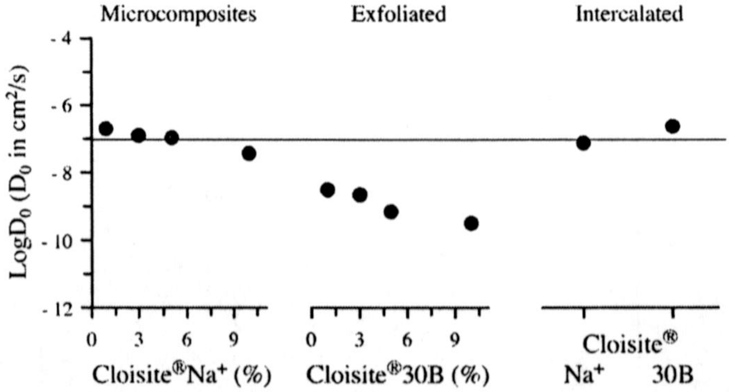

Figure 6. Dependence of water vapour diffusion log D_0 (D_0 = zero concentration diffusion coefficient) on the clay content of microcomposites, intercalated nanocomposites, and exfoliated nanocomposites.

This confirms the previous results and suggestions on polymeric exfoliated clay nanocomposites, showing that the morphological organization of the samples strongly affects the diffusion of vapours[2]. The nanometric dispersion of an inorganic component in the continuous polymeric phase is therefore very important to improve the barrier properties of the material. The presence of highly and finely dispersed clay platelets increases to a large extent the tortuousness of the system, leading to a large decrease of the diffusion coefficient.

3.2.5 Conclusions

The *in-situ* intercalative polymerization allows preparing PCL-based nanocomposites with either simple intercalation or complete exfoliation as well as intermediate morphology, that is to say semi intercalated/semi exfoliated nanocomposites. Interestingly, perfectly intercalated structures are obtained starting from Cloisite®Na^+, whereas direct melt blending of PCL with this natural clay only leads to microcomposites. All intercalated nanocomposites display improved properties by comparison to unfilled PCL but these properties remain in the same range whatever the technique used, i.e. melt blending or *in-situ* polymerization. Thus the major advantage of the *in-situ* intercalative polymerization process is the possibility to obtain fully exfoliated structures by the grafting of PCL chains at the clay surface.

As expected, these delaminated structures exhibit improved thermal and barrier properties due to the homogeneous high dispersion of individual clay platelets within the polymer matrix. Furthermore, the grafting of the polymer chains plays a key role in preventing the exfoliated silicate layers from aggregating during post-thermal and processing treatments of the composites and triggers a drastic enhancement of the aforementioned properties.

3.3 Masterbatch Process

As previously described, melt intercalation of preformed polymers and *in-situ* intercalative polymerization represent the most commonly used techniques to prepare polymer/clay nanocomposites. Both techniques display their own advantages. The major advantage of the melt blend intercalation is its ease of processing since it only requires the addition of the filler during the extrusion/kneading process, without making any important change on the industrial production line. On the other hand, the *in-situ* polymerization process allows for obtaining exfoliated nanocomposites (when using the appropriate functionalized organo-modified clay) displaying much enhanced properties.

Accordingly, a two-step method, named masterbatch process, has been approached for the preparation of PCL layered silicate nanocomposites by combining the *in-situ* intercalative polymerization and the melt blend intercalation process[15]. In such a process, a highly clay-filled (organo-modified) PCL is first prepared by *in-situ* intercalation polymerization of ε-CL, followed by its addition as masterbatch, that is blended with the molten polyester matrix (commercial PCL CAPA®650). As it will be shown, this method permits to prepare PCL-based nanocomposites with a high degree of exfoliation, which cannot be achieved by directly mixing PCL and clay.

3.3.1 Synthesis and Morphology

The polymerization of ε-CL was conducted in bulk at 100 °C by using dibutyltin dimethoxide as catalyst and in the presence of 25 to 50 wt% of (organo)clay. In order to collect PCL composites with high inorganic content, polymerizations were terminated at rather low monomer conversion. In agreement with observations made for PCL layered silicate nanocomposites prepared by the same technique and containing a small amount of clay (1-10 wt%), intercalated nanocomposites are recovered when the nanofiller is Cloisite®Na^+ or Cloisite®25A. In contrast, partially exfoliated/partially intercalated structures are formed in the presence of Cloisite®30B; this situation results from the grafting of PCL chains on the clay surface as already discussed. Then, the desired PCL-based nanocomposites were prepared by melt blending the PCL/clay masterbatch with the PCL matrix (M_n ~ 50,000 g mol^{-1}) in order to reach ultimate clay contents of 1, 3, 5, and 10 wt%.

The obtained composites have been analyzed by XRD and TEM in order to estimate the extent of the nanofiller dispersion in the PCL matrix. In all cases, intercalated nanocomposites were formed as evidenced by the significant increase in the interlayer distance. For instance, the interlayer distance increases from 1.17 nm to 1.79 nm for Cloisite®Na^+-filled PCL composite (Figure 7). The intercalated nanostructures have been confirmed by TEM analysis.

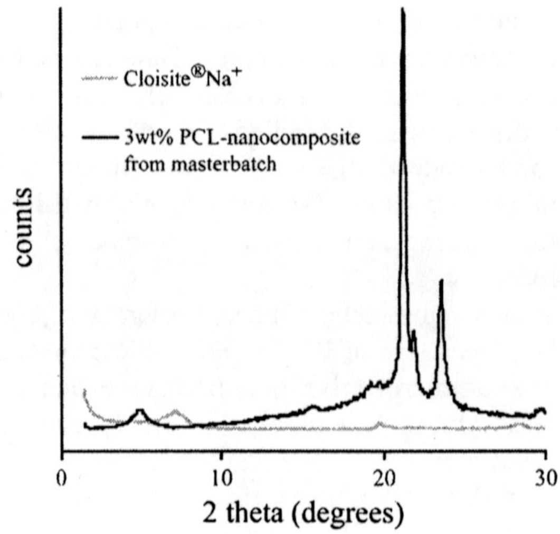

Figure 7. XRD patterns of Cloisite®Na^+ and of the corresponding PCL-based nanocomposite (3 wt%) prepared from the masterbatch containing 51 wt% of inorganics.

This situation is a remarkable progress as compared to the microcomposite structure that is formed when PCL is directly melt-blended with the natural sodium clay (Cloisite®Na^+). It appears that the intercalated structure found in the masterbatch (obtained by *in-situ* polymerization) is preserved upon melt blending with high molecular weight PCL. In the case of Cloisite®25A and Cloisite®30B, the interlayer distance increases from 1.86 nm to 2.87 nm and from 1.85 nm to 3.33 nm, respectively. In the latter case, a partially exfoliated structure can be observed by TEM.

3.3.2 Mechanical Properties

It can be interesting to compare the mechanical properties measured for PCL/Cloisite®Na^+ nanocomposites (obtained by the two-step masterbatch process) and for microcomposites prepared by direct melt blending of PCL and Cloisite®Na$^+$ under the same thermal and processing conditions. The difference in observed properties for the two series of composites is the expression of their morphology difference.

The elastic modulus is improved from 217 MPa for unfilled PCL up to 365 MPa for the nanocomposites (filled with 10 wt% Cloisite®Na$^+$) prepared by the masterbatch process. In contrast, the microcomposites prepared by direct melt blending with Cloisite®Na$^+$ showed only very limited increase in stiffness.

All PCL nanocomposites originated from the masterbatch remain rather ductile, with elongation at break higher than 300%. The stress at break slightly drops when the clay content is increased but remains at an acceptable level. Similar observations have been reported for the two organo-modified clays (Cloisite®25A and Cloisite®30B), that is increased stiffness and preserved ultimate tensile properties.

3.3.3 Thermal Properties

The thermal stability of PCL-based nanocomposites prepared by masterbatch process has been studied by TGA (at 20 °C/min.). All these nanocomposites have a degradation temperature higher than neat PCL and this improvement in thermal stability is slightly greater than for the composites obtained by direct melt blending.

Thus, the temperature at which 50% weight is lost shifts by about 50 °C to higher temperature when 3 wt% of inorganic filler is added. The thermal degradation has been studied also as function of the clay content in the final nanocomposites prepared via the masterbatch process. As far as Cloisite®30B is concerned, the temperature at which 50% weight loss is recorded shifts by about 65 °C to higher temperature at a clay content as low as 1 wt%. However and even if some enhanced thermal stability is still observed, this beneficial effect continuously decreases upon increasing the clay content. This thermal behaviour is very similar to that already observed for PCL-based nanocomposites prepared by direct melt intercalation.

3.3.4 Conclusions

The masterbatch process allows to obtain PCL intercalated silicate nanocomposites, even with natural Cloisite®Na$^+$, and it even leads to partially intercalated/partially exfoliated structure when using a masterbatch filled with Cloisite®30B. Stiffness of PCL nanocomposites is significantly improved as compared to microcomposites directly prepared by melt intercalation. As expected, the thermal stability of these nanocomposites is also enhanced as result of restricted permeation to oxygen and combustion gas by the finely dispersed silicate sheets. For instance, the temperature at which 50% weight loss is recorded shifts at higher temperature by at least 50 °C.

This process presents major advantages since it allows preparing intercalated nanocomposites or even exfoliated structures with high molecular weight polymer matrix. Furthermore, this synthetic pathway permits to bring compatibility between the inorganic filler and the polymer matrix by anchoring PCL chains on the clay surface. This strategy could be extended to all polymers known for being miscible with PCL.

Finally, by its easiness of processing (simple addition of the filler during the extrusion process), the masterbatch process presents a good potential for further industrial applications.

4 PLA – BASED NANOCOMPOSITES BY MELT INTERCALATION

In addition to the PCL matrix, preparation of biodegradable nanocomposites has been performed also with poly(L-lactide). Although PLA finds numerous applications in medical field, its thermal stability and gas barrier properties still need to be improved to enlarge its application domains, especially in food packaging. It is in this perspective that nanocomposite technology can bring some improvements. In addition, this polyester is a relatively stiff and brittle polymer with low deformation at break. In order to render PLA competitive with flexible commodity polymer, the matrix needs to be plasticized, e.g., with poly(ethylene glycol) 1000 (PEG 1000) before mixing in the molten state with the (organo)clay.

4.1 Influence of the Clay Organo-Modifier

Similarly to PCL layered silicate nanocomposites, melt intercalation of plasticized PLA with a constant amount of nanoclays (3 wt%) leads to an intercalated nanostructure[16,17], even for the unmodified natural montmorillonite (Cloisite®Na^+). This particularity can be explained by the sole intercalation of the plasticizer (PEG chains) into the interlayer spacing of the filler, leading to an increase of the interlayer distance from 1.21 to 1.77 nm, as already observed for simple blends of natural montmorillonite with PEG alone. Selective PEG intercalation was further confirmed by the impossibility to form a nanocomposite by melt blending (non plasticized) PLA with Cloisite®Na^+, only microcomposites could be recovered. XRD analysis performed on organoclay based blends (Figure 8) does not allow excluding a competition between the plasticizer and the polyester matrix for intercalation, since Cloisite®25A or Cloisite®30B and neat PLA can form intercalated structure.

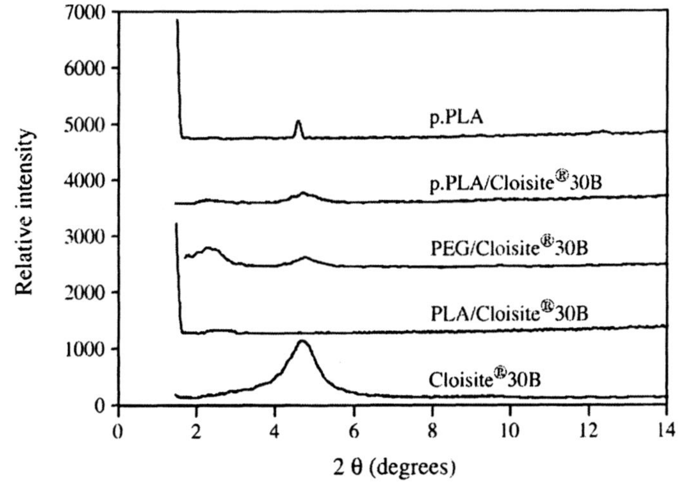

Figure 8. XRD diffractograms of Cloisite®30B, PLA/Cloisite®30B nanocomposite, PEG 1000/Cloisite®30B nanocomposite, plasticized PLA (p.PLA), and plasticized PLA/3 wt% Cloisite®30B nanocomposites.

As far as the thermal stability is concerned, TGA shows that similarly to PCL-based blends, Cloisite®30B triggers the largest thermal improvement, the main degradation process being shifted at higher temperature by 40 °C (Figure 9). On another hand, differential scanning calorimetry (DSC) analysis indicates that the (organo)clay do not influence either T_g or T_m of the plasticized PLA matrix, that are maintained around 15 and 170 °C, respectively. Visual tests of flame combustion were also performed. Compared to the unfilled plasticized PLA matrix, the intercalated nanocomposites based on organo-modified clays do not produce any burning droplets anymore. They are characterized by marked char formation. Interestingly, the blend prepared with plasticized PLA and Cloisite®Na^+ burns producing burning droplets, which seems to indicate that the sole intercalation of PEG chains into the interlayer space of the clay does not allow to promote fire retardant properties.

4.2 Influence of Clay Content

Different compositions have been prepared by incorporating 1, 3, 5, and 10 wt% of either Cloisite®30B or Cloisite®Na^+. When the filler content increases, no significant change in the interlayer distance can be detected as demonstrated for Cloisite®30B where $d_{(0,0,1)}$ remains close to 3.80 nm for loading from 1 to 10 wt% and for Cloisite®Na^+ with an almost constant interlayer distance close to 1.77 nm. Such an observation can be considered as an additional argument for the sole intercalation of PEG 1000. An increase of the thermal stability with the organo-modified clay content has been observed by TGA, a maximum being attained at 5 wt% loading, as also pointed out for PCL-based nanocomposites (Figure 10).

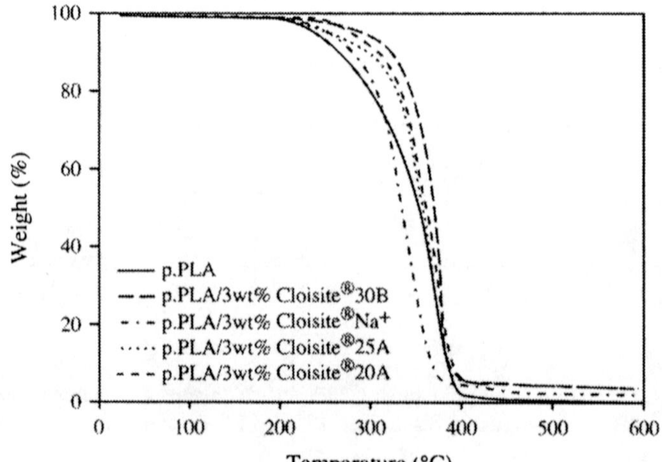

Figure 9. Thermogravimetric analysis of PLA plasticized with PEG 1000 and filled with 3 wt% of (organo-modified) montmorillonite. Cl. stands for Cloisite® and p.PLA for plasticized PLA. (Experiments carried out under air flow at 20 K/min heating rate).

On the other hand, filling the plasticized PLA matrix with increasing amount of unmodified Cloisite®Na^+ leads to a decrease of the thermal stability of the composite. It must be reminded that Na^+ cations in the interlayer space of the clay are highly hydrated ions. At high temperature, water release can be responsible of the partial hydrolysis of PLA matrix, leading to shorter chains, which can undergo degradation at lower temperature. As an evidence, the maximum found in TGA curve derivatives shifted from 371 to 325 °C by filling plasticized PLA with 1 to 10 wt% Cloisite®Na^+. Size exclusion chromatography analyses confirmed the degradation of PLA chains as the number average molecular weight of the matrix decreases from 33,200 to 8,900 upon increasing the clay content. DSC curves reveal that increasing the Cloisite®Na^+ content from 1 to 10 wt% results in a decrease of PLA melting temperature from 170 to 160 °C, very likely because of the formation of shorter PLA chains.

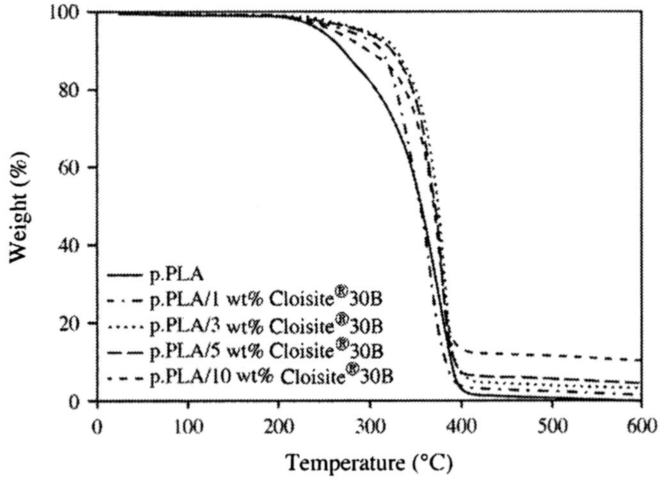

Figure 10. Thermogravimetric analysis of PLA plasticized with PEG 1000 (p.PLA) and filled with various amounts of Cloisite®30B. (Experiments carried out under air flow at 20K/min heating rate).

4.3 Influence of the Plasticizer Content

In order to shed some light on the actual role of the plasticizer, the effect of PEG 1000 content on the morphological and thermal properties of Cloisite®30B-based nanocomposites has been studied. XRD analysis reveals that increasing the PEG 1000 level from 5 to 20 wt% in a blend containing 3 wt% of clay does not trigger a significant morphological evolution in the intercalation. This observation tends to confirm that the intercalation is also partially due to PLA chains and not only to the plasticizer. As expected, PEG 1000 induces a higher PLA crystallinity, as shown by the increase of the melting and crystallinity enthalpy values (Table 4).

Table 4. Effect of the PEG 1000 content on the thermal properties of plasticized PLA filled with 3 wt% of Cloisite®30B[a]

PEG 1000 (wt%)	T_g (°C)	T_m (°C)	ΔH_m (J/g)	ΔH_c (J/g)
5	32	173	44.7	39.6
10	25	172	47.3	41.6
20	16	170	58.4	52.7

[a] Determined by DSC, under nitrogen flow; heating ramp 10 K/min. from –50 to 200 °C; cooling ramp 10 K/min. from 200 to –50 °C.

From the TGA curves, it can be concluded that the level of plasticizer has no influence on the thermal stability of the nanocomposites, indeed the maximum thermal degradation occurs at about 380 °C whatever the PEG 1000 content (ranging from 5 to 20 wt%).

4.4 Concluding Remarks

Plasticized poly(L-lactide) (organo)clay nanocomposites have been prepared by direct melt blending. X-ray diffraction has pointed out that all studied clays, including natural montmorillonite, lead to intercalated nanostructures. In the latter case, owing the polarity difference between the PLA matrix and the non modified clay, the intercalation is essentially provoked by the interlayer migration of poly(ethyleneglycol) chains used as plasticizer. Morphological analyses carried out on PLA/organo-modified montmorillonite nanocomposites have shown the possible competition between PLA chains and the plasticizer for intercalation between the aluminosilicate layers. From the thermogravimetric analyses, it is found that the more efficient organo-modified clay for forming intercalated and possibly partially exfoliated nanocomposites is the montmorillonite surface-treated with bis(2-hydroxyethyl) methyl (hydrogenated tallow alkyl) ammonium cations.

5 GENERAL CONCLUSIONS

Aliphatic polyester layered silicate nanocomposites based on poly(ε-caprolactone) (PCL) and on plasticized poly(L-lactide) (PLA) have been prepared first by melt blending of the respective polymer matrix with different (organo-modified) montmorillonites. It has been demonstrated that melt blending with organo-modified clay such as Cloisite®20A, 25A or 30B, yields intercalated nanocomposites with the possibility of partial exfoliation. Even at low organoclay content, substantial improvement of thermal stability, gas barrier properties and physical-mechanical performances have been noticed. However, melt blending of natural montmorillonite with PCL or neat (non plasticized) PLA leads to microphase-separated compositions.

As an alternative, *in-situ* intercalative polymerization of ε-caprolactone directly in between the silicate layers was also approached. In this way, intercalated nanocomposites could be obtained even with Cloisite®Na^+. Furthermore, exfoliated nanocomposites have been recovered starting from Cloisite®30B, thus a montmorillonite surface-modified by ammonium cations bearing two primary hydroxyl functions. Such a morphology results from the direct grafting of the growing polyester chains onto the hydroxyl functionalized silicate layers. As expected, clay delamination gives rise to an additional improvement of materials properties, in comparison with intercalated nanocomposites obtained by melt blending.

Interestingly enough, both synthetic pathways have been successively combined to each other, via the so-called masterbatch process. It consists of producing first a highly filled clay/PCL nanocomposite by *in situ* polymerization and then dispersing it within the preformed PCL matrix by melt blending. Either intercalated (with Cloisite®Na$^+$) or partially intercalated/partially exfoliated (with Cloisite®30B) nanocomposites have been prepared with much improved physical, mechanical, and thermal properties.

In-situ intercalative polymerization of lactides will be investigated in a near future. Biodegradation of these PCL or PLA-based nanocomposites by composting is under current investigation as well.

ACKNOWLEDGMENTS

B. Lepoittevin, N. Pantoustier, and M. Alexandre are warmly thanked for their collaboration. E. Pollet is grateful to Cécile Delcourt for efficient technical assistance, and to the Région Wallonne D.G.T.R.E. for support in the frame of the W.D.U. program TECMAVER. M.-A. Paul thanks the F.R.I.A. for her PhD grant. The authors thank Prof. A. Rulmont and C. Henrist from the General Chemistry and Physics Chemistry Department, University of Liège (Belgium) for their help in the XRD measurements. LMPC thanks the Services Fédéraux des Affaires Scientifiques, Techniques et Culturelles for general support in the frame of the PAI-5/03.

REFERENCES

1. (a) Vert, M., Schwarch, G., Coudane, J., 1995, Present and future of PLA polymers. *JMS Pure Appl. Chem.* **A32** (4): 787-796;
   (b) Hyon, S.-H., Jamshidi, K., Ikada Y., 1999, Synthesis of polylactides with different molecular weights. *Biomaterials* **18**: 1503-1508.
2. (a) Messersmith, P.B., Giannelis, E.P., 1993, Polymer-layered silicates nanocomposites: in situ intercalative polymerization of CL in layered silicates. *Chem. Mater.* **5**: 1064-1066;
   (b) Alexandre, M., Dubois, Ph., 2000, Polymer-layered silicate nanocomposites: preparation, properties, and uses of new class of materials. *Mater. Sci. Eng.* **R28** (1-2): 1-65.
3. Messersmith, P.B., Giannelis, E.P., 1995, Synthesis and barrier properties of poly(ε-caprolactone)-layered silicate nanocoposites. *J. Polym. Sci., Part A* **33**: 1047-1057.
4. Giannelis, E.P., Krishnamoorti, R., Manias, E., 1999, Polymer-silicate nanocomposites: model systems for confined polymers brushes. *Adv. Polym. Sci.* **118**: 108-147.
5. Lagaly, G., 1986, Interaction of alkylamines with different types of layered compounds. *Solid State Ionics* **22**: 43-51.
6. Vaia, R.A., Teukolsky, R.K., Giannelis, E.P., 1994, Interlayer structure and molecular environment of alkylammonium layered silicates. *Chem. Mater.* **6**: 1017-1022.

7. Pantoustier, N., Lepoittevin, B., Alexandre, M., Kubies, D., Calberg, C., Jérôme, R., Dubois, Ph., 2002, Biodegradable polyester layered silicate nanocomposites based on poly(ε-caprolactone). *Polym. Eng. Sci.* **42**(9):1928-1937.
8. Pantoustier, N., Alexandre, M., Degée, Ph., Kubies, D., Calberg, C., Jérôme, R., Henrist, C., Rulmont, A., Dubois, Ph., 2001, Poly(ε-caprolactone) layered silicate nanocomposites: effect of clay surface modifiers on the melt intercalation process. *e-Polymers* **009**: 1-9.
9. Lepoittevin, B., Devalckenaere, M., Pantoustier, N., Alexandre, M., Kubies, D., Calberg, C., Jérôme, R., Henrist, C., Cloots, R., Rulmont, A., Dubois, Ph., 2002, Poly(ε-caprolactone)/clay nanocomposites prepared by melt intercalation: mechanical, thermal and rheological properties. *Polymer* **43**: 4017-4023.
10. Tortora, M., Gorrasi, G., Vittoria, V., Pollet, E., Lepoittevin, B., Pantoustier, N., Alexandre, M., Dubois, Ph., 2003, Vapor barrier properties of polycaprolactone montmorillonite nanocomposites: effect of clay dispersion. *Polymer* **44**: 2271-2279.
11. Kubies, D., Pantoustier, N., Dubois, Ph., Rulmont, A., Jérôme, R., 2002, Controlled Ring-Opening Polymerization of ε-caprolactone in the presence of layered silicates and formation of nanocomposites. *Macromolecules* **35**: 3318-3320.
12. Lepoittevin, B., Pantoustier, N., Devalckenaere, M., Alexandre, M., Calberg, C., Jérôme, R., Dubois, Ph., 2002, Poly(ε-caprolactone)/clay nanocomposites by in-situ intercalative polymerization catalyzed by dibutyltin dimethoxide. *Macromolecules* **35**: 8385-8390.
13. Mecerreyes, D., Jérôme, R., Dubois, Ph., 1999, Novel macromolecular architectures based on aliphatic polyesters: relevance of the coordination–insertion ring opening polymerization. *Adv. Polym. Sci.* **147**: 1-59.
14. Huong, X., Brittain, W.J., 2001, Synthesis and characterization of PMMA nanocomposites by suspension and emulsion polymerization. *Macromolecules* **34**: 3255-3260.
15. Lepoittevin, B., Pantoustier, N., Devalckenaere, M., Alexandre, M., Calberg, C., Jérôme, R., Henrist, C., Rulmont, A., Dubois, Ph., 2003, Polymer/layered silicate nanocomposites by combined intercalative polymerization and melt intercalation: a masterbatch process. *Polymer* **44**: 2033-2040.
16. Pluta, M., Galeski, A., Paul, M.-A., Alexandre, M., Dubois, Ph., 2002, Polylactide/montmorillonite nanocomposites and microcomposites prepared by melt blending: structure and some physical properties. *J. Appl. Polym. Sci.* **86**: 1497-1506.
17. Paul, M.-A., Alexandre, M., Degée, Ph., Henrist, C., Rulmont, A., Dubois, Ph., 2003, New nanocomposite materials based on plasticized poly(L-lactide) and organo-modified montmorillonites: thermal and morphological study. *Polymer*, **44**: 443-450.

Organic Catalysis: A New and Broadly Useful Strategy for Living Polymerization

GREGORY W. NYCE, ERIC F. CONNOR, T. GLAUSER, ANDREAS MÖCK, and JAMES L. HEDRICK
IBM Almaden Research Center, San Jose, California 95120

1. INTRODUCTION

Significant effort has gone into the development of biodegradable polymers over the past few decades with the purpose of designing resorbable biomaterials, and, more recently, for designing commodity thermoplastics from renewable resources. Aliphatic polyesters, particularly polylactide, combine biocompatibility and biodegradability with remarkable physical properties and have the requisite thermal stability at processing temperatures. One of the most common synthetic routes to polyesters uses transition metal initiation compounds to affect the ring-opening polymerization (ROP) of the cyclic ester monomer. Advances in organometallic chemistry in the design and synthesis of single-site metal catalysts for ROP techniques[1] has enabled the preparation of well-defined functional polymeric materials with predictable molecular weights, narrow polydispersities, architectural and stereochemical control. The ring-opening (ROP) polymerization of lactide has been accomplished from a variety of metal catalysts including aluminium, tin, zinc and yttrium through a coordination-insertion mechanism.[2] Removal of the metal contaminant, bound to the chain-end, must be considered for many applications.

There are only a few reports on the ROP of lactides that do not use organometallic promoters[3]. The synthesis of biomacromolecules generally involves *in vivo* enzyme-catalyzed chain growth polymerization reactions within cells. Enzymes exhibit high, stereo-, reaction- and substrate specificity, and come from renewable resources that can be easily recycled. Recent advances in non-aqueous enzymology have permitted alternative reaction environments including organic solvents resulting in new biocatalytic methodologies for the controlled ROP of a wide variety of monomer sets.

Over the last few years, several groups have reported their ongoing efforts to improve classic organic synthetic reactions using, in many cases, simple organic compounds owing to the availability of enantiopure molecules. For example, proline has been used as a benign organocatalyst for the Mannich reaction (precursor to β-lactam) in place of the traditional organometallic catalysts[4]. Along similar lines, MacMillan has reported the first highly enantioselective organocatalytic inter- and intramolecular Diels-Alders reactions, 1,3-dipolar cycloadditions, and 1,4-conjugate Friedel-Crafts additions involving pyrroles as well as the first enantioselective organocatalytic alkylation of indoles catalyzed by imidazolidinone[4]. Several groups have reported effective non-enzymatic catalyst for the kinetic resolution of secondary alcohols using chiral phosphines[5] or amine catalysts[6]. These nucleophilic catalysts, particularly the "planar-chiral" heterocycles based on the tertiary phosphine and amine frameworks, provide good levels of enantiomeric excess. To some degree this strategy mimics that carried out by enzymes[7]. These trends towards environmentally sound organocatalysts have stimulated "greener" versions of classic synthetic asymmetric reactions[8]. The extension of organic catalysis to controlled polymerization procedures would be a highly desirable alternative to traditional organometallic approaches (Scheme 1).

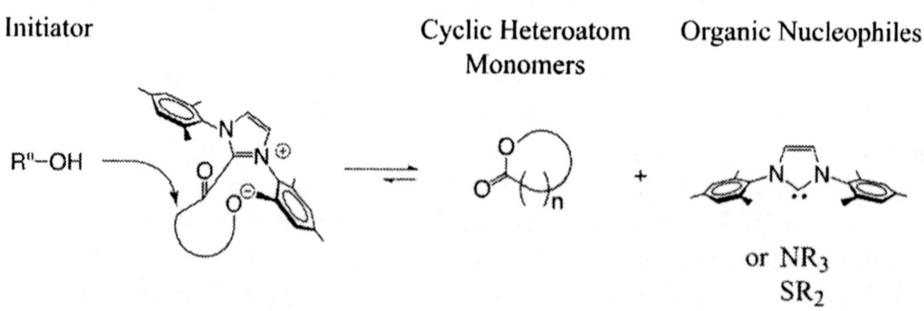

Scheme 1. General strategy for ROP of lactides using organic catalysts.

In designing an organic catalyst for the ROP of cyclic esters, we turned towards the likely active nucleophilic species of biocatalysis. Lewis bases including tertiary amines and phosphines, pyridines and imidazoles have been shown to be effective nucleophilic catalysts accelerating a wide variety of processes. For example, 4-(dimethylamino)pyridine (DMAP) and related bases such as 4-pyrrolidinopyridine (PPY) are widely used and extremely efficient reagents for acylation, alkylation, silylation, phosphorylation, condensation and transesterification reactions[9]. Likewise, other Lewis bases such as tertiary phosphines have been shown to be remarkable acylation catalysts that proceed through a nucleophilic-activation mechanism[10]. The use of nucleophilic phosphine ligands is pervasive in organometallic chemistry and homogeneous catalysis[11], and it is likely that their structure and function can be applied to other systems. Moreover, this class is particularly attractive owing to the widespread availability of potentially useful phosphines and phosphines in enantiopure form, which may function as enatioselective catalysts and provide an alternative to the organometallic standards[12-14]. Thiophines are another general class of nucleophilic compounds that were investigated for the ROP of lactide. As with the phosphines, the attractive features of these compounds are their commercial availability and/or ease of synthesis. N-heterocyclic carbenes are another class of possible nucleophilic compounds that have not yet been exploited as a catalyst to effect polymerizations. The resurgence in research on N-heterocyclic carbenes was stimulated by Arduengo's report on stable, isolable compounds. The use of N-heterocyclic carbenes has, in many cases, replaced the electron-rich phosphine ligands, producing organometallic complexes that exhibit superior catalytic performances to the phosphine analogue[15]. Moreover, nucleophilic N-heterocyclic carbene compounds are readily prepared with significant structural diversity including chiral derivatives[16], which are of interest to us for stereoselective polymerizations. These compounds will be compared with respect to their catalytic activity towards the ROP of lactide.

2. RESULTS AND DISCUSSION

2.1 Synthetic Strategy/Catalyst Evaluation

When the general approach for the ROP of lactide and evaluation of numerous catalysts is considered, it is apparent that a proliferation of variables be considered. For example, the type and concentration of catalysts, solution or bulk reaction conditions, polymerization temperature, undesirable transesterification side-reactions and the general classes of initiators all influence the polymerization. To survey these variables, a parallel approach to polymer synthesis, designed to enable rapid screening of optimal catalyst-initiator systems and associated polymerization conditions, was employed. The Quest 210 robotic reactor proved to be an excellent platform for performing up to 20 polymerizations in parallel providing the necessary environment, temperature control and agitation to accomplish the ROP[17]. The general strategy for the ROP of lactide in the presence of these nucleophilic organic catalysts requires a nucleophile such as an alcohol to initiate polymerization (Scheme 1). In this way, the α-chain end of the polylactide bears an ester functionality derived from the alcohol and polymerization proceeds when the terminal ω-hydroxyl group acts as a nucleophile to facilitate further chain growth. To demonstrate the initiating species and end-group control, pyrenylbutanol was used as the initiator since it is easily detected by ^{1}H NMR and GPC using the UV detector. Another important question with these catalysts is whether the weakly nucleophilic propagating chain-end in the presence of the catalysts is active not only to the cyclic monomer but the chain itself, leading to adverse transesterification reactions and loss of control. This is typically manifested in a broad or multimodal molecular weight distribution.

2.2 Tertiary Amine Catalyzed ROP

The catalytic behaviour of 4-(dimethylamino)pyridine (DMAP) and related bases such as 4-pyrrolidinopyridine (PPY) in the polymerization of lactide was studied using either ethanol or benzyl alcohol as initiators with 2.0 equivalents of amine relative to initiating alcohol (Scheme 1)[18]. Although a number of polymerization media were evaluated, most of the polymerizations were conducted in either methylene chloride at 35 °C, since it is a good solvent for both the lactide and poly(lactide) or in toluene at 100 °C. For a targeted degree of polymerization (DP) of 30 calculated from the monomer to initiator ratio, a measured value of 29 was obtained by ^{1}H NMR analysis of the end-groups in quantitative yields (36 h) with a polydispersity of 1.13[19]. A plot of molecular weight versus monomer conversion for the ROP of lactide initiated from ethanol in the presence of DMAP was linear, a correlation consistent with a living polymerization procedure. The versatility of the organocatalyzed ROP of lactide is demonstrated by the data shown in Table 1, where selected results of the poly(lactide) polymerization initiated from assorted alcohols in the presence of several amines for different targeted molecular weights in either methylene chloride, toluene or in bulk are shown. DMAP and PPY showed comparable catalytic activity, producing polymers with molecular weights that closely tracked the monomer-to-initiator ratio ([M]/[I]) (Samples **1-3**). The polydispersities are extremely low, and unlike most organometallic promoted polymerizations[20], remained invariant to high monomer conversions (100% conversion, 36 h). Bulk polymerization of D, L-lactide was also investigated and accomplished at 135 °C using benzyl alcohol as the initiator. Narrowly dispersed polylactides were obtained in about 5 to 20 minutes, depending on the targeted molecular weight, **4**.

Table 1. Characteristics of polylactide from a variety of organic catalysts

Sample Entry	Catalyst (Reaction Conditions)	Time (h)	Temp (°C)	Conv. (%)	[M]/[I]	DP	PDI
1	DMAP (CH_2Cl_2)	36	35	90	30	29	1.12
2	CMAP (CH_2Cl_2)	50	35	90	60	60	1.10
3	DMAP (toluene)	20	100	90	90	85	1.18
4	DMAP (bulk)	0.5	135	88	140	120	1.14
5	$P(Bu)_3$ (bulk)	0.5	135	88	60	52	1.12
6	$PPh(ME)_2$ (bulk)	20	135	78	60	47	1.23
7	$P(Ph)_2Me$ (bulk)	25	135	60	60	45	1.4
8	$P(Ph)_3$ (bulk)	30	135	60	60	51	1.21
9	Thiopine (toluene)	48	80	0	60	-	-
10	Carbene (THF)	2	RT	>95	60	59	1.09
11	Thiazolium (CH_2Cl_2)	64	RT	85	60	52	1.10

To substantiate the initiating species, reaction of excess benzyl alcohol with one equivalent of DMAP to lactide was investigated to generate a single turnover event. The lactide was ring opened with the formation of the benzyl ester, however due to the large excess of the benzyl alcohol and DMAP catalyst, transesterification with the diester product, afforded the monoester[21]. Conversely, *sec*-phenethyl alcohol ring-opened lactide quantitatively to the diester product with no evidence of adverse side reactions (Scheme 2)[22].

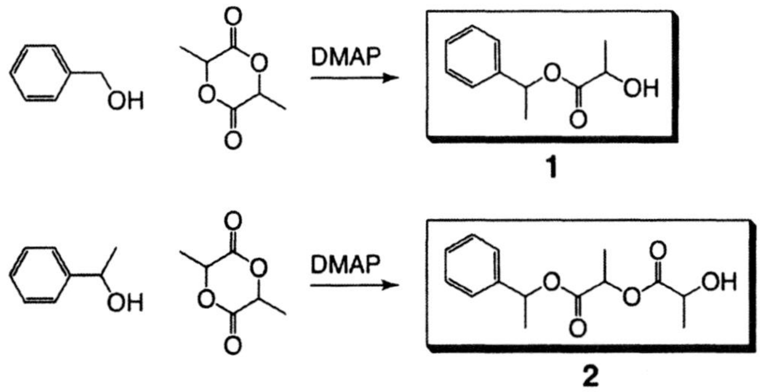

Scheme 2. Transesterification differences between primary and secondary alcohols.

These data clearly demonstrates the susceptibility of lactic acid derivatives towards selective transesterification with primary alcohols and confirms that secondary alcohols are dormant towards transesterification with the ring-opened products. Thus, the "living" polymerization character is a manifestation of the rapid initiation and the weakly nucleophilic propagating species (secondary alcohol) that is active only to the cyclic diester monomer, precluding undesirable transesterification reactions!

2.3 Organocatalytic Depolymerization

These results prompted us to survey a new depolymerization strategy based on a single transesterification reaction that should occur between a primary alcohol and poly(lactide) in the presence of DMAP. The primary alcohol should selectively cleave a PLA chain and produce an α-chain-end bearing the ester of the alcohol and a ω-chain-end having a secondary hydroxyl, which is dormant to subsequent reactions (Scheme 3)[23]. The feasibility of the organocatalytic chain scission of poly(lactide) was demonstrated with a commercially available high molecular weight poly(L-lactides) (Mn 50,000 g/mol, Mw/Mn = 1.60.

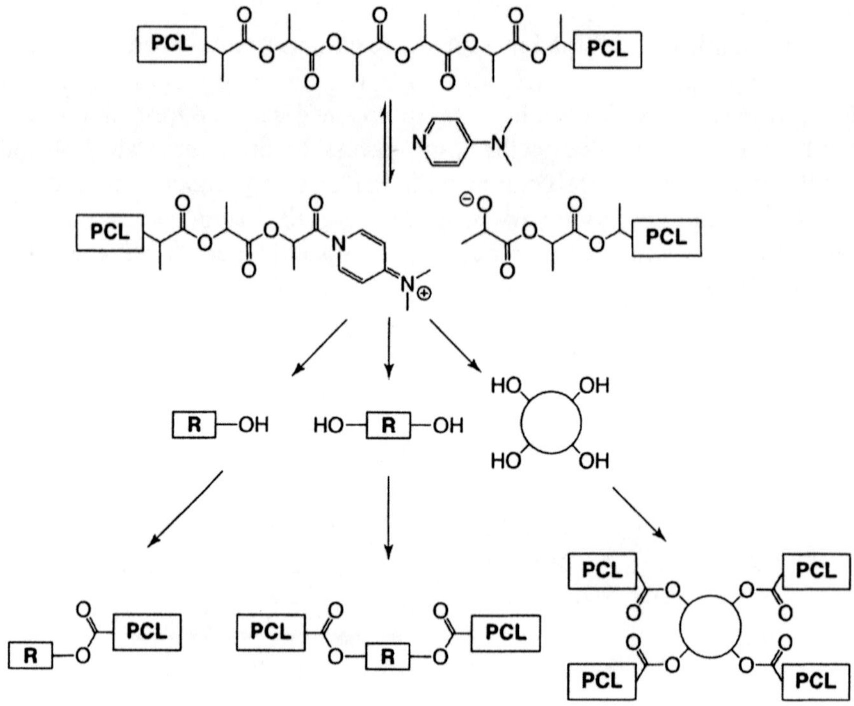

Scheme 3. Organocatalytic depolymerization.

The organocatalytic transesterification of polylactide using either benzyl alcohol, pentaerythritol or monohydroxyl functional poly(ethylene oxide) oligomers produced either linear (**12, 13**), star-shaped (**14, 15**) or block polymers (**16, 17**), respectively (Table 2). The reactions were performed in bulk (135 °C) and, at these temperatures, the homogeneous mixtures allowed effective transesterification for both of the high molecular weight polylactide samples investigated. In each case, the ^{1}H-NMR spectra clearly showed the resonances associated with the transesterification alcohol as well as the resonances associated with the hydroxyl chain end, allowing molecular weight determination. In each case, the molecular weight of the polylactide was comparable to the alcohol-to-polymer ratio and the polydispersity was monomodal with no evidence of either of the homopolymers. These combined data clearly demonstrate the versatility of the organic catalyzed chain scission approach to functional polylactide block copolymers and architectures in a single-step one-pot approach.

Table 2. Organocatalytic depolymerization: new route to functional polylactides

Sample Entry	Alcohol	Target DP	Exp. DP	PDI
12	$PHCH_2OH$	110	95	1.25
13	$PhCH_2OH$	18	10	1.35
14	Pentaerythritol	50	48	1.65
15	Pentaerythritol	25	16	1.35
16	Poly(ethylene oxide)	100	78	1.55
17	Poly(ethylene oxide)	40	31	1.61

2.4 Tertiary Phosphine Catalyzed ROP

The catalytic behaviour of a number of phosphines was investigated in bulk at 135 and 180 °C and in solution in a variety of polymerization media including THF, toluene and CH_2Cl_2 using benzyl alcohol as the initiator[24]. Polymerization in bulk at the temperatures surveyed produced polymer within 10 minutes to 50 hours depending on the catalyst type, concentration and targeted molecular weight (Scheme 1). The phosphine-catalyzed ROP of lactide in solution was considerably slower than those carried out at 135 and 180 °C in bulk (**5-8**)[25]. The catalyst concentration proved to be an important variable in the polymerizations, as phosphine contents above one equivalent relative to initiator tended to give somewhat broader molecular weight distributions (1.3-1.5), indicative of possible adverse transesterification reactions at these high polymerization temperatures. Shown in Figure 1 is a plot of molecular weight versus conversion for the ROP of lactide initiated with benzyl alcohol (target DP of 60) at 135 °C in the presence of either P(*n*-Bu)$_3$, P(*tert*-Bu)$_3$, $PhPMe_2$, Ph_2PMe or PPh_3. The correlation between molecular weight and conversion is consistent with a living or controlled polymerization. The molecular weight distributions are low and remain low to very high conversions (~95%). Comparison of phosphine activity in lactide polymerization decreases as the catalyst is varied according to the following order of reactivity: P(*n*-Bu)$_3$ > P(*tert*-Bu)$_3$ >$PhPMe_2$ > Ph_2PMe > PPh_3 > P(MeO)$_3$ (unreactive).

As expected with a nucleophilic catalyst mechanism, the substitution of the phosphine is a dominant feature in controlling reactivity. For example, in the $P(n\text{-}Bu)_3$ catalyzed polymerizations, polydispersity remained consistently below 1.2 throughout. However, it is important to note that prolonged polymerization times after monomer consumption, particularly at 180 °C, leads to a broadening of the polydispersity (~1.3–1.5), analogous to the many organometallic-promoted polymerizations[26]. Furthermore, it is well documented that phosphines are susceptible to oxidation and have suspect thermal stability: both of which could limit the catalytic activity. The polymerizations turned amber, particularly at the 180 °C reaction temperature or for the phenyl-substituted phosphines that required prolonged reaction times, suggesting catalyst degradation[11,13b]. Several chiral phosphine catalysts were investigated to possibly kinetically resolve the different entantiomers of *rac*-lactide and possible stereochemically controlled polymerizations (Scheme 4). However, for each of the commercial catalysts surveyed, no enhancement in either isomer was detected at low conversions.

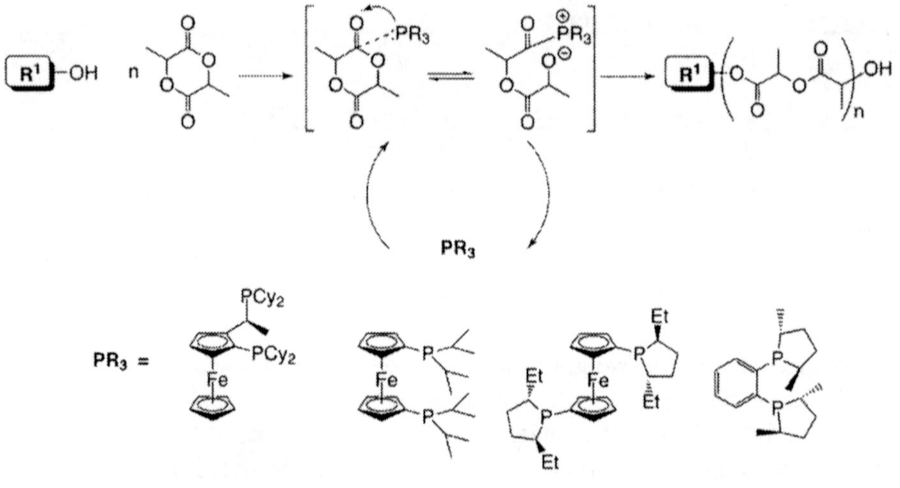

Scheme 4. Chiral phosphine catalysts.

2.5 N-Heterocyclic Carbene Catalyzed ROP

Two general classes of N-heterocyclic carbenes were investigated including the thiazolium and imidazolium carbenes (Scheme 5)[27]. The thiazolium carbenes were formed *in situ* from their respective salts with 5 equivalents of triethylamine, **11**. The ROP of lactide was accomplished in CH_2Cl_2 under mild conditions (2–4 days at 25 °C) producing polymers of controlled molecular weight and narrow polydispersity with ~85%

conversions. DP's in excess of 100 were difficult to obtain in reasonable times. The imadazolium carbenes prepared according to literature procedures were found to be extremely active catalysts such that extremely dilute conditions and low catalyst concentrations were required to produce narrow polydispersity. Remarkable turnovers were observed with these catalysts, even at low concentrations and temperatures below 0 °C, **10**. Polymers of predictable molecular weights from the initiator to monomer ratio with narrow polydispersity were obtained. Moreover, control of the end groups was also demonstrated.

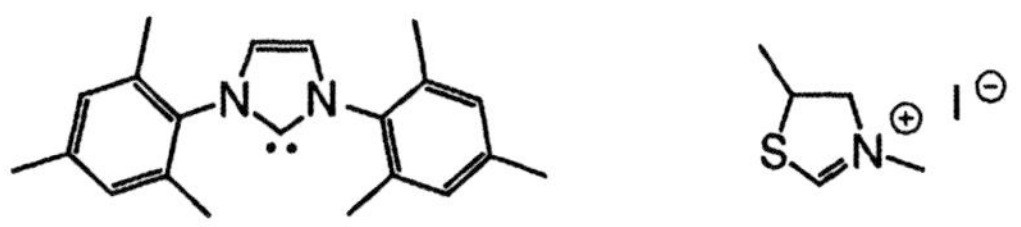

Scheme 5. N-heterocyclic carbene catalyst platform.

3. CONCLUSIONS

The ROP of lactide using nucleophilic catalysts including tertiary amines and phosphines and N-heterocyclic carbenes was demonstrated. The tertiary amines and phosphines were extremely active at elevated temperatures using bulk conditions, whereas the carbenes were active in solution using mild conditions. Narrowly dispersed polymers of predictable molecular weights were obtained. The polymerization is believed to occur through a monomer-activated mechanism, where initiation occurs when a nucleophile (alcohol) reacts with the lactide-catalyst complex to form the mono adduct and polymerization proceeds when the terminal ω- hydroxyl group continues the propagation. In addition, the organic catalyzed chain scission of polylactide was demonstrated as a general route to polylactides with controlled molecular weight, functionality and architecture. These developments will be incorporated to the library of building blocks for the construction of new macromolecular architectures based on dendritic architectures[28] for the generation of nanoporous materials for "on-chip" applications[29].

ACKNOWLEDGMENTS

The authors would like to thank IBM, the NSF Center for Polymeric Interfaces and Macromolecular Assemblies (CPMIA) and the National Institute of Standards and Technology through an ATP cooperative agreement (70NANB8H4013) for financial support.

REFERENCES

1. For example: Ovitt, T. M., and Coates, G. W. 1999, Stereoselective Ring-Opening Polymerization of meso-Lactide: Synthesis of Syndiotactic Poly(lactic acid). *J. Am. Chem. Soc.* **121**: 4072-4073; Chamberlain, B.M., Sun, Y., Hagadorn, J.R., Hemmesch, E.W., Young, V.G., Jr., Pink, M., Hillmyer, M.A., and Tolman, W.B., 1999, Discrete Yttrium(III) Complexes as Lactide Polymerization Catalysts. *Macromolecules* **32**: 2400-2402; Cheng, M., Attygalle, A. B., Lobkovsky, E., and Coates, G. W., 1999, Single-Site Catalysts for Ring-Opening Polymerization: Synthesis of Heterotactic Poly(lactic acid) from rac-Lactide. *J. Am. Chem. Soc.* **121**: 11583-11584; Chisholm, M. H., Eilerts, N., Huffman, J., Iyre S., Pacold, M., and Phomphrai, K., 2001, Molecular Design of Single-Site Metal Alkoxide Catalyst Precursors for Ring-Opening Polymerization Reactions Leading to Polyoxygenates. 1. Polylactide Formation by Achiral and Chiral Magnesium and Zinc Alkoxides, (η3-L)MOR, Where L = Trispyrazolyl- and Trisindazolylborate Ligands. *J. Am. Chem. Soc.* **122**: 11845-11854; O'Keef, B., Monnier, S., Hillmyer, M. A., and Tolman, W. B. 2001, Rapid and controlled polymerization of lactide by structurally characterized ferric alkoxides. *J. Am. Chem. Soc.* **123**: 339-40.
2. For example: Kowalski, A., Duda, A., and Penczek, S., 1998, Polymerization of L,L-Lactide Initiated by Aluminum Isopropoxide Trimer or Tetramer. *Macromolecules* **31**: 2114-2122; Dubois, Ph., Jacobs, C., Jerome, R., and Tessie, Ph., 1991, Macromolecular engineering of polylactones and polylactides. 4. Mechanism and kinetics of lactide homopolymerization by aluminum isopropoxide. *Macromolecules* **24**: 2266-2270; Kricheldorf, H. R., Lee, S., and Bush, S., 1996, Polylactones 36. Macrocyclic Polymerization of Lactides with Cyclic Bu2Sn Initiators Derived from 1,2-Ethanediol, 2-Mercaptoethanol, and 1,2-Dimercaptoethane. *Macromolecules* **29**: 1375-1381; Stevels, W. M., Dijkstra, and P. Feijen, J., 1997, New initiators for the ring-opening polymerization of cyclic esters. *Trends Polym. Sci.* **5**: 300-305.
3. For example: Kumar, A., and Gross, R. A., 2000, Candida antarctica Lipase B-Catalyzed Transesterification: New Synthetic Routes to Copolyesters. *J. Am. Chem. Soc.* **122**: 11767-11770; Kumar, A., and Gross, R. A., 2000, Candida antartica lipase B catalyzed polycaprolactone synthesis: effects of organic media and temperature. *Biomacromolecules* **1**: 133-138; Kobayashi, S., Uyama, H., Namekawa, S., and Hayakawa, H., 1998, Enzymic Ring-Opening Polymerization and Copolymerization of 8-Octanolide by Lipase Catalyst. *Macromolecules* **31**: 5655.
4. For example: Ahrendt, K. A., Borths, C. J., and MacMillan, D. W. C., 2000, New Strategies for Organic Catalysis: The First Highly Enantioselective Organocatalytic Diels-Alder Reaction. *J. Am. Chem. Soc.* **122**: 4243-4244; Jen, W. S., Wiener, J. J. M., and MacMillan, D. W. C., 2000, New Strategies for Organic Catalysis: The First Enantioselective Organocatalytic 1,3-Dipolar Cycloaddition. *J. Am. Chem. Soc.* **122**: 9874-9875; Dong, V. M., and MacMillan D. W. C., 2001, Design of a New Cascade Reaction for the Construction of Complex Acyclic Architecture: The Tandem Acyl-Claisen Rearrangement. *J. Am. Chem. Soc.* **123**: 2448-2449; Fu, G. C., 2000, Enantioselective Nucleophilic Catalysis with "Planar-Chiral" Heterocycles. *Acc. Chem. Res.* **33**: 412-420; Arai, S., Bellemin-Laponnaz, S., and Fu, G. C., 2001, Kinetic resolution of amines by a nonenzymatic acylation catalyst. *Angew. Chem. Int. Ed.* **40**: 234-236; Tao, B., Lo, M. C., and Fu, G. C., 2001, Planar-chiral pyridine N-oxides, a new family of asymmetric catalysts: exploiting an eta(5)-C(5)Ar(5) ligand to achieve high enantioselectivity. *J. Am. Chem. Soc.* **123**: 353-354; Cordova, A., Notz, W., Zhong, G., Betancort, J. M., and Barbas, C. F., 2002, A Highly Enantioselective Amino Acid-Catalyzed Route to Functionalized α-Amino Acids. *J. Am. Chem. Soc.* **124**: 1842-1843;

Cordova, A., Notz, W., Zhong, G., Betancort, J. M., and Barbas, C. F., 2002, A Highly Enantioselective Route to Either Enantiomer of Both α- and β-Amino Acid Derivatives. *J. Am. Chem. Soc.* **124**: 1866-1867.

5. For example: Vedejs, E., Daugulis, O., and Diver, S. T., 1996, Enantioselective Acylations Catalyzed by Chiral Phosphines. *J. Org. Chem.* **61**: 430-431; Vedejs, E., and Rozners, E., 2001, Parallel Kinetic Resolution under Catalytic Conditions: A Three-Phase System Allows Selective Reagent Activation Using Two Catalysts. *J. Am. Chem. Soc.* **123**: 2428-2429; Qiao, S., and Fu, G. C., 1998, The First Application of a Planar-Chiral Phosphorus Heterocycle in Asymmetric Catalysis: Enantioselective Hydrogenation of Dehydroamino Acids. *J. Org. Chem.* **63**: 4168-4169.
6. For example: Fu, G. C., 2000, Enantioselective Nucleophilic Catalysis with "Planar-Chiral" Heterocycles. *Acc. Chem. Res.* **33**: 412-420; Arai, S., Bellemin-Laponnaz, S., and Fu, G. C., 2001, Kinetic resolution of amines by a nonenzymatic acylation catalyst. *Angew. Chem. Int. Ed.* **40**: 234-236; Tao, B., Lo, M. C., and Fu, G. C., 2001, Planar-chiral pyridine N-oxides, a new family of asymmetric catalysts: exploiting an eta(5)-C(5)Ar(5) ligand to achieve high enantioselectivity. *J. Am. Chem. Soc.* **123**: 353-354; Fu, G.C., 2001, The chemistry of borabenzenes (1986-2000). *Adv. Organomet. Chem.* **47**: 101-119; Fu, G. C., 2001, Asymmetric catalysis with "planar-chiral" heterocycles. *Pure Appl. Chem.* **73**: 347-357; Hodous, B. L., and Fu, G. C., 2002, Enantioselective Staudinger Synthesis of β-Lactams Catalyzed by a Planar-Chiral Nucleophile. *J. Am. Chem. Soc.* **124**: 1578-1579.
7. Liu, L., and Breslow, R., 2002, A Potent Polymer/Pyridoxamine Enzyme Mimic. *J. Am. Chem. Soc.* 124: 4978-4979.
8. Borman, S., 2002, Improvin classics. Organocatalysts inspire 'greener' asymmetric versions of classic synthetic reactions. *C&E News* **80** (8): 33.
9. For example: Herrmann, W. A., and Kocher, C., 1997, N-heterocyclic Carbenes. *Angew. Chem. Int. Ed. Eng.* **36**: 2162-2187; Spassky, N., Wisniewski, M., Pluta, C., and LeBorgne, A., 1996, Highly stereoelective polymerization of rac-(D,L)-lactide with a chiral Schiff's base/aluminum alkoxide initiator. *Macromol. Chem. Phys.* **197**: 2627-2637.
10. Vedejs, E., Bennett, N. S., Conn, L. M., Diver, S. T., Gingras, M., Lin, S., Oliver, P. A., and Peterson, M. J., 1993, Tributylphosphine-catalyzed acylations of alcohols: scope and related reactions. *J. Org. Chem.* **58**: 7286-7288; Vedejs, E., and Diver, S. T., 1993, Tributylphosphine: a remarkable acylation catalyst. *J. Am. Chem. Soc.* **115**: 3358-3359.
11. For example: Parshall, G. W., and Ittel, S. D., 1992, *Homogeneous Catalysis*, Wiley-Interscience: New York; Sanford, M. S., Love, J. A., and Grubbs, R. H., 2001, A versatile precursor for the synthesis of new ruthenium olein metathesis catalysts. *Organometallics* **20**: 5314-5318; Fu, G.C., Nguyen, S.T., and Grubbs, R.H., 1993, Catalytic ring-closing metathesis of functionalized dienes by a ruthenium carbene complex. *J. Am. Chem. Soc.* **115**: 9856-9857.
12. Reetz, M. T., 2001, Combinatorial and evolution-based methods in the creation of enantioselective catalysts. *Angew. Chem. Int. Ed.* **40**: 284-310.
13. For example: Burk, M. J., Feaster, J. E., and Harlow, R. L., 1991, New chiral phospholanes; synthesis, characterization, and use in asymmetric hydrogenation reactions. *Tetrahedron: Asym.* **2**: 569-592; Gladiali, S., Dore, A., Fabbri, D., De Lucchi, O., and Manassero, M., 1993, Novel atropisomeric phosphorus ligands: 4,5-dihydro-3H-dinaphtho[2,1-c;1',2'-e]phosphepine derivatives. *Tetrahedron: Asym.* **5**: 511-514; McNeil, P. A., Roberts, N. K., and Bosnich, B., 1981, Asymmetric synthesis. Asymmetric catalytic hydrogenation using chiral chelating six-membered ring diphosphines. *J. Am. Chem. Soc.* **103**: 2273-2280.
14. For example: Ovitt, T. M., and Coates, G. W. 1999, Stereoselective Ring-Opening Polymerization of meso-Lactide: Synthesis of Syndiotactic Poly(lactic acid). *J. Am. Chem.*

Soc. **121**: 4072-4073; Chamberlain, B.M., Sun, Y., Hagadorn, J.R., Hemmesch, E.W., Young, V.G., Jr., Pink, M., Hillmyer, M.A., and Tolman, W.B., 1999, Discrete Yttrium(III) Complexes as Lactide Polymerization Catalysts. *Macromolecules* **32**: 2400-2402; Cheng, M., Attygalle, A. B., Lobkovsky, E., and Coates, G. W., 1999, Single-Site Catalysts for Ring-Opening Polymerization: Synthesis of Heterotactic Poly(lactic acid) from rac-Lactide. *J. Am. Chem. Soc.* **121**: 11583-11584; Chisholm, M. H., Eilerts, N., Huffman, J., Iyre S., Pacold, M., and Phomphrai, K., 2001, Molecular Design of Single-Site Metal Alkoxide Catalyst Precursors for Ring-Opening Polymerization Reactions Leading to Polyoxygenates. 1. Polylactide Formation by Achiral and Chiral Magnesium and Zinc Alkoxides, (η3-L)MOR, Where L = Trispyrazolyl- and Trisindazolylborate Ligands. *J. Am. Chem. Soc.* **122**: 11845-11854; Dove, A., Gibson, V. C., Marshall, E. L., White, A.J. P., and Williams, D., 2001, A well defined tin(II) initiator for the living polymerisation of lactide. *Chem. Comm.* 283-284.

15. For example: Arduengo, A. J., 1999, Looking for Stable Carbenes: The Difficulty in Starting Anew. *Acc. Chem Res.* **32**: 913-921; Arduengo, A. J., and Krfczyk, R., 1998, On the search of stable carbenes. *Chem. Z.* **32**: 6-14; Bourissou, D., Guerret, O., Gabbaie, F. P., and Betrand, G., 2000, Stable carbenes. *Chem. Rev.* **100**: 39-91; Herrmann, W.A., and Kocher, C., 1997, Essays on organometallic chemistry. 9. N-Heterocyclic carbenes. *Angew. Chem., Int. Ed. Eng.* **36**: 2162-2165.
16. For example: Herrmann, W. A., Goossen, L. Kocher, C., and Artus, G., 1996, Heterocyclic carbenes. 9. Chiral heterocyclic carbenes in asymmetric homogeneous catalysis. *Angew. Chem., Int. Edt. Eng.* **35**: 2805-2807; Lapert, M., 1988, The coordination chemistry of electron-rich alkenes (enetetramines). *J. Organomet. Chem.* **358**: 185-187.
17. Initial screening of different organic catalysts, solvents and a variety of polymerization conditions was performed on a Quest 210 reactor (Argonaut Technologies). This robotic reactor allowed up to 20 polymerizations to be performed in parallel under the appropriate environment. Polymers with targeted DP of 30 were prepared and assayed by SEC and ^{1}H NMR to optimize the polydispersity and the molecular weight control. For general description of ROP in Quest see: *Argonaut Application Note* **33**.
18. Nederberg, F., Connor, E. F., Möller, M., Glauser, T., and Hedrick, J. L., 2001, New paradigms for organic catalysts: The first organocatalytic living polymerization. *Angew. Chem. Int. Eng. Ed.* **40**: 2712-2715.
19. General procedure for D,L-lactide polymerization: a round-bottom flask equipped with a stirbar and sealed with a septum was flamed under vacuum and purged with nitrogen. The D,L-lactide (1.0 g, 6.94 mmol) and DMAP (0.56 g, 0.46 mmol, for DP = 30) was added in a glove box. Approximately, 5 ml of CH_2Cl_2 and ethanol (14 μl, 0.23 mmol) were added and the flask was heated to 35 °C. The polymer was isolated in cold methanol and dried to a constant weight, 100% yield. ^{1}H NMR (acetone-d_6) δ = 1.46-1.56 (d, 3H, -CH_3), 5.05-5.25 (q, H, -*CH*-). ^{13}C NMR (acetone-d_6) δ = 17.0 (CH_3-), 69.8 (-*C*H-), 169.8 (-*C*O-).
20. Möller, M., Kånge, R., and Hedrick, J. L., 2000, Sn(OTf)2 and Sc(OTf)3: efficient and versatile catalysts for the controlled polymerization of lactones. *J. Polym. Sci., Part A, Polym. Chem.* **38**: 2067-2074.
21. In a glove box, the D,L-lactide (1.0 g, 6.94 mmol) and the DMAP (1 eq to the D,L-lactide, 0.848 g, 6.94 mmol) was added to a round bottom flask. An excess of benzyl alcohol (10 eq to the D,L-lactide, 7.18 ml, 0.069 mmol) was charged together with CH_2Cl_2 and the reaction flask was slowly heated to 35 °C (5 h). The product was isolated by flash chromatography (9:1 dichloromethane/ethyl acetate). ^{1}H NMR (acetone-d_6) δ = 1.34 (d, 3H, -CH_3), 2.05 (s, H, -*OH*), 4.30 (q, H, -*CH*-), 5.16 (s, 2H, -CH_2-), 7.40-7.30 (m, 5H, C_6H_5-). ^{13}C NMR (acetone-d_6) δ = 175.4, 137.3, 129.3, 128.9, 128.8, 67.6, 66.8, 20.8.

22. Nederberg, F., Connor, E. F., Moller, M., Glauser, T., and Hedrick, J. L., 2001, New Paradigms for Organic Catalysts: The First Organocatalytic Living Polymerization. *Angew. Chem. Int. Eng. Ed.* **40**: 2712-2715.
23. Nederberg, F., Connor, E. F., Glauser, T., and Hedrick, J. L., 2001, Organocatalytic chain scission of poly(lactides): a general route to controlled molecular weight, functionality and macromolecular architecture. *Chem. Commun.* 2066-2067.
24. Myers, M., Connor, E., Glauser, T, Möck, A., Nyce, G., and Hedrick, J. L., 2001, Phosphines: nucleophilic organic catalysts for the controlled ring-opening polymerization of lactides. *J. Polym. Sci. Part A: Polym. Chem. Ed.* **40**: 844-851.
25. General procedure for lactide polymerization: a round-bottom flask equipped with a stirbar and sealed with a septum was flamed under vacuum and purged with nitrogen. The L-lactide (1.0 g, 6.94 mmol.) and $P(Bu)_3$ (0.023 g, 0.115 mmol, for DP = 60) were added in a glove box. Benzyl alcohol (12 μl, 0.115 mmol) was added and the flask was heated to 135 °C. The polymer was dissolved in THF and isolated in cold methanol and dried to a constant weight.
26. For example: Dubois, Ph., Barakat, I., Jerome, R. and Teyssie, Ph., 1993, Macromolecular engineering of polyactones and polyactides. 12. Study of the depolymerization reactions of poly(ε-caprolactone) with functional aluminum alkoxide end groups. *Macromolecules* **26**: 4407-4412.
27. Connor, E. F., Nyce, G. W., Möck, A., and Hedrick, J. L., 2002, First Example of N-Heterocyclic Carbenes as Catalysts for Living Polymerization: Organocatalytic Ring-Opening Polymerization of Cyclic Esters. *J. Am. Chem. Soc.* **124**: 914-915
28. Trollsås, M., and Hedrick, J. L., 1998, Dendrimer-like Star Polymers. *J. Am. Chem. Soc.* **120**: 4644-4651; Trollsås, M., Claesson, H., Atthoff, B., and Hedrick, J. L., 1998, Layered dendritic block copolymers. *Angew. Chem. Int. Ed. Engl.* **37**: 3132-3136; Trollsås, M., Hedrick, J. L., Mecerreyes, D., Dubois, Ph., Jérôme, R., Ihre, H., and Hult, A., 1998, Versatile Synthesis to Highly Functional Branched and Dendri-Graft Polyesters. *Macromolecules* **31**: 2756-2763; Trollsås, M., Kelly, M. A., Claesson, H., Siemens, R., and Hedrick, J. L., 1999, Highly branched block copolymers: design, synthesis, and morphology. *Macromolecules* **32**: 4917-4924.
29. Nguyen, C., Carter, K. R., Hawker, C. J., Hedrick, J. L., Jaffy, R., Miller, R. D., Remenar, J., Rhee, H., Rice, P., Toney, M., and Yoon, D., 1999, Low-Dielectric, Nanoporous Organosilicate Films Prepared via Inorganic/Organic Polymer Hybrid Templates. *Chem. Mater.* **11**: 3080-3085; Nguyen, C., Hawker, C. J., Miller, R., Hedrick, J. L., and Hilborn, J. G., 2000, Hyperbranched Polyesters as Nanoporosity Templating Agents for Organosilicates. *Macromolecules* **33**: 4281-4284; Heise, A., Nguyen, C., Malek, R., Hedrick, J. L., Frank, C. W., and Miller, R. D., 2000, Starlike Polymeric Architectures by Atom Transfer Radical Polymerization: Templates for the Production of Low Dielectric Constant Thin Films. *Macromolecules* **33**: 2346-2354; Mecerreyes, D., Huang, E., Magbitang, T., Volksen, W., Hawker, C. J., Lee, V., Miller, R. D., and Hedrick, J. L., 2001, Application of hyperbranched block copolymers as templates for the generation of nanoporous organosilicates. *High Perform. Polym.* **13**: S11-S19; Hedrick, J. L., Hawker, C. J., Trollsas, M., Remenar, J., Yoon, D. Y., and Miller, R. D., 1998, Templating nanoporosity in organosilicates using well-defined branched macromolecules. *Mat. Res. Symp. Proc.* **519**: 65-75.

Index

Printed in the United States
120588LV00004B/77/A

9 780306 478840